This clear and concise advanced textbook is a comprehensive introduction to power electronics. It considers the topics of analogue electronics, electric motor control and adjustable speed electrical drives, both a.c. and d.c.

In recent years, great changes have taken place in the types of semiconductor devices used as power switches in engineering applications. This book provides an account of this developing subject through such topics as: d.c. choppers, controlled bridge rectifiers, and the speed control of induction motors by variable voltage–variable frequency inverters. This being the second edition of this popular text, a further completely new chapter has been added, dealing with the application of pulse-width modulation (PWM) techniques in induction motor speed control. The chapters dealing with electronic switching devices and with adjustable speed drives have been entirely rewritten, to ensure the text is completely up-to-date.

With numerous worked examples, exercises, and the many diagrams, advanced undergraduates and postgraduates will find this a readable and immensely useful introduction to the subject of power electronics.

Power electronics and motor control
SECOND EDITION

Power electronics and motor control

SECOND EDITION

W. SHEPHERD

L. N. HULLEY

D. T. W. LIANG

Dept. of Electronic and Electrical Engineering
University of Bradford
England

Published by the Press Syndicate of the University of Cambridge
The Pitt Building, Trumpington Street, Cambridge CB2 1RP
40 West 20th Street, New York, NY 10011-4211, USA
10 Stamford Road, Oakleigh, Melbourne 3166, Australia

© Cambridge University Press, 1987, 1995

First published 1987
Second edition 1995

Printed in Great Britain at the University Press, Cambridge

A catalogue record for this book is available from the British Library

Library of Congress cataloguing in publication data

Shepherd, W. (William), 1928–
Power electronics and motor control / W. Shepherd, L. N. Hulley,
D. T. W. Liang. – 2nd ed.
  p. cm.
Includes bibliographical references and index.
ISBN 0-521-47241-5. – ISBN 0-521-47813-8 (pbk.)
1. Power electronics. 2. Electronic control. I. Hulley, L.N.
(Lance Norman) II. Liang, D. T. W. III. Title
TK7881.15.S54 1995
621.31'7–dc20   94-46438 CIP

ISBN 0 521 47241 5 hardback
ISBN 0 521 47813 8 paperback

KTS

CONTENTS

Preface to first edition — xv
Preface to second edition — xvii
List of principal symbols — xix

1 Power switching theory — 1
  1.1 Power flow control by switches — 1
  1.2 Attributes of an ideal switch — 2
  1.3 Sources of incidental dissipation in imperfect switches — 3
  1.4 Estimation of switching dissipation — 3
      1.4.1 Soft load – series resistance — 3
      1.4.2 Hard load – series resistance–inductance — 5
  1.5 Modification of switching dissipation – switching aids — 6
      1.5.1 Approximate calculations of switching loss reduction — 8
          1.5.1.1 Turn-on aid — 8
          1.5.1.2 Turn-off aid — 9
      1.5.2 Detailed calculation of switching loss reduction — 12
  1.6 Estimation of total incidental dissipation — 15
  1.7 Transfer of incidental dissipation to ambient – thermal considerations — 17
  1.8 Worked examples — 21
  1.9 Review questions and problems — 28

2 Switching devices and control electrode requirements — 32
  2.1 Rating, safe operation area and power handling capability of devices — 32
      2.1.1 Power handling capability (PH) — 32
      2.1.2 Principles of device fabrication — 33
      2.1.3 Safe operation area (SOA) — 33
      2.1.4 Ratings and data sheet interpretation — 34
  2.2 Semiconductor switching devices — 35
      2.2.1 Bipolar junction transistor (BJT) — 36
          2.2.1.1 Forward current transfer ratio — 37

		2.2.1.2 Switch-on and switch-off characteristics	40
		2.2.1.3 Construction and properties of some types of power bipolar transistors	41
		2.2.1.4 Switching properties of bipolar devices	43
	2.2.2	Metal–oxide–semiconductor field-effect transistor (MOSFET)	48
2.3	Compound devices		52
	2.3.1 Cascade connected devices		52
		2.3.1.1 Power Darlington transistor	52
		2.3.1.2 Insulated gate bipolar transistor (IGBT)	53
	2.3.2 Cumulative feedback connected devices (thyristors)		57
		2.3.2.1 Basic thyristor theory	58
		2.3.2.2 Triac (bidirectional SCR)	73
		2.3.2.3 Gate turn-off thyristor (GTO)	75
		2.3.2.4 Metal–oxide controlled thyristor (MCT)	82
2.4	Device selection strategy		84
	2.4.1 Voltage and current ratings		84
	2.4.2 Switching frequency (slew rate)		84
	2.4.3 Ruggedness against abuse		85
	2.4.4 Ease of triggering		85
	2.4.5 Availability and cost		86
	2.4.6 Incidental dissipation (ID)		86
	2.4.7 Need for aids and/or snubbers		87
2.5	Review questions and problems		87

3	System realisation		**94**
3.1	Introduction		94
3.2	Preventive protection circuitry		95
	3.2.1 Voltage and current snubber circuits		95
		3.2.1.1 Requirement for snubber circuits	95
		3.2.1.2 Design of snubber circuits	95
		3.2.1.3 Worked examples on snubber circuits	102
	3.2.2 Ancillary environmental protection		105
		3.2.2.1 Current surge protection	105
		3.2.2.2 Time cut strategies	106
		3.2.2.3 Electromagnetic interference (EMI)	106
3.3	Abuse protection circuitry		107
	3.3.1 Overcurrent protection		107
	3.3.2 Overvoltage protection – crowbar		108
3.4	Isolation circuitry		108
	3.4.1 Pulse isolation transformer		109
	3.4.2 Opto-isolator		111
3.5	System realisation strategy		112
3.6	Prototype realisation		114
	3.6.1 Principles		114
	3.6.2 Example – single-phase voltage control circuit		114

	3.7	Device failure – mechanisms and symptoms	115
	3.8	Review questions and problems	118

4 Adjustable speed drives — 121
- 4.1 Basic elements of a drive — 121
- 4.2 Load torque–speed characteristics — 122
- 4.3 Stability of drive operations — 123
  - 4.3.1 Steady-state stability — 123
  - 4.3.2 Transient stability — 127
- 4.4 Principal factors affecting the choice of drive (reference TP1) — 129
  - 4.4.1 Rating and capital cost — 130
  - 4.4.2 Speed range — 130
  - 4.4.3 Efficiency — 130
  - 4.4.4 Speed regulation — 134
  - 4.4.5 Controllability — 134
  - 4.4.6 Braking requirements — 135
  - 4.4.7 Reliability — 135
  - 4.4.8 Power-to-weight ratio — 136
  - 4.4.9 Power factor — 136
  - 4.4.10 Load factor and duty cycle — 136
  - 4.4.11 Availability of supply — 137
  - 4.4.12 Effect of supply variation — 137
  - 4.4.13 Loading of the supply — 137
  - 4.4.14 Environment — 138
  - 4.4.15 Running costs — 138
- 4.5 Types of electric motor used in drives — 139
  - 4.5.1 D.c. motors — 139
  - 4.5.2 Synchronous motors — 139
    - 4.5.2.1 Wound-field synchronous motors — 140
    - 4.5.2.2 Permanent magnet synchronous motors — 141
    - 4.5.2.3 Synchronous reluctance motors — 142
    - 4.5.2.4 Self-controlled (brushless) synchronous motors — 142
    - 4.5.2.5 Stepping (stepper) motors — 143
    - 4.5.2.6 Switched reluctance motors — 145
  - 4.5.3 Induction motors — 146
- 4.6 Different options for an adjustable speed drive incorporating an electric motor — 147
- 4.7 A.c. motor drives or d.c. motor drives? — 147
- 4.8 Trends in the design and application of a.c. adjustable speed drives — 149
  - 4.8.1 Trends in motor technology and motor control — 149
  - 4.8.2 Trends in power switches and power converters — 149
- 4.9 Problems — 150

5 D.c. motor control using a d.c. chopper — 152
- 5.1 Basic equations of motor operation — 152
- 5.2 D.c. chopper drives — 157
  - 5.2.1 Basic (class A) chopper circuit — 158
    - 5.2.1.1 Analytical properties of the load voltage waveform — 160
    - 5.2.1.2 Analytical properties of the load current waveform — 164
    - 5.2.1.3 Average current, r.m.s. current and power transfer — 167
  - 5.2.2 Class A transistor chopper — 170
  - 5.2.3 Class B chopper circuits (two-quadrant operation) — 171
- 5.3 Worked examples — 174
- 5.4 Problems — 187

6 Controlled bridge rectifiers with d.c. motor load — 190
- 6.1 The principles of rectification — 190
- 6.2 Separately excited d.c. motor with rectfied single-phase supply — 191
  - 6.2.1 Single-phase semi-converter — 192
  - 6.2.2 Single-phase full converter — 195
    - 6.2.2.1 Continuous conduction — 196
    - 6.2.2.2 Discontinuous conduction — 200
    - 6.2.2.3 Critical value of load inductance — 202
    - 6.2.2.4 Power and power factor — 202
  - 6.2.3 Worked examples — 203
- 6.3 Separately excited d.c. motor with rectified three-phase supply — 210
  - 6.3.1 Three-phase semi-converter — 211
  - 6.3.2 Three-phase full converter — 212
    - 6.3.2.1 Continuous conduction — 213
    - 6.3.2.2 Critical value of load inductance — 217
    - 6.3.2.3 Discontinuous conduction — 217
    - 6.3.2.4 Power and power factor — 220
    - 6.3.2.5 Addition of freewheel diode — 220
  - 6.3.3 Three-phase double converter — 221
  - 6.3.4 Worked examples — 222
- 6.4 Problems — 233

7 Three-phase naturally commutated bridge circuit as a rectifier or inverter — 236
- 7.1 Three-phase controlled bridge rectifier with passive load impedance — 236
  - 7.1.1 Resistive load and ideal supply — 237
    - 7.1.1.1 Load-side quantities — 240
    - 7.1.1.2 Supply-side quantities — 243

		7.1.1.3 Operating power factor	245
		7.1.1.4 Shunt capacitor compensation	246
		7.1.1.5 Worked examples	250
	7.1.2	Highly inductive load and ideal supply	254
		7.1.2.1 Load-side quantities	254
		7.1.2.2 Supply-side quantities	256
		7.1.2.3 Shunt capacitor compensation	259
		7.1.2.4 Worked examples	261
7.2	Three-phase controlled bridge rectifier–inverter		265
	7.2.1 Theory of operation		265
	7.2.2 Worked examples		271
7.3	Problems		275

8 Single-phase voltage controllers — 280

8.1	Resistive load with symmetrical phase-angle triggering	281
	8.1.1 Harmonic properties	281
	8.1.2 R.m.s. voltage and current	286
	8.1.3 Power and power factor	288
	8.1.3.1 Average power	288
	8.1.3.2 Power factor	291
	8.1.3.3 Reactive voltamperes and power factor correction	292
	8.1.4 Worked examples	296
8.2	Series R–L load with symmetrical phase-angle triggering	303
	8.2.1 Analysis of the instantaneous current variation	304
	8.2.2 Harmonic properties of the current	309
	8.2.3 R.m.s. current	312
	8.2.4 Properties of the load voltage	313
	8.2.5 Power and power factor	314
	8.2.6 Worked examples	316
8.3	Resistive load with integral-cycle triggering	323
	8.3.1 Harmonic and subharmonic properties	324
	8.3.2 R.m.s. voltage and current	327
	8.3.3 Power and power factor	327
	8.3.4 Comparison between integral-cycle operation and phase-controlled operation	328
	8.3.4.1 Lighting control	328
	8.3.4.2 Motor speed control	329
	8.3.4.3 Heating loads	329
	8.3.4.4 Electromagnetic interference	330
	8.3.4.5 Supply voltage dip	330
	8.3.5 Worked examples	331
8.4	Problems	337

xii Contents

9	**Three-phase induction motor with constant frequency supply**	**346**
	9.1 Three-phase induction motor with sinusoidal supply voltages	346
	9.1.1 Equivalent circuits	348
	9.1.2 Power and torque	350
	9.1.3 Approximate equivalent circuit	353
	9.1.4 Effect of voltage variation on motor performance	356
	9.1.5 M.m.f space harmonics due to fundamental current	358
	9.2 Three-phase induction motor with periodic nonsinusoidal supply voltages	359
	9.2.1 Fundamental spatial m.m.f. distribution due to time harmonics of current	359
	9.2.2 Simultaneous effect of space and time harmonics	360
	9.2.3 Equivalent circuits for nonsinusoidal voltages	361
	9.3 Three-phase induction motor with voltage control by electronic switching	362
	9.3.1 Approximate method of solution for steady-state operation	369
	9.3.1.1 Theory of operation	369
	9.3.1.2 Worked examples	370
	9.3.2 Control system aspects	378
	9.3.2.1 Representation of the motor	378
	9.3.2.2 Representation of the SCR controller	381
	9.3.2.3 Closed-loop operation using tachometric negative feedback	383
	9.3.2.4 Worked examples	386
	9.4 Three-phase induction motor with fixed supply voltages and adjustable secondary resistances	393
	9.4.1 Theory of operation	393
	9.4.2 Worked examples	396
	9.5 Problems	398
10	**Induction motor slip-energy recovery**	**404**
	10.1 Three-phase induction motor with injected secondary voltage	404
	10.1.1 Theory of operation	404
	10.1.2 Worked example	405
	10.2 Induction motor slip-energy recovery (SER) system	406
	10.2.1 Torque–speed relationship	408
	10.2.2 Current relationships	413
	10.2.3 Power, power factor and efficiency	416
	10.2.4 Speed range, drive rating and motor transformation ratio	419
	10.2.5 Filter inductor	422
	10.2.6 Worked examples	424
	10.3 Problems	433

11	**Induction motor speed control by the use of adjustable voltage, adjustable frequency step-wave inverters**	**435**
	11.1 Three-phase induction motor with controlled sinusoidal supply voltages of adjustable frequency	435
	11.1.1 Theory of operation	435
	11.1.2 Worked examples	440
	11.2 Three-phase, step-wave voltage source inverters with passive load impedance	444
	11.2.1 Stepped-wave inverter voltage waveforms	447
	11.2.1.1 Two simultaneously conducting switches	447
	11.2.1.2 Three simultaneously conducting switches	451
	11.2.2 Measurement of harmonic distortion	456
	11.2.3 Harmonic properties of the six-step voltage wave	457
	11.2.4 Harmonic properties of the optimum twelve-step waveform	458
	11.2.5 Six-step voltage source inverter with series R–L load	459
	11.2.5.1 Star-connected load	459
	11.2.5.2 Delta-connected load	460
	11.2.6 Worked examples	465
	11.3 Three-phase, step-wave voltage source inverters with induction motor load	471
	11.3.1 Motor currents	471
	11.3.2 Motor losses and efficiency	473
	11.3.3 Motor torque	475
	11.3.4 Worked examples	476
	11.4 Problems	482
12	**Induction motor speed control by the use of adjustable frequency PWM inverters**	**487**
	12.1 Properties of pulse-width modulated waveforms	487
	12.1.1 Single-pulse modulation	487
	12.1.2 Multiple-pulse modulation	489
	12.1.3 Sinusoidal modulation	491
	12.1.3.1 Sinusoidal modulation with natural sampling	491
	12.1.3.2 Overmodulation in sinusoidal PWM inverters	496
	12.1.3.3 Sinusoidal modulation with regular sampling	499
	12.1.4 Optimal pulse-width modulation (harmonic elimination)	500
	12.1.5 PWM voltage waveforms applied to three-phase inductive load	503
	12.1.6 Worked examples	505
	12.2 Three-phase induction motor controlled by PWM voltage source inverter (VSI)	512

		12.2.1 Theory of operation	512
		12.2.2 Worked example	514
	12.3	Three-phase induction motor controlled by PWM current source inverter (CSI)	516
		12.3.1 Current source inverter with passive load	516
		12.3.2 Current source inverter with induction motor load	516
	12.4	Secondary frequency control	518
	12.5	Problems	520

Appendix   General expressions for Fourier series   523
Answers to problems   525
References and bibliography   531
Index   536

PREFACE

to the first edition

This book is intended as a teaching textbook for advanced undergraduate and postgraduate courses in power electronics. The reader is presumed to have a background in mathematics, electronic signal devices and electric circuits that would be common in the early years of first degree courses in electrical and electronic engineering. It is the writers' experience that engineering students prefer to learn by proceeding from the particular to the general and that the learning route be well illustrated by many worked examples. Both of these teaching practices are followed here and a lot of problems are also included for attempt and solution at the ends of most chapters.

About one half of the text was written while the principal author (W.S.) was on study leave at the Department of Electrical and Computer Engineering, University of Wisconsin, Madison, Wisconsin, USA. His grateful thanks are acknowledged to the stimulating company of Professor Donald Novotny and Professor Tom Lipo during this period of sabbatical scholarship, sponsored by the Fulbright Commission.

It has become evident in recent years that the reign of the silicon controlled rectifier member of the thyristor family, as the universal semiconductor power switch, is drawing to a close. Except in very high power applications the technology of the immediate future lies with three-terminal, control electrode turn-off devices such as the gate turn-off thyristor (GTO), the bipolar power transistor and the field effect power transistor (FET). An important implication of this is that the complicated and expensive commutation (turn-off) circuits that are now necessary in many thyristor applications will not be needed. Accordingly, commutation circuits are not covered in this textbook.

For the specialised sections dealing with electronic engineering, namely Chapter 1 and Chapter 2, the authorship is mainly due to Mr L. N.

Hulley, Senior Lecturer in Electronic Engineering at the University of Bradford.

The brunt of the typing has been undertaken by Mrs Marlyn Walsh. Her good nature and forbearance have helped us greatly.

Most of the drawings and diagrams were prepared in fair copy by Mr David Jowett, Chief Technician to the Schools of Electrical and Electronic Engineering. Our grateful thanks to him also.

The most significant source of information for any university teacher is his/her students. Much of the material in this book has been included as part of relevant taught courses or of research projects over the past twenty years. Our grateful thanks are due to the several hundred undergraduate students and fifty or so postgraduate research students and colleagues who have worked with us in the 'thyristor business'. It is to be hoped that some of this benefit has been reproduced in the present book.

W. Shepherd
L. N. Hulley
Bradford, England
1987

PREFACE

to the second edition

The advances in power electronics since this book was first published in 1987 have chiefly been in the development of more effective semiconductor switching devices. In particular, the future of high power switching applications will involve reduced use of the silicon controlled rectifier (SCR) and gate turn-off thyristor (GTO) and increased use of the metal–oxide–semiconductor (MOS) controlled thyristor (MCT). The most influential development, however, is likely to be due to increased ratings of metal–oxide–semiconductor field-effect transistor (MOSFET) devices and, in particular, widespread use of the insulated gate bipolar transistor (IGBT). Design data of these switching devices is widely available from manufacturers. Increases in the range of semiconductor switches and their nonlinear nature has influenced practitioners to move towards computer based design rather than analytical studies. Simulation techniques are widespread and expert systems are under development.

Chapters 1–3 have been extensively revised, compared with the original text, to incorporate much new material, especially concerned with modern semiconductor power switches. With regard to the switching properties of semiconductor devices the authors have adopted an analytically fundamental approach rather than the current industrial standard. This is educationally easier to understand and more conservative in solution than accepted industrial practice.

Some re-organisation of the original text has permitted expansion of the section on *Adjustable Speed Drives*, now in Chapter 4, to include a brief treatment of various types of synchronous motor.

The previous work on step-wave inverters has been concentrated into the new Chapter 11 and an additional chapter has been included on pulse-width modulation (PWM) control.

Revision of the entire text has provided opportunity to eliminate the errors and obscurities of the original text. Any that remain or are newly introduced are the sole responsibility of the principal author (WS).

The authors are grateful to the students and teachers who have used the previous book, found it helpful and profitable, and have written to say so. They are also grateful to the University of Bradford, England, for permission to use classroom examples and examination materials from courses taught there.

Much of the book revision was undertaken while WS was on study leave as Visiting Stocker Professor at Ohio University, Athens, Ohio, USA. His gratitude is expressed to the Dean of Engineering, Richard Robe, and to his colleagues in the Department of Electrical and Computer Engineering. Dr D. A. Deib, formerly of Ohio University, was kind enough to check some of the analysis of Chapter 6.

The retyping has been carried out by Suzanne Vazzano of Athens, Ohio, and by Beverley Thomas of Bradford, England. Our thanks to them both for their patient forbearance.

W. Shepherd
L. N. Hulley
D. T. W. Liang
Bradford, England
1995

PRINCIPAL SYMBOLS

a_0	Fourier coefficient of zero order
a_n, b_n, c_n	Fourier coefficients of order n
e	instantaneous e.m.f., V
e_s	instantaneous supply e.m.f., V
e_L	instantaneous load e.m.f., V
e_{aN}, e_{bN}, e_{cN}	instantaneous supply phase voltages in a three-phase system, V
e_{ab}, e_{bc}, e_{ca}	instantaneous supply line voltages in a three-phase system, V
e_{AB}, e_{BC}, e_{CA}	instantaneous terminal voltages of a three-phase load, V
f	frequency, Hz
f_1, f_2	frequency in the primary and secondary windings of an induction motor, Hz
g_m	transconductance coefficient, mho
h_{FE}	current gain of common emitter connection
i	instantaneous current, A
i_a	instantaneous armature current in a d.c. motor, A
$i_a, i_b, i_c, i_{AN}, i_{BN}, i_{CN}$	instantaneous currents in the lines or phases of a three-phase system, A
i_s, i_L	instantaneous supply and load currents, respectively, A
i_A, i_B, i_G, i_K	instantaneous anode, base, gate and cathode currents, respectively, A
i_{CBO}	instantaneous collector–base current on open-circuit A
j	phasor operator, $(1\angle\frac{\pi}{2})$
k	order of Fourier harmonic
m	harmonic order of carrier wave sideband, in Chapter 12
n	instantaneous motor speed (Chapter 5), r.p.m. or rad/s
n	order of Fourier harmonic
n, p	designation of negative and positive semiconductor materials, respectively
n_p, n_s	number of effective turns on the primary and secondary windings of an induction motor, respectively

p	number of pole pairs in a motor
p	frequency ratio in PWM
p	differential operator d/dt
r	resistance, ohms
rfi	radio-frequency interference
s	Laplacian operator
t	time, s
t_d, t_f, t_r, t_s	time intervals during transistor switching, s
t_{on}, t_{off}	'on' and 'off' switching times of a transistor, respectively, s
v	instantaneous voltage, V
v_{AB}, v_{BC}, v_{CA}	instantaneous line–line voltages in a three-phase system, V
v_{AN}, v_{BN}, v_{CN}	instantaneous line–load neutral voltages in a three-phase system, V
$v_{AN_s}, v_{BN_s}, v_{CN_s}$	instantaneous line–supply neutral voltages in a three-phase system, V
v_{AO}, v_{BO}, v_{CO}	instantaneous line–pole neutral voltages in a three-phase system, V
v_{NN_s}	instantaneous voltage between load and supply neutral points, V
v_{AC}, v_{AK}	instantaneous voltage drop (anode–cathode) across a diode, triac or SCR, V
$v_{BE}, v_{CB}, v_{CE}, v_{DS}$	instantaneous voltage drops between designated electrodes on a transistor, V
x	SCR extinction angle, rad
A, B, C, D, E, G, K, S	designation of semiconductor switch electrodes
A_{pc}	power–control ratio
A_{V_0}	amplifier voltage gain at no-load
ASIC	application specific integrated circuit
B	viscous friction coefficient, Nm/rad/s
BJT	bipolar junction transistor
BV_{CEO}	breakdown collector–emitter voltage on open circuit, V
C	capacitance, F
CSI	current source inverter
D	diode
D_L	distortion component of input voltamperes, VA
E	r.m.s. value of voltage, V
E_{av}	average value of voltage, V
E_{av_0}	average voltage at no-load, V
E_b	battery voltage, V
E_L, E_s	r.m.s. values of load and supply voltages, respectively, V
E_m	peak value of sinusoidal voltage, V
E_n	r.m.s. value of nth harmonic voltage, V
E_1, E_2	r.m.s. values of the e.m.f.s induced in the primary and secondary windings, respectively, of an induction motor, V
E_T, E_{Th}	r.m.s. voltage across an SCR or transistor, V

FET	field-effect transistor
$G(s)$	open-loop transfer function
HF	high frequency
I	r.m.s. current, A
I_a, I_b, I_c	r.m.s. currents in the lines or phases of a three-phase system, A
I_a	average value of armature current, A
I_{av}	average current, A
I_f	field current or filter circuit current, A
I_{dc}	d.c. current level, A
I_{inv_1}	r.m.s. value of fundamental component of inverter current, A
I_m	peak value of sinusoidal current, A
I_n	r.m.s. value of nth harmonic current, A
I_{on}, I_{off}	maximum values of state currents in transistor switching, A
I_B, I_C, I_E	steady-state value of current in the base, collector and emitter electrodes, respectively, of a transistor
I_{CE}	steady-state value of collector–emitter current, A
$I_{CE_{sat}}$	steady-state value of collector–emitter current under saturated conditions, A
I_L, I_s	r.m.s. values of load and supply currents, respectively, A
I_{s_n}	r.m.s. value of nth harmonic component of the supply current, A
I_1, I_2	r.m.s. currents in primary and secondary windings, respectively, of an induction motor, A
I_{max}	maximum value of current, A
IC	integrated circuit
ID	incidental dissipation
IGBT	insulated gate bipolar transistor
J	polar moment of inertia, kg/m^2
J_1, J_2, J_3, J_4	junctions of semiconductor structure
JFET	junction field-effect transistor
L	self-inductance coefficient, H
L_a	armature inductance of a d.c. motor, H
L_f	field inductance or filter circuit inductance, H
L_m, L_1, L_2	magnetising, primary and secondary per-phase inductances in an induction motor, H
L_s	supply inductance, per-phase, H
LASCR	light-activated, silicon controlled rectifier
M	modulation ratio, V_m/V_c
MCT	metal–oxide controlled thyristor
MOS	metal–oxide–semiconductor
MOSFET	metal–oxide–semiconductor field-effect transistor
N	motor speed, r.p.m. or rad/s
N	number of 'on' cycles in integral-cycle control
N_0	motor no-load speed, r.p.m. or rad/s

Principal symbols

N_1	motor synchronous speed, r.p.m. or rad/s
P	average power, W
P_g	average power in the air-gap of an induction motor, W
P_{link}	average power in d.c. link of a d.c. link inverter, W
P_{max}	maximum value of average power, W
P_{mech}	average value of mechanical power, W
$P_{in}, P_{losses}, P_{out}$	average values of input, loss and output power, W
P_L	load power, W
PF	power factor
PF_s	power factor at the supply point
PH	power handling capability
PWM	pulse-width modulation
Q_L	reactive component of voltamperes, VA
R	resistance, ohm
R_a	armature circuit resistance in a d.c. motor, ohm
R_f	field resistance or filter circuit resistance, ohm
R_L	load resistance, ohm
R_1, R_2	per-phase resistances of the primary and secondary windings, respectively, of an induction motor, ohm
R_{2_n}	value of R_2 in the presence of secondary motor currents of nth harmonic frequency, ohm
RF	ripple factor
RSOA	safe operating area under conditions of reverse bias
S	per-unit slip
S_m	per-unit slip at which peak torque occurs
S_L	apparent load voltamperes, VA
SCR	silicon controlled rectifier
SCS	silicon controlled switch
SOA	safe operating area
T, Th	designations of a transistor or an SCR
T	number of 'on' plus 'off' cycles with an integral cycle waveform (Chapter 8)
T	torque, Nm
T_{case}	case temperature of an SCR, °C
T_j	junction temperature of an SCR, °C
T_m	maximum value of torque, Nm
T_s	static friction (i.e. stiction) torque, Nm
T_L, T_{FW}	load and friction and windage torques in a motor, Nm
T_{on}, T_{off}	time intervals in a chopper circuit (Chapter 5), s
T_1, T_2	electrode designations of a triac
V	voltage, V
V_{av}	average value of voltage, V
V_{bus}	bus or rail voltage, V
V_c, V_m	peak values of carrier and modulating voltages, respectively (Chapter 12), V
V_{dc}	d.c. voltage level, V

V_m	peak value of sinusoidal voltage, V
V_{on}, V_{off}	maximum values of state voltages in transistor switching, V
V_s	supply voltage level, V
V_{CC}	collector rail voltage in a transistor circuit, V
VSI	voltage source inverter
V_1	primary (applied) voltage in an induction motor, V
V_1, V_{L_1}	fundamental components of modulated phase and line voltages, respectively, (Chapter 12), V
W	energy, J
W_{on}, W_{off}	energy dissipation during transistor switching 'on' and 'off' respectively, J
X	reactance, ohm
X_C, X_L	capacitive and inductive reactance, respectively, ohm
X_m	per-phase magnetising reactance of an induction motor, ohm
X_1, X_2	per-phase leakage reactances for the primary and secondary windings, respectively, of an induction motor, ohm
Z	impedance, ohm
Z_L, Z_{L_1}	impedance of load to sinusoidal currents of fundamental frequency, ohm
Z_{L_n}	impedance of load to sinusoidal currents of nth harmonic frequency, ohm
α	firing-angle of SCR, rad or degrees
α_p, α_n	base current gains of a bipolar transistor (Chapter 2)
$\alpha_1, \alpha_2, \ldots, \alpha_n$	switching points on an optimally modulated wave (Chapter 12)
β	conduction interval, rad or degrees
γ	ratio of time intervals in a chopper circuit (Chapter 5)
δ	pulse-width (Chapter 12), rad or degrees
ε	Napierian logarithmic base
ε	electrical error signal, V, (Chapter 9)
ϕ	instantaneous flux, weber
τ	time constant, s
θ_c	conduction interval, rad or degrees
ψ	phase-angle, rad or degrees
ψ_s, ψ_{s_1}	phase-angle of fundamental current at the supply point, rad or degrees
ψ_{s_n}	phase-angle of nth harmonic current at the supply point, rad or degrees
μ	overlap angle, rad or degrees
ω	angular supply frequency, rad/s
Φ	flux/pole for steady-state operation of an induction motor, weber
Φ	phase-angle to sinusoidal currents of supply frequency, rad or degrees

1
Power switching theory

1.1 POWER FLOW CONTROL BY SWITCHES

The flow of electrical energy between a fixed voltage supply and a load is often controlled by interposing a controller, as shown in Fig. 1.1. Viewed from the supply, the apparent impedance of the load plus controller must be varied if variation of the energy flow is required. Conversely, seen from the load, the apparent properties of the supply plus controller must be adjusted. From either viewpoint, control of power flow can be realised by using a series-connected controller with the desired properties. If a current source supply is used instead of a voltage source supply, control can be realised by the parallel connection of an appropriate controller. For safety reasons the latter technique is rarely adopted.

The series-connected controller in Fig. 1.1 can take many different forms. In a.c. distribution systems where variability of power flow is a secondary requirement, transformers are often the prevalent interposing elements. The insertion of reactive elements is inconvenient because variable inductors and capacitors of appropriate size are expensive and bulky. It is easy to use a series-connected variable resistance instead, but at the expense of a considerable loss of energy. Viewing from the load side, loads that absorb significant electric power usually possess some form of energy 'inertia'. This allows amplitude variations created by the interposed controller to be effected in an efficient manner.

Amplitude variations of the controller may be exchanged for a fractional time variation of connection and disconnection from the supply. If the frequency of such switching is so rapid that the load cannot track the switching events due to its electrical inertia, then no energy is expended in an ideal controller. The higher the load electrical inertia and the switching frequency, the more the ripple is reduced in significance.

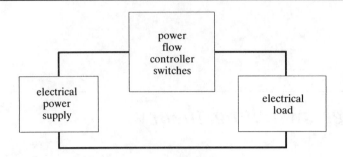

Fig. 1.1 Generalised representation of a controller.

With modern semiconductor devices the switching operation of a series-connected controller can be implemented with high efficiency. For this reason, controllers are almost exclusively realised with power electronic switches. Inefficiency in the switching operation causes wasted energy in the switching devices. This wastage usually appears as heat and contributes to the 'incidental dissipation', which has to be removed from the controller in order to ensure safe operation.

1.2 ATTRIBUTES OF AN IDEAL SWITCH

The attributes of an ideal switch are summarised as follows:
1 *Primary attributes*
   (a) the switching times of the state transitions between 'on' and 'off' should be zero,
   (b) the 'on' state voltage drop across the device should be zero,
   (c) the 'off' state current through the device should be zero,
   (d) the power–control ratio, A_{pc} (i.e. the ratio of device power handling capability to the control electrode power required to effect the state transitions) should be infinite,
   (e) the 'off' state voltage withstand capability should be infinite,
   (f) the 'on' state current handling capability should be infinite,
   (g) the power handling capability of the switch should be infinite ($PH_{max} \to \infty$).
2 *Secondary attributes*
   (a) complete electrical isolation between the control function and the power flow,
   (b) bi-directional current and voltage blocking capability.

Every switching device, from a manual switch to a fast metal–oxide–semiconductor field-effect transistor (MOSFET), is deficient in all of the above features. Different devices possess particular features in which their performance excels. It is the job of power electronics engineers to select the form of switch most suited to a particular application. The number and range of semiconductor switches available increases all the time. An awareness of the weak and strong features of the many options is as much part of the design task as is the knowledge of power electronics circuits. This selection process is covered in Section 2.4 of Chapter 2, below.

1.3 SOURCES OF INCIDENTAL DISSIPATION IN IMPERFECT SWITCHES

Practical semiconductor switches are imperfect. They possess a very low but finite on-state resistance which results in a conduction voltage drop. The off-state resistance is very high but finite, resulting in leakage current in both the forward and reverse directions depending on the polarity of the applied voltage. Furthermore, the switching-on and switching-off (i.e. commutation) actions do not occur instantaneously. Each transition introduces a finite time delay. Both switch-on and switch-off are accompanied by heat dissipation, which causes the device temperature to rise. In load control situations where the device undergoes frequent switchings, the switch-on and switch-off power losses may be added to the steady-state conduction loss to form the total 'incidental dissipation' loss, which usually manifests itself as heat. Dissipation also occurs in the device due to the control electrode action.

1.4 ESTIMATION OF SWITCHING DISSIPATION

1.4.1 Soft load – series resistance

A resistive load R_L in a semiconductor switching circuit is sometimes referred to as a soft load. A typical switching waveform is shown in Fig. 1.2. The on-state voltage drop and off-state leakage current are neglected, and the voltage v and current i both change linearly with time during each transition. During turn-off, the current and voltage undergo simultaneous transitions:

$$i = I_{max}\left(1 - \frac{t}{t_{off}}\right) = \frac{V_{bus}}{R_L}\left(1 - \frac{t}{t_{off}}\right) \qquad (1.1)$$

$$v = V_{bus}\frac{t}{t_{off}} \qquad (1.2)$$

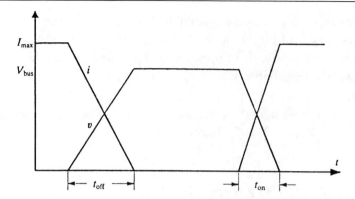

Fig. 1.2 Simplified principal electrode waveform trajectories under soft load conditions.

The switching energy loss W_{off} during such transitions can be evaluated by integrating the product of the voltage and current waveform over the time interval t_{off} as follows:

$$W_{\text{off}} = \int_0^{t_{\text{off}}} vi\,dt \tag{1.3}$$

Substituting (1.1) and (1.2) into (1.3) gives

$$W_{\text{off}} = \int_0^{t_{\text{off}}} V_{\text{bus}} \frac{t}{t_{\text{off}}} I_{\text{max}}\left(1 - \frac{t}{t_{\text{off}}}\right) dt = \frac{V_{\text{bus}} I_{\text{max}}}{t_{\text{off}}} \int_0^{t_{\text{off}}} \left(t - \frac{t^2}{t_{\text{off}}}\right) dt$$

$$= \frac{V_{\text{bus}} I_{\text{max}}}{t_{\text{off}}} \left[\frac{t^2}{2} - \frac{t^3}{3t_{\text{off}}}\right]_0^{t_{\text{off}}} = \frac{V_{\text{bus}} I_{\text{max}} t_{\text{off}}}{6} \tag{1.4}$$

Similarly, the turn-on switching energy loss can be found by interchanging the current and voltage terms in (1.1) and (1.2), giving rise to

$$W_{\text{on}} = \int_0^{t_{\text{on}}} V_{\text{bus}}\left(1 - \frac{t}{t_{\text{on}}}\right) I_{\text{max}} \frac{t}{t_{\text{on}}}\, dt = \frac{V_{\text{bus}} I_{\text{max}} t_{\text{on}}}{6} \tag{1.5}$$

At a switching frequency f, the incidental dissipation due to switching ID_S is

$$ID_{S(\text{soft})} = \frac{V_{\text{bus}} I_{\text{max}}(t_{\text{on}} + t_{\text{off}})f}{6} \tag{1.6}$$

1.4.2 Hard load – series resistance–inductance

In practice, even with R loads, stray elements arise such that greater overlap occurs between the current and voltage waveforms, resulting in greater switching dissipation than that of (1.5). In order to assess this distortion of the switching trajectories, simplified linear approximations of state transitions are depicted in Fig. 1.3. In contrast with the resistive load case, voltage transitions occur while the current is finite and constant, while current transitions occur with the voltage remaining constant at the bus value.

At turn-off, it is seen that

$$W_{\text{off}} = \int_0^{t_{\text{off}}} V_{\text{bus}} I_{\text{max}} \left(1 - \frac{t}{t_{\text{off}}}\right) dt \qquad (1.7)$$

$$= V_{\text{bus}} I_{\text{max}} \left[t - \frac{t^2}{2 t_{\text{off}}}\right]_0^{t_{\text{off}}} = \frac{V_{\text{bus}} I_{\text{max}} t_{\text{off}}}{2}$$

Similarly, the turn-on switching energy loss can be found as

$$W_{\text{on}} = \int_0^{t_{\text{on}}} I_{\text{max}} V_{\text{bus}} \left(1 - \frac{t}{t_{\text{on}}}\right) dt = \frac{V_{\text{bus}} I_{\text{max}} t_{\text{on}}}{2} \qquad (1.8)$$

At a switching frequency f, the incidental dissipation due to switching is

$$ID_{S(\text{hard})} = \frac{V_{\text{bus}} I_{\text{max}} (t_{\text{on}} + t_{\text{off}}) f}{2} \qquad (1.9)$$

Comparing hard load switching with soft load switching, the incidental dissipation due to switching is increased by a factor of 3 for each switching

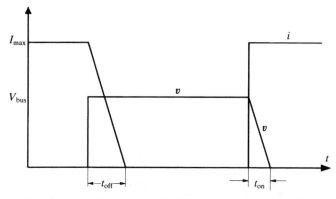

Fig. 1.3 Simplified principal electrode waveform trajectories under hard load conditions.

transition. Due to the high slew rates (i.e. fast switching times) of the switching of modern power devices, the switching trajectories of the current and voltage waveforms are often distorted by the presence of stray inductances and capacitances, which may be small compared with the load. To evaluate whether the presence of such stray elements constitute a significant effect on the switching waveform, it is convenient to relate the quoted switching time (on or off) of the device to the probable size of the stray elements. Assuming a first-order resistance–capacitance (RC) or resistance–inductance (RL) circuit transient, it will take roughly 4 times the time constant for a complete state transition (i.e. from on to off, or vice-versa). This will establish the equivalent of the half-power frequency of the stray elements for a given load resistance

$$L_{\text{stray}} \leq \frac{t_{\text{off}} R_L}{4} \tag{1.10}$$

$$C_{\text{stray}} \leq \frac{t_{\text{on}}}{4 R_L} \tag{1.11}$$

$$f_{3\,\text{dB(off)}} = \frac{2}{\pi t_{\text{off}}} ; \quad f_{3\,\text{dB(on)}} = \frac{2}{\pi t_{\text{on}}} \tag{1.12}$$

Hence, if the stray capacitance is of the order of picofarads and the stray inductance is of the order of microhenries, it can be seen that the frequency will be in the range of MHz. Also, the slowest power semiconductor switches, like the SCR, typically switch in less than 20 µs while the faster devices such as the MOSFET switch in less than 50 ns. Based on the above equations, one can calculate the effect of small parasitic values of L and/or C in the circuit. For example, if t_{on} or t_{off} is of the order of 1 µs, then f_3 is of the order of 600 kHz. Hence stray radiation occurs from such elements as well as from the lead wires and load. Unless suitable steps are taken, such radiation will cause interference problems in trigger and processing circuits. A further discussion of this phenomenon is given in Section 3.2.2.3 of Chapter 3, below.

1.5 MODIFICATION OF SWITCHING DISSIPATION – SWITCHING AIDS

Much confusion exists in the literature between 'snubbers' and 'switching aids'. This is made worse by the fact that they both enjoy the same topological location and the same circuit elements may serve both purposes. However, conceptual ambiguity should not exist. The purpose of a snubber is to protect a device against a weak feature inherent in its construction. For example, semiconductor and mechanical switches attempt to turn off too rapidly for their own good. The reasons for these effects are discussed in detail in Chapter 2 and protective snubber design is presented in Chapter 3. If the current is allowed to rise too rapidly during turn-on, beyond the device design limit, the device will be destroyed. If the voltage is allowed to rise too rapidly during turn-off, the device is likely to be spuriously triggered. Both effects can have disastrous consequences on the remaining elements of the whole system. Therefore snubbers are mandatory to prevent abuse, while switching aids are not.

Switching aids are components which are included in main electrode circuits to reduce switching dissipation in the device because the active region of the device-controlling junction is then allowed to operate at a lower temperature. The thermal stress on localised regions across the junction is thereby reduced. The use of switching aids has the advantages of

(a) improved reliability,
(b) reduced enclosure size (since this is often dominated by thermal considerations).

In some instances, significant increase in power handling capability can be realised. It must be noted that the use of switching aids does not result in an improvement of power transfer efficiency. The device energy loss associated with switching transitions in unaided operation is diverted away from the switch to other external circuit elements which are more able to cope with it.

It was shown in Section 1.4 above that the presence of inductance and capacitance in semiconductor devices and circuits can significantly increase the incidental dissipation during switching transitions. In particular, the presence of capacitance has a detrimental effect during switch-on and the presence of inductance has a detrimental effect during switch-off. The situation can be greatly improved by the use of properly designed switching aids where the reactive components are used conversely. Inductance (usually in the form of a saturable reactor) and capacitance are placed in electrical proximity to the principal electrodes of the switch so as to modify the switching performance. An appropriate use of inductance reduces the turn-on dissipation, while the use of capacitance reduces the turn-off dissipation. Switching aid design is usually aimed to ameliorate one transition only, not both switch-on and switch-off. With bipolar transistors, for example, the aim is to aid turn-off, whereas for MOSFETS it is to aid turn-on.

1.5.1 Approximate calculations of switching loss reduction

1.5.1.1 Turn-on aid

The inclusion of inductor L, with or without the clamp diode D, in series with the load resistor R_L in Fig. 1.4(a) limits the rate of rise of current after switch-on. If the switch voltage decreases linearly with time as shown in Fig. 1.4(b), then

$$v = V_{bus}\left(1 - \frac{t}{t_{on}}\right) \qquad (1.13)$$

$$i = I_{max}(1 - \varepsilon^{-t/\tau_{on}})$$
$$\text{where } I_{max} = \frac{V_{bus}}{R_L} \quad \text{and} \quad \tau_{on} = \frac{L}{R_L} \qquad (1.14)$$

Now initially, for the t_{on} interval, the current $i(t)$ is presumed to ramp up linearly in time,

$$i = I_{max}\frac{t}{\tau_{on}} = \frac{V_{bus}}{R_L}\frac{t}{\tau_{on}} \qquad (1.15)$$

Hence, the switching energy losses can be evaluated from the voltage and current relationships

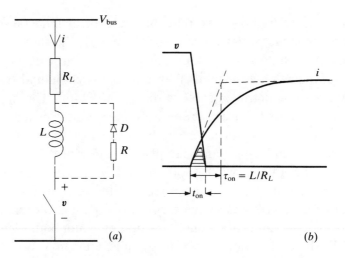

Fig. 1.4 Switch-on action. (a) Equivalent circuit, (b) transition waveforms.

$$W_{on} = \int_0^{t_{on}} vi\,dt = \int_0^{t_{on}} V_{bus}\left(1 - \frac{t}{t_{on}}\right) I_{max} \frac{t}{T_{on}}\,dt$$

$$= \frac{V_{bus}I_{max}}{T_{on}} \int_0^{t_{on}} \left(t - \frac{t^2}{t_{on}}\right) dt = \frac{V_{bus}I_{max}}{T_{on}}\left(\frac{t_{on}^2}{6}\right) \qquad (1.16)$$

$$= \frac{t_{on}}{T_{on}} \frac{V_{bus}I_{max}t_{on}}{6} = k_{on}\left[\frac{V_{bus}I_{max}t_{on}}{6}\right]$$

It was shown in (1.4) above that the square-bracketed term is the unaided energy loss with resistive load. In the presence of the switching aid, therefore

$$W_{on} = k_{on} \times [\text{Unaided turn-on loss with soft load}] \qquad (1.17)$$

where $k = t_{on}/T_{on}$, which is the fractional reduction in turn-on switching loss. The reduced loss in the presence of the inductance is illustrated by the shaded area as shown in Fig. 1.4(b).

In practice, a diode may have to be included in parallel with the inductor as indicated by the dashed line in Fig. 1.4(a) to prevent the inductance from causing a transient overvoltage exceeding V_{bus} across the switch during turn-off, which would otherwise cause destruction of the device. This is often, erroneously, referred to as a free wheel diode. The diode D merely clamps the device to no more than a diode volt-drop above the d.c. rail voltage V_{bus} as the transient overvoltage occurs. The energy stored in the inductance has to be dissipated during the off period, and a long inductor current path time constant may result due to the small intrinsic resistance of the inductor. To this end, a compromise situation of the inclusion of series resistance R is sometimes adopted together with the endurance of some excess voltage due to the volt-drop across the resistor. In this way, the rating of the diode can also be reduced. The inductor L often takes the form of a saturable reactor.

1.5.1.2 Turn-off aid

In Fig. 1.5(a), a capacitor C is inserted in parallel with the switching device as a turn-off aid. During the turn-off interval t_{off}, the switch current is presumed to reduce linearly in time,

$$i(t) = I_{max}\left(1 - \frac{t}{t_{off}}\right) = \frac{V_{bus}}{R_L}\left(1 - \frac{t}{t_{off}}\right) \qquad (1.18)$$

The voltage across the open switch is presumed to increase with the simplest form of curved characteristic, namely a parabola,

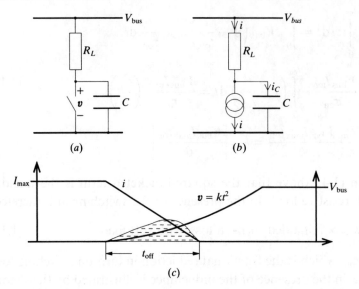

Fig. 1.5 Turn-off action (a), (b) equivalent circuits, (c) transition waveforms.

$$v(t) = k_v t^2 \tag{1.19}$$

so that

$$\frac{dv}{dt} = 2k_v t \tag{1.20}$$

In the circuit of Fig. 1.5, it is seen that

$$V_{bus} = v + \left(i + C\frac{dv}{dt}\right) R_L \tag{1.21}$$

Combining (1.19), (1.20) and (1.21) gives

$$V_{bus} = k_v t^2 + \left[\frac{V_{bus}}{R_L}\left(1 - \frac{t}{t_{off}}\right) + 2k_v Ct\right] R_L \tag{1.22}$$

Re-arranging (1.22) results in

$$V_{bus} = k_v(2\tau_{off} + t)t_{off} \quad \text{where} \quad \tau_{off} = R_L C \tag{1.23}$$

Since t is relatively small compared with $2\tau_{off}$,

$$k_v \approx \frac{V_{bus}}{2\tau_{off} t_{off}} \tag{1.24}$$

1.5 Modification of switching dissipation – switching aids

and

$$v(t) = \frac{V_{bus} t^2}{2\tau_{off} t_{off}} \tag{1.25}$$

The switching energy loss based on the v–i characteristics of Fig. 1.5(c) is

$$W_{off} = \int_0^{t_{off}} vi\, dt = \int_0^{t_{off}} \frac{V_{bus} t^2}{2\tau_{off} t_{off}} I_{max}\left(1 - \frac{t}{t_{off}}\right) dt$$

$$= \frac{V_{bus} I_{max}}{2\tau_{off} t_{off}} \int_0^{t_{off}} \left(t^2 - \frac{t^3}{t_{off}}\right) dt = \frac{V_{bus} I_{max}}{2\tau_{off} t_{off}} \left[\frac{t^3}{3} - \frac{t^4}{4 t_{off}}\right]_0^{t_{off}} \tag{1.26}$$

$$= \frac{V_{bus} I_{max}}{2\tau_{off} t_{off}} \frac{t_{off}^3}{12} = \frac{V_{bus} I_{max} t_{off}}{24}\left[\frac{t_{off}}{\tau_{off}}\right]$$

Comparing (1.26) with (1.6) shows that

$$W_{off} = \left[\frac{V_{bus} I_{max} t_{off}}{2}\right]\left[\frac{t_{off}}{12\tau_{off}}\right]$$

$$= [\text{Unaided turn-off loss with hard load}] \times \left[\frac{k_{off}}{12}\right] \tag{1.27}$$

where $k_{off} = t_{off}/\tau_{off}$.

Equation (1.27) shows that the presence of the capacitance C ensures a 1/12 reduction of turn-off energy even if the turn-off time is as slow as $t_{off} = \tau_{off}$. With faster turn-off, a still greater reduction of W_{off} can be realised. Rigorous analysis gives a 1/14.5 reduction of switching loss so that the 1/12 figure of (1.27) is conservative.

In practice, the presence of a capacitor as a turn-off aid will result in increased incidental dissipation during the turn-on period, as the capacitor charged energy must be discharged before the next turn-off. A discharge resistor R_{dis} may be added, with a shorting diode D, as shown in Fig. 1.6.

Again, it should be noted that the effect of the turn-off switching aid is to re-distribute the energy loss from the device to the discharge resistor R_{dis} and hence excessive heat dissipation in the device is alleviated. The resistance value for R_{dis} is determined by two factors. The capacitor current path time constant must be chosen so that the capacitor completes its discharge within the minimum device on-state time $t_{on\text{-state}}$. Normally $t_{on\text{-state}} = 4 R_{dis} C$ is sufficient to ensure proper operation and forms the basis for the minimum on-state time of the device. However, R_{dis} must have a minimum value so as to dissipate the capacitor energy of $\frac{1}{2} C V_{bus}^2$ with a limited initial capacitor

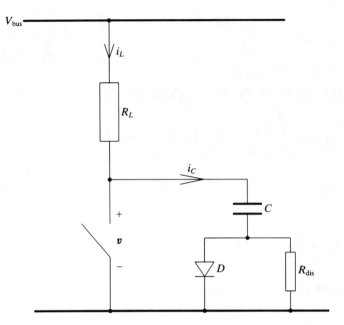

Fig. 1.6 Switching aid for semiconductor turn-off.

discharge current of V_{bus}/R_{dis}. Normally, if the initial capacitor current at turn-on, $i_{C_{max}}$ is limited to 20% of I_{max}, the device should operate within its safe operation area, as defined in Section 2.1, of Chapter 2, below. A numerical example of the design of a turn-off aid is given in Example 1.1, below.

1.5.2 Detailed calculation of switching loss reduction

The design criteria described in the preceding section are adequate for most first-order approximation calculations. In many instances, only a single aid is used either for turn-on or, more often, for turn-off. In some applications, however, a turn-on inductor L and a turn-off capacitor C need to be used simultaneously and they then interact with each other. A more detailed result of the circuit action is given here which can be used if such a procedure is required.

By the inclusion of a capacitor and associated auxiliary components the switch voltage v can be allowed to rise gradually. The rise of v is controlled not only by the switch current, but by the network shown in Fig. 1.7(a). Because of the gradual rise of voltage v, the incidental dissipation in the switch may be calculated using the equivalent circuit of Fig. 1.7(b).

1.5 Modification of switching dissipation – switching aids

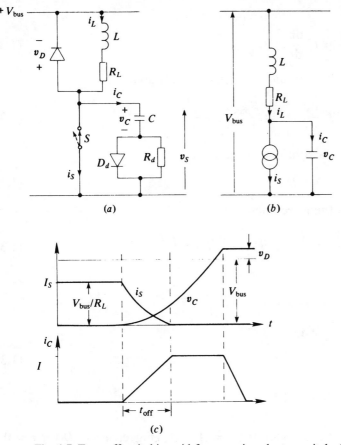

Fig. 1.7 Turn-off switching aid for a semiconductor switch: (*a*) circuit arrangement, (*b*) equivalent circuit during turn-off, (*c*) waveforms.

Switch current $i_S(t)$ is given by

$$i_S = I_S\left(1 - \frac{t}{t_{\text{off}}}\right) = i_L - i_C \tag{1.28}$$

where I_S is the steady-state switch current.

Now

$$V_{\text{bus}} = v_S + (i_S + i_C)R + L\frac{\mathrm{d}}{\mathrm{d}t}(i_S + i_C) \tag{1.29}$$

But

$$I_S = \frac{V_{\text{bus}}}{R} \tag{1.30}$$

and

$$i_C = C\frac{dv_C}{dt} = C\frac{dv}{dt} \tag{1.31}$$

Substituting (1.28), (1.30) and (1.31) into (1.29) gives

$$LC\frac{d^2v}{dt^2} + RC\frac{dv}{dt} + v = \frac{V_{bus}}{t_{off}}\left(t + \frac{L}{R}\right) \tag{1.32}$$

Now let (1.32) have a solution of the form

$$v = A\varepsilon^{Bt} + C't + D \tag{1.33}$$

The LHS of (1.32) then becomes

$$LCA\varepsilon^{Bt}\left(\frac{1}{LC} + \frac{R}{L}B + B^2\right) + C't + CRC' + D \tag{1.34}$$

where

$$\begin{aligned} B &= -\frac{R}{2L} \pm \sqrt{\frac{R^2}{4L^2} - \frac{1}{LC}} \\ C' &= \frac{V_{bus}}{t_{off}} \\ D &= \frac{V_{bus}}{t_{off}}\left(\frac{L}{R} - CR\right) = -A \end{aligned} \tag{1.35}$$

At $t = 0$, from (1.33), it is seen the $v = A + D$. Hence, $A = -D$ for $v = 0$ at $t = 0$. For conditions where the circuit is less than critically damped (i.e. when $4\tau_2 > \tau_1$) it is found that

$$v = \frac{V_{bus}}{t_{off}}\left[\varepsilon^{-t/2\tau_2}\left\{\frac{\pm 2\tau_2 \cos(\omega t - \theta)}{\sqrt{4 - \tau_1/\tau_2}}\right\} + t + \tau_2 - \tau_1\right] \text{ for } 0 \leq t \leq t_{off}$$

$$\tag{1.36}$$

where

$$\tau_1 = R_L C, \quad \tau_2 = \frac{L}{R_L} \tag{1.37}$$

$$\tan\theta = \frac{(\tau_1 - 3\tau_2)\sqrt{\tau_1}}{(\tau_1 - \tau_2)\sqrt{4\tau_2 - \tau_1}} \tag{1.38}$$

In (1.36), the positive sign is used when $\tau_1 \geq \tau_2$, otherwise the negative sign should be used. The energy in joules stored in the switching device during turn-off is, in general,

$$W_{\text{off}} = \int_0^{t_{\text{off}}} v i_S \, dt = \int_0^{t_{\text{off}}} \frac{V_{\text{bus}} I_S}{t_{\text{off}}} \left(1 - \frac{t}{t_{\text{off}}}\right) f(t) \, dt \qquad (1.39)$$

For the condition of underdamping, which is the practical condition of prime concern, the turn-off energy is obtained when $f(t)$ in (1.39) is equated to the square-bracketed term of (1.36). Evaluation of the energy equation (1.39) is very tedious, but gives the result

$$\begin{aligned} W_{\text{off}} = V_{\text{bus}} I_S &\left[\pm \frac{\sqrt{\tau_1 \tau_2}\, \tau_2 \varepsilon^{-t_{\text{off}}/2\tau_2}}{\omega t_{\text{off}}^2} \cos(\omega t_{\text{off}} - \alpha) \right] \\ &+ V_{\text{bus}} I_S \left[\frac{t_{\text{off}} + 3(\tau_2 - \tau_1)}{6} \right. \\ &\left. - \frac{\tau_1}{t_{\text{off}}^2} \{\tau_1^2 + \tau_2^2 - 3\tau_1\tau_2 + t_{\text{off}}(2\tau_2 - \tau_1)\} \right] \end{aligned} \qquad (1.40)$$

where

$$\omega = \sqrt{\frac{1}{\tau_1 \tau_2} - \frac{1}{4\tau_2^2}} \qquad (1.41)$$

and

$$\tan \alpha = \frac{(5\tau_2^2 - 5\tau_1\tau_2 + \tau_1^2)\sqrt{\tau_1}}{(\tau_1^2 + \tau_2^2 - 3\tau_1\tau_2)\sqrt{4\tau_2 - \tau_1}} \qquad (1.42)$$

In (1.40), the positive sign is applicable when $(\tau_1 - \tau_2)^2 \geq \tau_1\tau_2$ and the negative sign applies when $(\tau_1 - \tau_2)^2 < \tau_1\tau_2$.

The energy loss at turn-off is highly dependent on the damping state of the circuit. The condition of critical damping (i.e. $4\tau_2 = \tau_1$) is satisfied when $L = R^2 C/4$. If τ_1 is made equal to t_{off}, the turn-off energy is $W_{\text{off}} \approx (V_{\text{bus}} I_S)/26$. A summary of the use of turn-off aids is given in Fig. 1.8.

1.6 ESTIMATION OF TOTAL INCIDENTAL DISSIPATION

Power switching devices dissipate power in the form of heat at all times when connected to an external load and supplied with a voltage supply source V_{bus}. Let δ be the duty cycle of the switching device. The on-state and off-state dissipation are given in (1.43), where I_{max} is the maximum load current when the device is turned on, I_{leakage} is the leakage current when the device is turned off and V_{on} is the on-state voltage drop across the device.

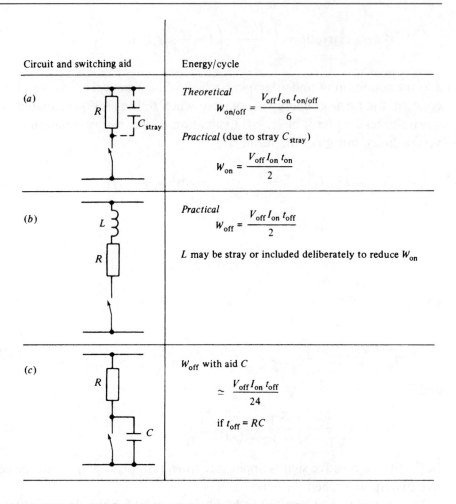

Fig. 1.8 Use of turn-off aids to reduce switching losses.

$$ID_{\text{on-state}} = \delta V_{\text{on}} I_{\text{max}}$$
$$ID_{\text{off-state}} = (1 - \delta) V_{\text{bus}} I_{\text{leakage}} \qquad (1.43)$$

The control electrode dissipation, ID_{CE}, varies enormously for different devices, but may be represented, in general, by

$$ID_{\text{CE}} = \frac{\text{Power handling capability }(PH)}{\text{Power–control ratio }(A_{pc})} \qquad (1.44)$$

The total incidental dissipation ID may therefore be expressed as a summation

$$ID = ID_S + ID_{\text{on-state}} + ID_{\text{off-state}} + ID_{\text{CE}} \qquad (1.45)$$

If this dissipation is not removed by heat transference, the temperature rises and, in consequence, the device is destroyed. The simple formula (1.44) is based on the following assumptions:
   (a) the switching times are an insignificant fraction of the periodic on or off time of a device,
   (b) negligible leakage current during switching,
   (c) negligible on-state voltage drop during switching.

If either $v = 0$ or $i = 0$ for the whole of either (or both) switching transitions the ID and stress on the chip is drastically reduced. Such switching conditions can be realised by circuit design or by control electrode signal processing but could result in loss of power and intolerable waveform distortion.

1.7 TRANSFER OF INCIDENTAL DISSIPATION TO AMBIENT – THERMAL CONSIDERATIONS

Figure 1.9(a) depicts the attachment of the encapsulated switch to its heat sink. A detailed description of the internal structure of a silicon controlled rectifier (SCR) switch is given in Section 2.3.2.1(D). It should be noted, in particular, that the semiconductor crystal must be secured by a good thermal bond to the attachment header. Even when an epoxy encapsulation is used in conjunction with lug mounting, the principal objective remains good thermal bonding. This form of bonding provision is not made in signal-level devices, in which physical handling is the prime consideration.

The incidental dissipation which arises in the device occurs mainly from normal operation and is expressed empirically in (1.45). This results in temperature increase at a rate determined by the thermal capacity of the various elements of the device and its associated attachments. The temperature rise in the semiconductor is moderated or limited within safe operation if the heat dissipated can be transferred to the ambient medium in which the device is situated.

In silicon semiconductor devices the main controlling junction can rise in temperature to such a level that device failure results. The precise mechanisms of failure are varied and complex. Further, when device destruction occurs, the physical evidence of the failure is nearly always destroyed so that diagnosis of the exact cause is often impossible. This also applies for most other forms of device abuse.

Simplified thermal equivalent circuits, such as shown in Fig. 1.9(b), may be used in the thermal management of the device, where $T(°C) \equiv$ voltage,

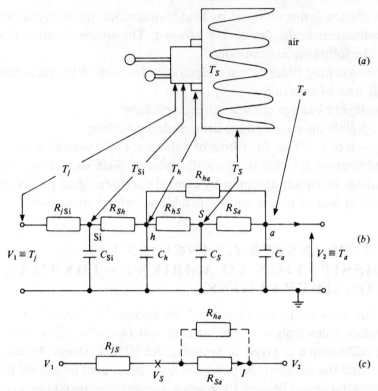

Fig. 1.9 Electrical analogue of heat sink operation: (*a*) finned heat sink layout, (*b*) equivalent circuit, (*c*) simplified equivalent circuit (*jSi* = junction → bulk silicon, *Sih* = silicon → header, *hs* = header → sink, *Sa* = sink → air, *ha* = header (case) → air, T_S = sink temperature).

$\sum ID$ (watt-second) ≡ current, $C_T \equiv$ thermal capacity (watt-second/°C), $R_T \equiv$ thermal resistance(°C/watt). In practice, it is not possible to separate C_T and R_T. They both represent transient effects that are continuous in some parts and discontinuous in others. Also, such transient storage effects must be considered in empirically derived form. For thermal steady-state operation all of the capacitor C_T may be eliminated, resulting in a simpler equivalent circuit Fig. 1.9(*c*).

The frequently used term 'heat sink' is a partial misnomer. Whilst it is capable of absorbing transient energy the main purpose of a heat sink is to enhance the steady heat flow path from the encapsulation of the device to the ambient medium. Heat sinks are usually made of finned, extruded aluminium, as in Fig. 1.9(*a*), and usually operate by natural convection of the ambient air. Blown air or water cooling is used where a very large heat flow is required. Equivalent circuits can be used to model the heat flow and calculate the sizes of heat sink required. Because of the relative thermal

1.7 Transfer of incidental dissipation to ambient – thermal considerations

conductivities the thermal resistance R_{Sa} from sink to air is very large compared with the thermal resistance R_{hS} from the header to the sink. Also the unaided resistance $R_{ha} \gg (R_{hS} + R_{Sa})$ for a power device. Therefore R_{hS} and R_{ha} are normally neglected. In Fig. 1.9 the voltages are analogues of the temperature while the currents are analogues of the heat power flow. Referring to Fig. 1.9(c)

$$\frac{V_j - V_S}{I} = R_{jS} = \frac{T_j - T_S}{P} = \text{thermal resistance} \qquad (1.46)$$
$$= \text{case derating factor in } °C/W$$

Under transient conditions C_{Si} and C_h play a very significant role and often permit the device to have a relatively high surge rating. Under steady-state thermal conditions R_{jSi}, R_{Sih} and R_{hS} are lumped in series and their combined value R_{jS} for a given device is quoted in the literature for $T_{j\max}$, often in the form of a derating curve, Fig. 1.10. The only parameter over which the user normally has any control is the thermal resistance R_{ha} from the case to air. This must be arranged to be sufficiently small to prevent the junction temperature T_j exceeding its maximum permissible value $T_{j\max}$. Furthermore, a temperature difference must exist between the sink and the ambient medium otherwise a transfer of heat energy will not take place.

Power devices have either a lug or stud attached to the silicon supporting structure (header) to facilitate good thermal contact. No such provision is made for a signal-level device. Heat sinks are commercially available and the R_{Sa} values for them are usually quoted per unit length. A very simple method of mounting a device is to use a vertical square plate of copper or aluminium. An empirically derived value for the heat sink–air thermal resistance R_{Sa} under conditions of natural convection is then found to be

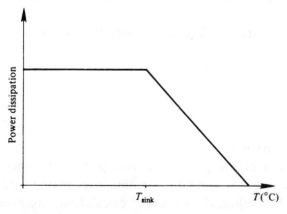

Fig. 1.10 Typical case derating characteristic.

Table 1.1 *Thermal resistivities.*

Substance	Resistivity (SI)
Al	0.48
Cu	0.26
Mica	100–400
Silicone grease	400
PTFE	400
Air	3800

$$R_{Sa} = \frac{0.5}{L}, \text{ where } L \text{ is the side length in metres} \tag{1.47}$$

The corresponding surge impedance, for a 10 cm square plate, is

$$Z_{Sa} \simeq 5(1 - \varepsilon^{-t/200}) \tag{1.48}$$

where $t =$ time in seconds of the heat pulse.

The thermal resistivity of some relevant materials is given in Table 1.1.

The area of contact between the device and the heat sink must be very clean and also be as large as possible. If the device and the sink are in electrical contact the heat sink then possesses the electrical potential of the stud anode (or cathode). If electrical isolation is needed between the device and the sink this is usually achieved by the use of mica or PTFE washers. In addition to being electrical insulators, both mica and PTFE have high thermal resistivity, Table 1.1. The heat insulation effect of this is mitigated, to some extent, by the fact that the area of contact is large compared with a simple metal-to-metal contact.

If forced convection is used, by blowing air over the heat sink, (1.47) becomes modified to

$$R_{bSa} = R_{Sa}\sqrt{\frac{2}{3V}} \tag{1.49}$$

where V is the net air velocity in m/s.

With these equations and knowing the maximum permissible operating temperature and the value of the incidental dissipation, the minimum dimensions of a heat sink can be evaluated. Numerical calculations are given in Examples 1.6 and 1.7 below.

1.8 WORKED EXAMPLES

Example 1.1
A power switch has turn-off times of 1 μs when switching a 10A highly inductive load at 1 kHz on a 100V d.c. rail. Design the turn-off aid such that it completes its function as the switch completes its switching action. The duty cycle of the switch varies between 5% to 95%.

Solution. The device switching waveforms and the capacitor current waveform are shown in Fig. 1.11.

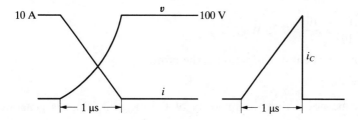

Fig. 1.11 Switching characteristics in Example 1.1

The capacitance required can be found by

$$C = \frac{1}{V_{bus}} \int_0^{t_{off}} i_C \, dt = \frac{1}{V_{bus}} \left[\frac{I_{max} t_{off}}{2} \right] = \frac{10 \times 10^{-6}}{100 \times 2} = 50 \text{ nF}$$

To ensure that C has discharged completely, from Section 1.5.1.2 above,

$$t_{\text{on-state}} = 4 R_{dis} C$$

Hence,

$$R_{dis} = \frac{t_{\text{on-state}}}{4C} = \frac{5}{100} \times \frac{1}{1000} \times \frac{1}{4 \times 50 \times 10^{-9}} = 250 \, \Omega$$

To evaluate the power rating P of R_{dis},

$$P = \frac{1}{2} C V_{bus}^2 f = \frac{1}{2} \times 50 \times 10^{-9} \times 100^2 \times 10^3 = 0.25 \text{ W}$$

Example 1.2
A semiconductor switch with a resistive load switches at 10 kHz taking 1 μs to switch on and 5 μs to switch off. The on-state current is 10 A and the switch saturation voltage is 150 mV when the supply rail is 100 V. If the device conducts for one-half of the periodic time, calculate the switching and on-state dissipations.

22 Power switching theory

Solution. The circuit is shown in Fig. 1.8(a). Using the practical equations (1.7), (1.8) it is found that the switching energy dissipated per cycle is

$$W_{on} = \frac{V_{bus}I_{sat}t_{on}}{2} = \frac{100 \times 10}{2} \times 10^{-6} = 500\,\mu J$$

$$W_{off} = \frac{V_{bus}I_{sat}t_{off}}{2} = \frac{100 \times 10 \times 5 \times 10^{-6}}{2} = 2500\,\mu J$$

The total rate of energy dissipation or switching power, at 10 kHz, is

$$P_{on} + P_{off} = P_{switch} = \frac{3000}{10^6} \times 10^4 = 30\,W$$

When the collector–emitter voltage saturates, for the steady-state 'on' condition,

$$P_{sat} = \frac{150}{10^3} \times \frac{10}{2} = 0.75\,W$$

The total incidental dissipation is therefore

$$P_{total} = 30.75\,W$$

It is to be noted that the majority of the incidental loss is generated by switching.

Example 1.3
A switching aid is to be used in the circuit of Example 1.2 to drastically reduce the switching loss. Deduce suitable component values if the load resistance is 10 Ω.

Solution. Let the circuit of Fig. 1.7(a) be used as a switching aid. The load current is

$$I_S = \frac{100}{10} = 10\,A$$

Using the criterion of (1.23) the capacitor C has the value

$$C = \frac{\tau}{R} = \frac{t_{off}}{R} = \frac{5}{10^6 \times 10} = 0.5\,\mu F$$

In the turn-on switching interval the rise of the switch current and the fall of the switch voltage can both be approximated by linear characteristics, as shown in Fig. 1.2. From (1.5),

$$W_{on} = \frac{V_{off}I_{on}}{t_{on}^2} \int_0^{t_{on}} t(t_{on} - t)\,dt$$

$$= \frac{V_{off}I_{on}}{t_{on}^2} \left[\frac{t^2 t_{on}}{2} - \frac{t^3}{3}\right]_0^{t_{on}} = \frac{V_{off}I_{on}t_{on}}{6}$$

The on-current is given by

$$I_{on} = \frac{Vt}{L}$$

Let the added inductor L have the arbitrary value defined by the following relationship (but note that almost any value of L will suffice):

$$\frac{L}{R} = t_{on}$$

In this case, therefore,

$$L = 10 t_{on} = 10 \times 10 \times 1 \times 10^{-6} = 100\ \mu H$$

After 1 μs the value of I_{on} has risen from zero to

$$I_{on|1\mu s} = \frac{100 \times 10^{-6}}{100 \times 10^{-6}} = 1\ A$$

The switch-on energy loss is therefore

$$W_{on} = \frac{100 \times 1 \times 10^{-6}}{6} = 16.7\ \mu J$$

which represents a great reduction from the unaided value of 500 μJ. At a switching frequency of 10 kHz,

$$P_{on} = \frac{16.7}{10^6} \times 10^4 = 0.167\ W$$

The critical damping value of inductance is found, Section 1.5.2, from

$$L_{crit} = \frac{R^2 C}{4} = \frac{100 \times (0.5)}{4 \times 10^6} = 12.5\ \mu H$$

The inserted value of 100 μH is obviously much greater than the critical value (which is typically true) and justifies the use of the practical equation (1.7) for turn-off, rather than the theoretical equation (1.3). The use of the turn-off aid circuit results in considerable reduction of the turn-off switching loss. If the combination of L, C and R gives (say) a tenfold reduction of W_{off} then the total power becomes

$$P_{total} = P_{sat} + P'_{on} + P_{off}$$
$$= 0.75 + 0.167 + 2.5$$
$$= 3.42\ W$$

This is roughly one-tenth of the unaided incidental loss.

Example 1.4
A MOSFET used as a switch, Fig. 1.12, has parameters $V_S = 30\ V$, $I_D = 30\ A$, $R_{DS} = 25\ m\Omega$, $V_{GS} = 15\ V$, $t_{d(on)} = 25\ ns$, $t_r = 80\ ns$, $t_{d(off)} = 70\ ns$, $t_f = 30\ ns$ and $f_S = 25\ kHz$. The drain-source leakage current

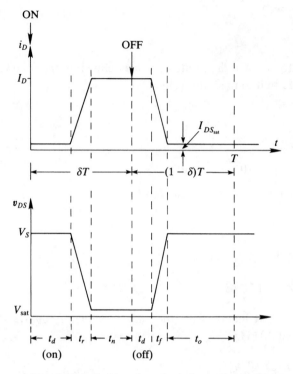

Fig. 1.12 Switching characteristics in Example 1.4.

$I_{DS_{sat}} = 200$ μA and the duty cycle $\delta = 50\%$. Determine the power dissipation due to the drain current, (a) during turn-on, $t_{on} = t_{d(on)} + t_r$, (b) during the conduction period t_n, (c) during turn-off, $t_{d(off)} + t_f$, (d) after turn-off t_o, and (e) the total average power loss during the state transitions.

Solution.

$$T = \frac{1}{f_S} = \frac{1}{25 \times 10^3} = 40 \times 10^{-6} \text{ s}$$

$\delta = 0.5$ so that $\delta T = 20 \times 10^{-6}$ s

Referring to Fig. 1.12,

$t_n = 20 - 0.025 - 0.08$
$\quad = 19.895$ μs

$t_o = (1 - \delta)T - t_{d(off)} - t_f$
$\quad = 20 - 0.07 - 0.03$
$\quad = 19.9$ μs

(a) During turn-on
For the delay period $t_{d(\text{on})}$

$$P_d = \frac{1}{T}\int_0^{t_d} I_{DS_{\text{sat}}} V_S \, dt$$

$$= I_{DS_{\text{sat}}} V_S t_d f_s$$
$$= 200 \times 10^{-6} \times 30 \times 25 \times 10^{-9} \times 25 \times 10^3$$
$$= 3.75 \, \mu W$$

For the turn-on rise time t_r,

$$P_d = \frac{1}{T}\int_0^{t_r}\left[V_S\left(1-\frac{t}{t_r}\right)+V_{\text{sat}}\right]\left[I_D\frac{t}{t_r}+I_{DS_{\text{sat}}}\right]dt$$

$$= \frac{1}{T}\int_0^{t_r}\left[V_S I_D\left(1-\frac{t}{t_r}\right)\frac{t}{t_r}+V_S I_{DS_{\text{sat}}}\left(1-\frac{t}{t_r}\right)+V_{\text{sat}}I_D\frac{t}{t_r}+V_{\text{sat}}I_{DS_{\text{sat}}}\right]dt$$

$$= \frac{1}{T}\left\{V_S I_D\left[\frac{t^2}{2t_r}-\frac{t^3}{3t_r^2}\right]_0^{t_r}+V_S I_{DS_{\text{sat}}}\left[t_r-\frac{t^2}{2t_r}\right]_0^{t_r}+V_{\text{sat}}I_D\left[\frac{t^2}{2t_r}\right]_0^{t_r}+V_{\text{sat}}I_{DS_{\text{sat}}}[t]_0^{t_r}\right\}$$

$$= \frac{1}{T}\left[\frac{V_S I_D t_r}{6}+\frac{V_S I_{DS_{\text{sat}}} t_r}{2}+\frac{V_{\text{sat}}I_D t_r}{2}+V_{\text{sat}}I_{DS_{\text{sat}}} t_r\right]$$

If V_{sat} is negligibly small,

$$P_d = 25 \times 10^3\left[\frac{30 \times 30 \times 80 \times 10^{-9}}{6}+\frac{30 \times 200 \times 10^{-6} \times 80 \times 10^{-9}}{2}\right]$$
$$= 25 \times 10^3[12 \times 10^{-6}+240 \times 10^{-12}]$$
$$= 300.006 \times 10^{-3} = 0.3 \, W$$

$\therefore P_{\text{turn-on}} = 3.75 \times 10^{-6}+0.3 = 0.3 \, W$

(b) During the conduction period t_n,

$i_{DS} = I_D$

$v_{DS} = V_{\text{sat}} = 0$

$\therefore P_{\text{loss}} = 0$

(c) During turn-off
For the delay period $t_{d(\text{off})}$

$$P_d = I_D V_{\text{sat}} t_{d(\text{off})} f_s = 0$$

because $V_{\text{sat}} = 0$

For the fall period t_f,

$$P_d = \frac{1}{T} \int_0^{t_f} \left[I_D \left(1 - \frac{t}{t_f}\right) + I_{DS_{sat}} \right] \left[V_S \frac{t}{t_f} + V_{sat} \right] dt$$

$$= \frac{1}{T} \left[\frac{V_S I_D t_f}{6} + \frac{V_{sat} I_D t_f}{2} + \frac{V_S I_{DS_{sat}} t_f}{2} + V_{sat} I_{DS_{sat}} \right]$$

Neglecting the V_{sat} terms,

$P_d = 25 \times 10^3 [12 \times 10^{-6} + 240 \times 10^{-12}]$

$= 0.3$ W

(d) *During the off period t_o,*

$P_d = V_S I_{DS_{sat}} t_o f_s$

$= 30 \times 200 \times 10^{-6} \times 19.9 \times 10^{-6} \times 25 \times 10^3$

$= 0.003$ W

(e) *Total transitional power loss*

$P_{\text{total}} = (a) + (b) + (c) + (d)$

$= [3.75 \times 10^{-6} + 0.3] + [0] + [0.3] + [0.003]$

$= 0.603$ W

Example 1.5

Calculate the switch-off power loss for a switch operating at 1 kHz with the turn-off waveform shown in Fig. 1.13.

Solution. In Fig. 1.13 the projection of the falling current intersects the time axis at a value t_1, where

$$t_1 = 0.33 \times \frac{100}{90} = 0.36 \ \mu s$$

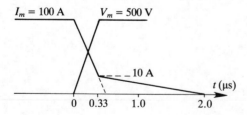

Fig. 1.13 Turn-off waveform in Example 1.5.

Then

$$W_{\text{off}} = \int_0^{0.33} \frac{V_m t}{0.33} \times I_m\left(1 - \frac{t}{0.36}\right) dt + \int_0^{1.67} V_m \times 10\left(1 - \frac{t}{1.67}\right) dt$$

$$= \frac{V_m I_m}{0.33} \int_0^{0.33} \left(t - \frac{t^2}{0.36}\right) dt + V_m \times 10 \int_0^{1.67} \left(1 - \frac{t}{1.67}\right) dt$$

$$= \frac{500 \times 100}{0.33 \times 10^{-6}} \left[\frac{t}{-2} - \frac{t^3}{3 \times 0.36 \times 10^{-6}}\right]_0^{0.33 \times 10^{-6}}$$

$$+ 500 \times 10 \times \left[t - \frac{t^2}{2 \times 1.67 \times 10^{-6}}\right]_0^{1.67 \times 10^{-6}}$$

$$= (3.2 + 4.17) \text{ mJ} = 7.38 \text{ mJ}$$

$$P_{\text{off}} = W_{\text{off}} \times f$$
$$= (7.38 \times 10^{-3})(1 \times 10^3) = 7.38 \text{ W}$$

Example 1.6
A power transistor develops a power loss of 3.42 W and is mounted on a square heat sink. This transistor is linearly case derated from 20 °C to 200 °C at 40 °C/W. Calculate the sink temperature. Evaluate the minimum dimensions of the heat sink to suitably enhance the thermal conductivity to the 20 °C ambient medium.

Solution. An electrical equivalent analogue of the heat sink is given in Fig. 1.9. In this example
$V_1 \equiv 200 \,°\text{C}$
$V_2 \equiv 20 \,°\text{C}$
$V_S \equiv$ sink temperature
$I \equiv$ device power dissipation, P
Then, from the device to the sink, in (1.46),

$$\frac{I}{V} = \frac{P}{T_1 - T_S} = \frac{3.42}{200 - T_S} = \frac{1}{40} = \frac{1}{R_{jS}}$$

which gives

$T_S = 63.2 \,°\text{C}.$

From the sink to the air, in Fig. 1.9,

$$\frac{T_s - 20}{3.42} = \frac{43.2}{3.42} = 126 \,\Omega = R_{Sa}$$

Using the empirical relationship (1.47) gives

$$L = \frac{0.5}{R_{Sa}} = \frac{0.5}{12.6} \times 100 = 3.97 \simeq 4\,\text{cm}$$

Example 1.7
An SCR device has a thermal resistance of 1.5 °C/W from junction to air (including the heat sink). The ambient temperature is 25 °C.
(a) If the specified maximum junction temperature $T_{j_{\text{max}}}$ is 125 °C calculate the maximum power loss of the device.
(b) If the transient thermal impedance is 0.05 °C/W for a surge of 10 ms duration, what is the maximum power dissipation that the device will withstand without exceeding its $T_{j_{\text{max}}}$?

Solution.
(a) From (1.46), the steady-state power loss is

$$P = \frac{T_j - T_a}{R_{ja}} = \frac{125 - 25}{1.5} = 66.7\,\text{W}$$

(b) The transient thermal impedance Z_{ja} is defined as

$$Z_{ja} = \frac{\text{temperature rise}}{\text{power loss in the specified time}}$$

$$\therefore P = \frac{125 - 25}{0.05} = 2000\,\text{W}$$

Note that the value of Z_{ja} is often quoted in a manufacturer's specification for a given device.

1.9 REVIEW QUESTIONS AND PROBLEMS

Switching action and incidental losses
1.1 List the attributes of an ideal switch.

Derive the formula for the switching loss in an imperfect switch and give an expression for the incidental dissipation.
 Show how the formula is usually modified for high-speed switching when stray and encapsulation reactances cause degradation of the switch.
1.2 With the aid of main electrode waveforms, derive formulae for the switching dissipation which occurs in an imperfect switch and give an approximate formula for the total incidental dissipation.
 Show how, and briefly state why, this formula becomes modified when the slew rates are high.

1.3 Show that the theoretical unaided turn-off switching loss for a semiconductor switch with resistive load is given by (1.3).

1.4 A semiconductor switch with resistive load R is shunted by an ideal capacitor C. Show that the turn-off switching loss is given by (1.26).

1.5 If a practical switch has linear decrements and increments of current and voltage during finite actuation times, t, derive an expression for the incidental dissipation during each state change.

By means of waveform diagrams representing the behaviour of a fast power bipolar transistor switching a resistive load, show that degradation results and give a more realistic expression for the incidental dissipation. Why is the turn-off loss greater than the turn-on loss?

Switching aids

1.6 Explain why switching aids are advantageous in solid-state switching circuits.

Show where they are connected, including the ancillary components which are additionally required.

With the aid of electrode waveform diagrams of the loci before and after the inclusion of switching aids, illustrate how they function.

1.7 A bipolar transistor circuit has a resistive load $R = 10\,\Omega$ and uses a rail voltage of 100 V. The collector–emitter saturation voltage is negligible. The transistor turn-off time is 6 μs and the switching frequency is 20 kHz. An ideal capacitor C is connected across the transistor as a switching aid. Calculate an appropriate value for C and the corresponding value of the switch-off dissipation compared with unaided operation. Assume $\tau_{off} = t_{off}$.

1.8 A power bipolar transistor switches a 15 Ω load to a 150 V supply. It has a saturation voltage of 1 V. The duty cycle of the 20 kHz operation is 0.6. Evaluate the incidental dissipation in the device and its efficiency if the two-state transition times are 1 and 4 μs.

1.9 A bipolar transistor is used to control power from a 100 V supply to a 10 Ω resistance and is heavily saturated, so that when 'on' the voltage across the device is 200 mV. It is switched at 1 kHz with a duty cycle of 50%. The turn-on time is 2 μs and the turn-off time is 10 μs. Calculate the incidental dissipation.

Show how extra components could be connected to drastically reduce the turn-off dissipation, and, with the aid of voltage and current waveforms, explain their purposes.

Thermal considerations and heat sinks

1.10 Explain what is meant by the term 'thermal derating'. A device is case rated at 50 W and is linearly derated from 50 to 200 °C, the latter being the maximum junction temperature. What is the thermal resistance to case of the device?

The device is to be screwed directly to commercially available extruded aluminium heat sinking which has a thermal conductivity of 3.0 W/°C per metre. If the device has to dissipate 15 W, find the minimum length of extrusion which must be purchased and what the case temperature will be if the ambient temperature is 20 °C.

1.11 A case rated device of 50 W maximum dissipation at 25 °C has a maximum permissible junction temperature of 175 °C. Obtain a value of thermal resistance, junction to sink, of the device.

The device is to be operated at 35 W. If the stud can be assumed to be at the same temperature as the heat sink, use any empirical formula you know to evolve the dimensions of a heat sink which will avoid the device being overheated, if the ambient temperature of the air is 20 °C.

What is the temperature of the heat sink?

1.12 A power transistor has a thermal resistance of 20 °C/W (device to sink) for steady-state operation following initial switch-on. The ambient temperature is 25 °C. What size of square heat sink is required to dissipate 4 W if the maximum permitted junction temperature is 120 °C? What is the steady-state power dissipation at the maximum permitted temperature, with and without the heat sink?

This power transistor has a surge thermal impedance of 1.5 °C/W for 100 ms surge pulses. What is the maximum permitted power of the pulses? Comment on this value, compared with the steady-state value, and explain why the difference arises.

1.13 A 200 V, 10 A SCR device has a thermal resistance of 1 °C/W. What is the steady-state power loss if the working junction temperature is 100 °C and the ambient temperature is 15 °C?

1.14 An SCR has a maximum junction temperature rating of 120 °C and works in an ambient temperature of 30 °C. After initial switch-on it is expected to experience a current surge that will cause 1000 W of dissipation in 5 ms. What surge impedance rating must the system possess to avoid overheating? Comment on the contribution of a heat sink. How must the fabrication of the device be arranged.

1.15 Distinguish between case and ambient temperature ratings for a solid-state semiconductor device. An intermediate power bipolar device has ratings of 1 W and 10 W at 20 °C. Explain the meaning of the two ratings. Why is it unlikely, in practice, that the device will dissipate 10 W?

1.16 Sketch a family of curves which represent the thermal behaviour of a controlled rectifier operating under various phase-angle firing conditions and explain why such curves exist.

Explain why thermal derating is necessary and distinguish between ambient and case temperature rating.

1.17 A triac is to be used in conjunction with a 50 Hz mains supply for either of two purposes:

(a) proportional continuous power control by means of phase-angle firing, or

(b) isolated integral cycle bursting firing for pulse periods of 10 s.

If the formula for the dynamic impedance of a heat sink to which the device is firmly mounted is given by $Z = 3(1 - \varepsilon^{-t/200})$ evaluate the thermal impedance under the two conditions and explain why they are different.

If the thermal resistance of the triac is $1\,\Omega$, establish the maximum junction temperature in each situation, if the ambient temperature is 20 °C for a thermal power flow of 20 W. What would happen if the device 'worked loose' on the heat sink?

1.18 A gate turn-off SCR (GTO) is used in a situation where its incidental dissipation is 20 W, its maximum junction temperature is 200 °C and it is case rated at 2 °C/W. If the ambient temperature is 20 °C and the heat sinking has a thermal resistance of $1\,\Omega/m$, calculate the minimum length required. What will be the heat sink temperature?

1.19 Distinguish between the ambient and the case derating curve of a semiconductor power device and explain the significance of the latter in evaluating heat sink dimensions.

Draw the steady-state thermal equivalent circuit of such a device, indicating where the difference exists.

A power bipolar transistor is used as a high-speed switch and it is found that the switching dissipations are 0.1 and 5 W and the 'on-state' dissipation is 2 W. It has a case derating curve slope of 5 °C/W. If commercially extruded heat sinking is available with a thermal conductivity of 0.5 W/°C per metre, calculate the minimum length required if the maximum storage temperature of the device is 180 °C and the ambient temperature is 30 °C. If a capacitor is available, which has a load-associated time constant equal to the turn-off time, calculate the minimum length now required.

1.20 A power metal–oxide device with $R_{DS_{on}} = 0.05\,\Omega$ is used to switch energise a $10\,\Omega$ resistance from a 100 V bus.

It is found that the transition times are 50 and 200 ns. Evaluate the incidental dissipation generated in the device if the duty cycle is 0.6 and it is actuated at 20 kHz.

The device is case derated at 0.1 W/°C from a maximum storage temperature of 150 °C. If a commercial heat sink has a thermal conductance of 1.0 W/°C per metre, find the minimum length required at its operational temperature.

2
Switching devices and control electrode requirements

2.1 RATING, SAFE OPERATION AREA AND POWER HANDLING CAPABILITY OF DEVICES

If the modern power control engineer is to make the correct choice of switching device for a given application it is necessary to be aware of the characteristics and limitations of the devices that are currently available. The obvious limitations to any device are its maximum voltage and current ratings, which must not be exceeded. It is hence essential to have some knowledge of device construction and fabrication.

2.1.1 Power handling capability (PH)

The maximum power handling capability PH_{max} of a semiconductor switch is related to the product of V_{bus} and I_{max}. For any semiconductor device the maximum voltage to which the device can be subjected is related to the avalanche breakdown value of the silicon p–n junction, while the maximum device current is limited to the chip current density. The maximum current density is affected by various factors, including temperature and mechanical stress.

Both the maximum voltage and maximum allowable current are affected by the impurity levels present in the chip. Moreover, they are affected inversely in such a way that any increase of one parameter will result in reduction of the other. When endeavours are made to increase a device power handling capability, limiting factors arise as with most engineering problems. A compromise has to be made as invariably the load current flows through the same junctions that have to withstand the supply voltages without unintentional avalanching. To explain the inter-dependency of device parameters, a brief review of the fabrication process is developed.

2.1.2 Principles of device fabrication

In semiconductor device fabrication the two main methods used are diffusion and epitaxial deposition. In the diffusion method the impurities are deposited into an existing relatively pure crystalline structure covering the surface with impurities (e.g. donor) and elevating the temperature for a considerable time. Surplus impurity is then removed and the process repeated with opposite (i.e. acceptor) impurity to form a submerged junction. Using this convenient process the thickness of the silicon does not change but the penetration variation in junction depth does vary statistically throughout the batch and across the chip.

With the epitaxial deposition method, the various regions are 'grown' by thermal deposition, going directly from the gaseous phase to the solid-state phase without the intermediate liquid phase. Using this epitaxial crystal-growing technique, abrupt changes in the doping concentrations and much greater control of the thicknesses of silicon can result. The deposition of silicon dioxide (SiO_2) on the surface can serve several functions:

(i) it provides masking to prevent diffusion in selected areas. Where this is etched away windows are formed and diffusion can occur. The process is repeated to create acceptor and donor regions.

(ii) junction edge contamination is minimised by the junction being formed under the silicon dioxide glass layer – a process known as surface passivation.

(iii) the thickness of the surface glass can be accurately controlled. By depositing aluminium as a contact an MOS capacitance is formed.

The sandwich structure of the metal–oxide–semiconductor provides the basis for the formation of depletion and carrier capacitance described, for example, in Section 2.2.1 below. Both the on and off times of the switching action arise from two principal sources of effective capacitance:

(i) charge storage and recovery current flow associated with the principal and control electrode currents

(ii) displacement currents which flow in the junctions by virtue of their depleted states.

2.1.3 Safe operation area (SOA)

In the case of proportional devices the power handling capability is defined in terms of the breakdown voltage maximum value $(BV)_{max}$ and the average value of the maximum allowable current.

$$PH_{max} = (BV)_{max} \times I_{max_{av}} \tag{2.1}$$

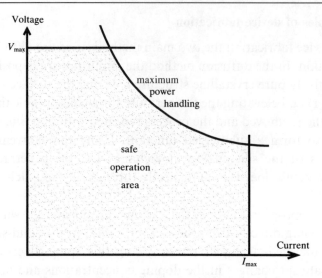

Fig. 2.1 Generalised diagram of power semiconductor demonstrating the safe operation area (SOA).

The terms on the RHS of (2.1) are both specified in the device data. Static characteristics of the principal electrode, Fig. 2.1, define limiting values of PH_{max} by two bounds of the safe operating area (SOA). The magnitude of this area is an indication of the general usefulness of the device.

A similar static characteristic also exists for the control electrode. With the exception of bipolar devices, however, where A_{pc} is relatively low, the contribution towards the total device dissipation can be ignored, although the limits of its voltage/current bounds must be satisfied.

Another performance indicator is that of the maximum usable frequency. As t_{on} and t_{off} become more significant fractions of the periodic actuation time, the switches progressively inhibit the performance of the controller. In the limit situation, when the on- and off-state times approach zero, the switches have become merely coarse signal analogue amplifiers. The times are controlled by impurity levels, the dimensions of the silicon and the control electrode signal characteristic.

2.1.4 Ratings and data sheet interpretation

Manufacturers' data sheets serve two basic purposes:
(1) they advertise the device by telling the user what it can do, particularly in comparison with its competitors (i.e. an advisory function). Typical parameter values are often listed, indicating the maximum and minimum values for the type.

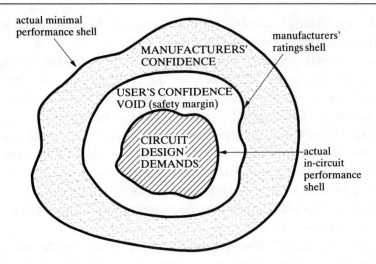

Fig. 2.2 Conceptual shells illustrating design demands and the ratings offered by the manufacturers of devices.

(2) they inform the user as to the limits, including absolute maximum ratings, outside which the manufacturer does not wish the user to operate the device. This is a legalistic function within the meaning of the Trades' Description Act in the United Kingdom. Usually the manufacturer is trying to prevent the device from being abused so that its life is not endangered, not to define precise limits of malfunction.

The user/supplier interface situation is best thought of as three concentric irregular shells with two interspaced voids, as illustrated in Fig. 2.2.

2.2 SEMICONDUCTOR SWITCHING DEVICES

Many power semiconductor devices are now (1995) available. Two-terminal uncontrolled switches are usually classified under the term 'rectifiers'. These are widely described in many existing books and are not discussed further in the present text.

Devices listed under the headings bipolar devices, field-controlled devices and current-controlled devices are all three-terminal controlled switching devices. The switching actions of semiconductor power devices fall into two categories:

(a) those in which the operation is basically analogue but the principal current is controlled in a proportional manner by the control electrode, by being overdriven (e.g. bipolar junction transistor, Darlington combination, MOSFET).

(b) those in which the state changes are or can be instigated by trigger action at the control electrode, but the control does not continue proportionally. Examples are the SCR, GTO and MCT, which are fabricated so as to internally generate positive feedback.

All of the various controlled switches now available incorporate the actions of one or other of the two basic power devices; namely the bipolar junction transistor (BJT) and the metal–oxide–semiconductor field-effect transistor (MOSFET). Individual devices have limitations in power control ratio and in power handling capability which may be overcome by compounding. The compounding takes one of two forms, described below:

(i) cascade connection – a signal level device is connected to the main power device, within a single encapsulation, and often on a single chip (e.g. Power Darlington, IGBT).

(ii) positive feedback connection – such as the constituent transistors of an SCR and, more latterly, the MCT.

Improvements that occur in certain aspects of a device specification usually do so at the expense of increased complexity and result in reduced limits on other features of the specification.

2.2.1 Bipolar junction transistor (BJT)

The safe operating area of a bipolar transistor occurs in the static collector current–voltage plane, corresponding to the form of Fig. 2.1. Operation is defined within specified boundaries that include the hyperbolic power characteristic.

The parameters controlling the performance of bipolar transistors are essentially the same regardless of chip size and power handling capacity. The basic theory of operation is well documented elsewhere. At the signal power level the triple diffused, or epitaxial silicon planar, surface passivated deposition device has existed for many years and probably represents the ultimate product, i.e. further development is unlikely. As power handling level and chip size are increased it is not possible to achieve in one device an all embracing performance representing economy of production, high power control ratio (current gain, h_{FE}), low 'on' resistance, high operating voltage and high-speed operation. Devices are constructed differently to achieve the optimum performance of particular parameters, which usually means accepting a reduced performance of other parameters.

2.2.1.1 Forward current transfer ratio

The power control ratio in the case of a bipolar transistor is indicated by the forward current gain, h_{FE}, where E indicates the common emitter configuration. It is defined by the expression

$$h_{FE} = \left.\frac{\partial i_C}{\partial d_B}\right|_{V_{CE}=\text{constant}}$$

$$= \frac{I_C}{I_B}, \text{ under steady conditions} \qquad (2.2)$$

where subscript 'C' in (2.2) refers to the collector current and subscript 'B' refers to the base current. Parameter h_{FE} is sensitive to changes of collector voltage, collector current and temperature. It varies with frequency in a simple, first-order, low pass manner. At low frequency reactive elements in the circuit may be ignored.

To achieve a high value of parameter h_{FE} a very thin base width together with relatively high doping levels is required. Thin bases are difficult to fabricate over large chip areas due to irregularities in the two diffusion processes required to produce the junctions J_1 and J_2 of Fig. 2.6 below. High doping levels result in low 'on' resistances but poor maximum voltage ratings. Simple equivalent circuits which may be used to represent the variation with frequency of h_{FE} are depicted in Fig. 2.3(b). The corresponding Bode diagram representing the loss of gain as the frequency (plotted on a log scale) increases is given in Fig. 2.3(c)). Capacitance C_T in Fig. 2.3 is given by

$$C_T = C_\pi + C_u(1 + g_m R_L) \qquad (2.3)$$

where C_π is the base charging capacitance, C_u is the J_2 depletion layer capacitance and $g_m R_L$ is the low frequency voltage gain. When $R_L = 0$ then $g_m R_L$ is also zero and from the definition of two-port circuit parameters, in terms of the complex variable $s(= \sigma + j\omega)$,

$$h_{FE} = \frac{g_m}{s(C_\pi + C_u) + g_\pi} = \frac{g_m/(C_\pi + C_u)}{s + g_\pi/(C_\pi + C_u)} \qquad (2.4)$$

Putting $s = j\omega$ in (2.4) demonstrates that there are two significant values of frequency ω.

(i) When $20\log_{10}|h_{FE}|$ is 3 dB down from its low frequency value in the Bode (attenuation/frequency) diagram, Fig. 2.3(c), the straight-line approximation has its cut-off value ω_B.

(ii) When $|h_{FE}| = 1.0$, $20\log_{10}|h_{FE}| = 0$ and the Bode characteristic cuts the frequency axis at ω_T. Putting $|h_{FE}| = 1.0$ in (2.4) gives

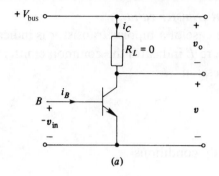

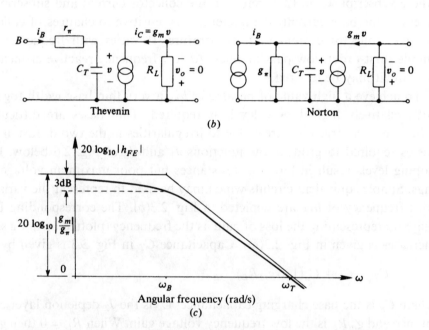

Fig. 2.3 Variation of forward current gain for a bipolar transistor: (a) circuit diagram, (b) Thevenin and Norton equivalent circuits, (c) Bode (attenuation–frequency) diagram.

$$1.0 = \left| \frac{g_m}{\omega_T(C_\pi + C_u) + g_\pi} \right| \tag{2.5}$$

But $C_\pi \gg C_u$ and $\omega_T C_\pi \gg g_\pi$. Therefore

$$\omega_T = 2\pi f_T \simeq \frac{g_m}{C_\pi} \tag{2.6}$$

When $\omega > \omega_T$ in Fig. 2.3(c) the gain in decibels (dB) is negative so that $|h_{FE}| < 1.0$.

Where bipolar transistors are used in variable frequency power applications it is important to know the frequency for which the current gain becomes small. The frequency f_T is usually established indirectly by extrapolation from the gain–bandwidth product because device-modelling is inaccurate.

In the equivalent circuit of Fig. 2.3(b) resistor r_π is the resistance of the bulk material of the emitter and represents the resistor across which forward (on-state) constant voltage falls. Admittance g_π is the reciprocal of r_π.

The time constant of the base circuit current in Fig. 2.3, during conduction, is

$$\tau = r_\pi(C_\pi + C_u) \simeq r_\pi C_\pi \qquad (2.7)$$

Combining (2.6), (2.7) gives

$$\tau \simeq \frac{r_\pi g_m}{\omega_T} \qquad (2.8)$$

The base charging capacitance C_π is proportional to the square of the physical base width and so, therefore, is the frequency response.

As a device increases in size the parameters C_π, C_u and g_π increase so that r_π decreases. But the most significant feature of large area (i.e. high current) bipolar transistors is the physical phenomenon of charge spreading. Even with interdigitation structure, discussed in Section 2.3.2.1(c) below, the base current has to migrate in a direction perpendicular to the principal (collector–emitter) current flow for relatively great distances. This phenomenon is represented in the equivalent circuit, Fig. 2.4, by the extrinsic resistance r_x. In addition to the extrinsic resistance effect the device has also to drive the load R_L. These two factors reduce the useful frequency range. A device should not be used within two octaves of the transition frequency ω_T, for the highest spectral component, otherwise distortion will be generated. Also, ω_T varies considerably with collector voltage and current. Furthermore, account should be taken of the additional drop caused by the non-zero value of R_L.

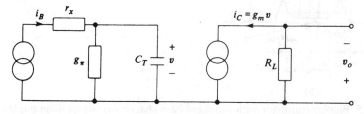

Fig. 2.4 Equivalent circuit for a large-area bipolar transistor.

40 Switching devices and control electrode requirements

The practical effect of extrinsic resistance is that f_T for a typical signal device will be 0.1–1 GHz and with specially made devices it may extend to 5 GHz. With power devices 1–10 MHz is more typical. Quoting ω_B could be more useful but is unsatisfactory as this varies with h_{FE} in batch production of a single type.

It can be anticipated from the foregoing that the types of square-wave voltages generated by inverters may be considerably distorted if the designer selects power bipolar devices that are too slow acting for the circuit being created.

2.2.1.2 Switch-on and switch-off characteristics
When a bipolar power transistor is switched on and off the times of commutation are usually different, Fig. 2.5(b). The transition times t_r or t_f in

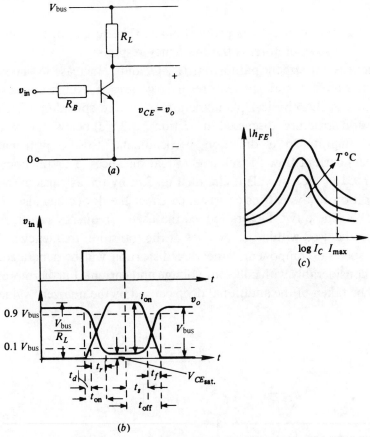

Fig. 2.5 Bipolar transistor with saturated base charge: (a) circuit diagram, (b) switching characteristics, (c) gain $|h_{FE}|$ versus current.

which the output voltage $v_o(t)$ falls from $0.9\,V_{CC}$ to $0.1\,V_{CC}$ (or vice-versa, respectively) are known as slew rates. The characteristics of Fig. 2.5 show the gain $|h_{FE}|$ to be a parameter which varies considerably with temperature and collector current. Also it is a parameter over which the producer has little control. Devices of the same batch even after selection may vary by at least 2:1 in gain. In designing a switching circuit the power electronics engineer must design for the worst condition and the worst device. Therefore most circuits of a batch, including one containing the worst device, will inevitably pass more current into the base of the device than is necessary for it to just saturate, i.e. when $v_{CB} = 0$. In saturation the collector–base junction becomes forward biased and $v_{CE} < v_{BE}$. There is then an excess of charge in the base which must be extracted before the switch-off action during t_f, associated with state change, occurs. The resulting saturation time constant τ_s can be several times greater than this slew time, which is loosely related to the transition frequency, ω_T. The excess charge can be extracted relatively quickly and τ_s reduced by returning the bases of such devices via resistors or inductors to appropriate negative potentials.

When $v_{CB} < 0$ the device is said to be heavily saturated, $v_{CE} < v_{BE}$ and the dissipation in the 'on' state is low. It may well then contribute less heat generation in the device than the switching loss. An analysis of switching operation is given in Sections 1.4, 1.5 of Chapter 1.

With power devices which possess relatively little chip thermal inertia it is essential to limit the dissipation so that the device operates within its safe operating area. If a power switching device comes out of saturation, even for a small fraction of time or only very slightly, by $v_{CE} > v_{BE}$, damage is likely especially if the device was previously operating at its rating limits.

2.2.1.3 Construction and properties of some types of power bipolar transistors

The important parameters of a power bipolar transistor are the current gain h_{FE}, the current rating $I_{C_{max}}$, the device breakdown voltage on no-load BV_{CEO}, the dissipation during conduction, which depends on the on-state saturated resistance $R_{CE_{sat}}$ and the switching speed, indicated by f_T.

At low currents the current gain is affected by the usual factors of emitter efficiency and base lifetime but at full load current the device geometry and base width become important. Breakdown voltage, BV_{CE}, is related to the material resistivity on either side of the collector–base J_2 junction. The 'on' resistance, $R_{CE_{sat}}$, is related to the bulk resistance of the collector. Special fabrication is required if low $R_{CE_{sat}}$ and high BV_{CE} values are to be achieved in the same device.

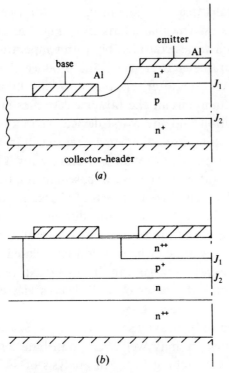

Fig. 2.6 Construction of power bipolar transistors: (a) single diffused device of mesa form, (b) triple diffused device with low R_{CEsat} or double diffused device with the lower n region formed by epitaxial deposition.

Some modes of construction are depicted in Fig. 2.6.

(A) Single diffused device

A high resistivity wafer has the same 'impurity' diffused into both sides. Acid etching produces a mesa emitter. The resultant base width, $(20-30) \times 10^{-6}$ m, cannot be very thin. A rugged device results with a comparatively large SOA as charge redistribution in the relatively thick base permits uniform current densities and avoidance of hot spots. Typical values are that transition frequency f_T is low at 2 MHz, $BV_{CEO} = 200$ V and $I_{C_{max}} = 30$ A. Such devices are virtually obsolete but usually form the basis of thyristor realisation.

(B) Triple diffused and epitaxial devices

By resorting to a three-diffusion process, or epitaxial deposition, it is possible to achieve a structure which possesses high resistance collector material near

J_2 but low resistance material for the bulk remainder of the collector supporting material. This preserves a high value of breakdown voltage yet results in a low value of resistance $R_{CE_{sat}}$ at the full rated current of the device.

The triple diffused device and the planar devices achieve just this effect. For the former $BV_{CEO} = 400\,\text{V}$ and $I_{C_{max}} = 15\,\text{A}$ with $f_T = 35\,\text{MHz}$ being typical of what can be achieved. This f_T improvement over the single diffused device is at the expense of a significantly smaller SOA. With triple diffused devices better control of epitaxy results in $f_T = 100\,\text{MHz}$ being achieved and $I_{C_{max}}$ is improved to 50 A. A very good feature of these devices is the opportunity to be able to fabricate both npn and pnp versions of similar characteristics. Punch-through is likely with thin bases and hence rapid action protection is required. Fast acting fuses are usually unsatisfactory.

By fabricating the base structures and the emitter with epitaxial deposition any diffusion is avoided. It is then possible to make very abruptly graded junctions. A resultant $f_T = 10\,\text{MHz}$ is only reasonable, $BV_{CEO} = 160\,\text{V}$ is low and $I_{C_{max}} = 50\,\text{A}$ but because of the even abruptness of the junctions the SOA is large and the device is correspondingly rugged.

Although interdigitation is used on high power devices to reduce the base spreading resistance (extrinsic resistance r_x) the best technique is that of multiple emitter construction. A separate contact has to be made to each emitter – the bases are 'fingered' together. In high power handling devices the costs involved in any such project are such that the extra expense involved with this technique is justified and, due to improvements to uniformity of current density across the chip, improvements to $I_{C_{max}}$ to 50 A result for f_T values of 30 MHz.

2.2.1.4 Switching properties of bipolar devices

The immediate application of a voltage between the collector and emitter of a power device when the base is open circuit or returned to the emitter via a resistor is not inconsequential, even if the amplitude does not exceed the BV_{CEO} rating. In fact, if the action is preceded by the existence of stored charge in high switch-rate circuits, the localised heating that occurs can be sufficient to destroy the device.

(A) Depletion layer capacitance

The depletion layer capacitance of a junction is given by the expression $C_{dep} = KV^{1/2}$ (for an abrupt junction). The constant K depends upon material permittivity and junction surface area. Because many modern power bipolars are constructed by successive overdoping, the emitter–base junction capacitance C_{EB} at 0 volts may be 5 nF whilst the collector–base junction capacitance C_{CB} may be 1 nF.

The 'hold-off' facility of the collector–base is thus shunted by C_{CB}. If a rise of voltage dV_{CE}/dt is applied to the device a current

$$i_d = C_{CB} \frac{dV_{CB}}{dt} \simeq C_{CB} \frac{dV_{CE}}{dt} \tag{2.9}$$

will be injected to the base of device. The route of this current flow greatly affects the total current flow in the collector circuit. Even if an external resistance R_B is strapped between base and emitter a sizeable component of current i_d will flow through the transistor on account of the extrinsic resistance r_x. This flows through C_{EB}. The effect is to generate a current flow $(h_{FE}+1)i_d$ by transistor action. If the whole of the current flows through junction EB (because B is open circuit) the following amplification of i_d occurs in the form of a steady collector current

$$I_C = (h_{FE}+1)C_{CB} \frac{dV_{CB}}{dt} \tag{2.10}$$

A current i_d, say equal to 0.1 A maximum, could well appear as $I_c = (50+1)0.1 = 5.1$ A.

The distribution of charge across the surface of the chip is fairly uniform – if the device is well made – and experience shows that local heating from this mechanism is minimal. Thermal runaway and subsequent failure are unlikely. If, however, sizeable residual charge is present in the base at the initiation of the rise a redistribution of charge takes place which can generate serious local heating. Such situations are unlikely to occur in amplifier circuits where only unidirectional voltages occur. In a.c. power circuits transistors are subjected to reverse voltages which can leave this residual charge in the base as the supply to the device alternates.

(B) The effect of reverse conduction on $(dv/dt)_{max}$ rating

Before considering the application of the forward voltage it is necessary to consider the effects of reversing the collector–emitter voltage V_{CE}.

If the collector and emitter are interchanged in a circuit, transistor action still takes place. Moreover, if the device is symmetrical it will have the same current gain. However, modern multiple diffused devices are deliberately constructed asymmetrically to improve this parameter in one preferred direction. Therefore, under most circumstances, reversing the applied voltage results in the transistor still operating as an active device, but with a vastly reduced current gain (50 one way, say, and 5 the other).

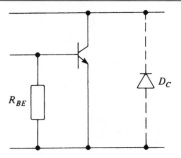

Fig. 2.7 Use of a thermal stabilising resistor to minimise forward leakage current.

The problem is caused by a thermal stabilising resistor R_{BE} connected as shown in Fig. 2.7 to minimise leakage in the forward direction. This now acts as a bias resistor to hold the device 'on' in the reverse direction. The inclusion of R_{BE} has an advantage in the reverse direction as its low value causes the transistor to saturate, minimising reverse dissipation. If this did not occur, V_{CE} would avalanche at 10–50 volts (say) with considerable reverse dissipation. If R_{BE} is not included a reverse acting clamp diode D_c has to be included.

If reverse saturation occurs immediately prior to the voltage across the device alternating to the positive direction, the rate of rise of voltage in the positive direction combined with any residual base charge which has not had a chance to recombine causes dangerous localised heating.

In the reverse direction the CB junction becomes forward biased and therefore it has a high capacitive value and will contain much charge in an exceedingly thin region of base. If the change to a high forward voltage occurs so rapidly as to prevent recombination this charge passes to the emitter and causes the current gain to generate a very large amplitude spike which may well be in the region of 50 A peak. Furthermore, the spike occurs at a time when V_{CE} is large.

There are two simple ways in which this phenomenon can be avoided. Fig. 2.8 indicates the insertion of diodes D either directly in series with the transistor or in series with the biassing resistor. It can be seen that the diode which is in series with the collector has to be rated at the same maximum current level as the device whereas the one used in the base protection configuration need only be rated at a lower level, dependent on the maximum average current flow through R_{BE}. If the device is transformer coupled at its input, Fig. 2.9, R_{BE} may be omitted and the diode included in series with the transformer. A further parallel circuit may also be needed into which the current from the inductance of the transformer can be routed and dissipated.

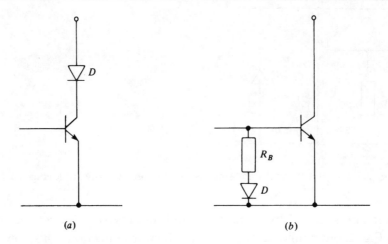

Fig. 2.8 Methods of protecting a power bipolar transistor from the effects of reverse current flow.

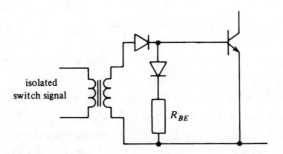

Fig. 2.9 Modified use of a thermal stabilising resistor with transformer coupled input.

An alternative is to include inductance in series with the collector merely to limit dv/dt rise. It is shown in Chapter 1 above that an inductance connected in this position is desirable to minimise the switch-on loss and it may well be that its value is adequate to limit the rate of rise of applied voltage without the inclusion of diodes.

Dependent upon the nature of the load, a capacitance connected between the emitter and earth could well restrict the rate of rise to a suitably low level. The capacitance is also helpful as a turn-off aid.

(C) Turn-off of power bipolars when subjected to high voltages
A power bipolar transistor may be extinguished from its control electrode by the following methods:

(a) open circuiting the base in order to remove the source of bias current,
(b) returning the base to earth via (i) a high value resistor, or (ii) a low value resistor,
(c) returning the base to a negative potential via an impedance.

For power applications and devices, method (a) is usually unsatisfactory as time t_s, Fig. 2.5(b), becomes too long. The result is a high value of switching dissipation as the collector current initially falls only very gradually, because the collector has to wait until the excess carriers in the base recombine. Method (b)(i) is similar in effect.

On account of voltage V_{BE} methods (b)(ii) and (c) are similar in action and allow values higher than the voltage BV_{CEO} to be used. This may be exceeded without reverse bias if the duty cycle of any pulses is so low that the chip thermal inertia prevents the maximum junction temperature $T_{j_{max}}$ from being exceeded. If the device is heavily saturated – which is usually the case – excess holes exist in the base and these may be extracted rapidly. But the charge destined for the collector cannot decrease until this extraction has taken place. If charge is initially extracted from the base too rapidly it leaves regions of the base depleted with excess charge concentrations still occurring in other areas remote from the base contact. As the device comes out of saturation hot-spots are created with the result that current concentrations occur in these regions which are under the emitter fingers. It can be seen that the higher the degree of interdigitation of overlaying, the less this effect is likely to occur. Therefore the rate at which charge is extracted from the base cannot be too large or too small, as in cases (a) and (b)(i) above.

In order to avoid device damage the forward voltage rating V_{CE} has to be limited to a value less than its normal rating BV_{CEO}. Devices are accordingly given a reverse bias safe operating area (RBSOA) rating. Also, a parallel combination of a diode and inductor (a few µH) is recommended for insertion in series with the base lead, Fig. 2.10, so as to optimise the rate of discharge and the RBSOA.

A resistance has to be included in parallel with the inductance in situations where the base–emitter capacitance forms a less than critically damped condition. If this component is not included the circuit can parasitically oscillate causing the transistor to switch on spuriously. Later on in the fall time the base–emitter voltage starts to fall and the resistance of the junction starts to rise. In order to sustain the rapid extraction of charge the negative returning voltage of the circuit must be adequate, i.e. greater than $-2\,\text{V}$. Even if the junction avalanches temporarily, with excess voltage, the transistor will not be damaged.

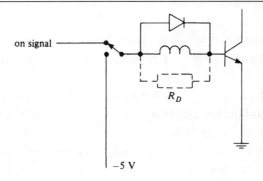

Fig. 2.10 Basic charge extraction circuit for bipolar transistor turn-off.

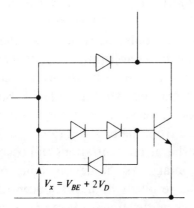

Fig. 2.11 The Baker clamp circuit for bipolar transistor saturation prevention.

One way of reducing the 'off' time is to prevent the device from being heavily saturated. The Baker (anti-saturation) clamp circuit depicted in Fig. 2.11 prevents V_{CE} becoming less than V_{CB}. If $V_x = V_{BE} + 2V_D = V_{CE} + V_D$ then $V_{CE} = V_{BE} + V_D$. Unfortunately, this action detracts from one of the main advantages of a triple diffused device, namely its very low 'on' state dissipation. Instead of $V_{CE_{sat}} = 0.1$ V during the 'on' state it is now 1.2–1.5 V, having a tenfold increase.

Such circuit aids can reduce t_f, Fig. 2.5(b), from 500 ns to 50 ns and virtually eliminate t_s, which was 5 μs. The turn-off time t_{off} then becomes 50 ns.

2.2.2 Metal–oxide–semiconductor field-effect transistor (MOSFET)

Historically the field effect was initially realised at signal level on a junction basis. After a very short time the MOS form of fabrication appeared, in which the gate is formed by a deposition of aluminium on top of the silicon

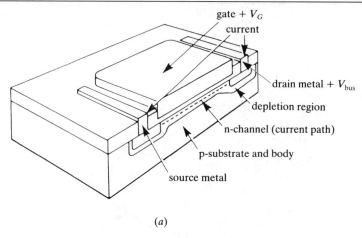

Fig. 2.12 Small-signal MOSFET construction (ref. 52).

dioxide, Fig. 2.12. This horizontal form of fabrication is, however, quite unsuited to handling large currents because the conduction channel has a relatively small cross-sectional area A and a large length L, such that resistance $R_{DS_{on}}$ is high. This is indicated by a large area of the I_D/V_{DS} plane static characteristic being unusable – a situation that prevailed for over a decade, with MOS devices being used mainly for signal processing.

By comparison with the horizontal construction of signal level devices, power level MOSFETS are fabricated in a vertical mode, Fig. 2.13(a). The aspect ratio L/A of the conducting channel is lower so that a low value of $R_{DS_{on}}$ can be achieved. Length L is formed by the base width of a fabricated bipolar chip. Also, in a single encapsulation of (say) a 10–25 A device, at least 1000 cells are paralleled to achieve the desired low on-state resistance. The positive temperature coefficient of resistance of the channel enhances current sharing. Epitaxial layers have to be added, as with bipolar devices, to enhance the value $BV_{DS_{max}}$.

The change of conducting channel aspect ratio in power devices has a further advantage in that it significantly increases the usable area of the I_D/V_{DS} plane and almost linearises the transfer characteristic.

Fig. 2.13(a) shows a vertical section through such a device. For convenience the gate is realised in polysilicon silicon. Its resistance does, in the limit, affect switching slew rates and aluminium deposition has to be alternatively adopted in HF applications.

A typical set of characteristics is given in Fig. 2.13(b), in which the knee of the characteristics can be seen to be relatively close to the i_D axis. Fig. 2.13(c) shows a dynamic equivalent circuit of a much simplified device,

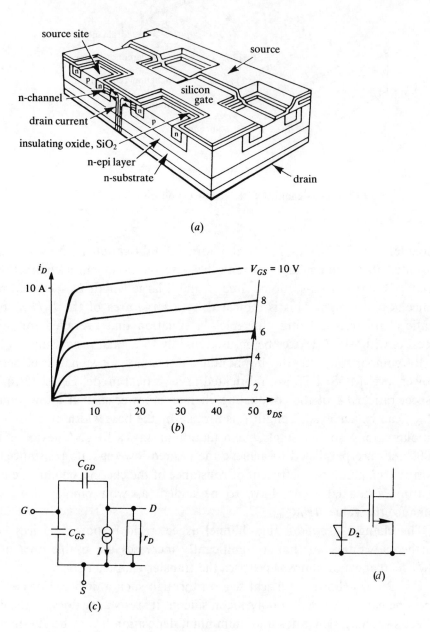

Fig. 2.13 Vertical etch groove, power MOSFET: (*a*) construction (horizontally compressed presentation), (*b*) static characteristics ($g_m = 1$ s, $C_{GD} = 300$ pF, $C_{GS} = 5000$ pF, $r_D = \infty$), (*c*) incremental equivalent circuit ($I = g_m V_{GS}$), (*d*) gate overvoltage protection by Zener diode.

incorporating capacitively coupled parasitic elements. It can be seen that the device works in enhancement mode. By comparison, in the recent past, depletion mode junction field-effect transistor (JFET) power devices have been fabricated and incorporated into designs. During state change the channel edge changes are depletion mode activities.

Exceeding the maximum voltage rating of a power MOSFET does not result in immediate device failure. Provided that thermal considerations are adhered to, avalanching is allowed. This is convenient as bus-limiting diodes, normally required for inductive loads, are not cheap and can then be omitted. It is possible to obtain types of power MOSFETs rated for repetitive avalanching.

A device circuit representation, including parasitic elements, Fig. 2.14, incorporates a parasitic bipolar transistor. A feature of the circuit is that rapid positive change of V_{DS} can excite the bipolar into conduction – a condition that has to be avoided. The need to, in effect, prevent the parasitic bipolar from becoming excited can create the need for a dV/dt snubber requirement. For a 25 A device the input capacitance is about 3000 pF, such that a low resistance or compensated voltage source is necessary if fast switching is required. This need not be sustained when the switching transitions have been effected. The actual voltage required is at least 10 V because $g_m \approx 1\text{--}10\,\text{A/V}$.

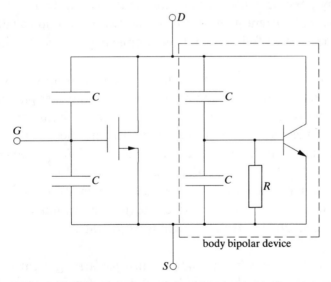

Fig. 2.14 Equivalent circuit of power MOSFET representing the possibility of avalanching.

In the case of a 'high side' controller, i.e. one with earth referred load, the bus voltage has to be exceeded by that same amount at the gate in order to switch the device on. This is inconvenient as bootstrapping or transformer coupling from earth referred control sources is then required.

With modern devices the production yields are so high that some of the chip can be devoted to intelligent protection activity. For example, temperature, overcurrent or overvoltage may be sensed. The gate signal can be removed, if required, and the principal electrodes isolated.

Specially configured devices are available which contain on-chip pump circuits. These can be switched from logic level (5V) sources but possess only lower $V_{DS_{max}}$ ratings.

2.3 COMPOUND DEVICES

2.3.1 Cascade connected devices

2.3.1.1 Power Darlington transistor

Power bipolar transistors consume considerable input control power. An $I_{C_{max}} = 50\,\text{A}$ device with a low current level h_{FE} of 50 may only possess a gain of between 15 and 20 at this level. It can be seen that an intermediate power driver device must be cascaded with this transistor before a unit may be considered for operation from a signal power level source.

There are many possible configurations for compounding two transistors to improve the overall current gain. Probably the two most popular are the Darlington double emitter follower and its complement, as shown in Fig. 2.15.

The complementary connection is particularly useful as it uses a low level pnp driver to convert a high power npn bipolar into a high power pnp unit. It is found that npn units can be constructed on an integrated basis, sometimes with three devices in cascade, with protecting diodes and thermal stabilising resistors also included. Such devices achieve values of $I_{C_{max}} = 500\,\text{A}$ for $BV_{CEO} = 400\,\text{V}$ with $|h_{FE}| > 100$ and $P_{max} = 3\,\text{kW}$ dissipation limit. External connections to the power device bases are made in order that these may be returned to negative voltages so that saturation charge can be extracted to speed up the device switching (t_{off} rating $= t_s + t_f$ in Fig. 2.5(b)).

It can be seen that the power Darlington transistor arrangement forms a Baker clamp. This is to ensure that the principal device cannot saturate fully, with the effect that the turn-off time is improved. External connections are

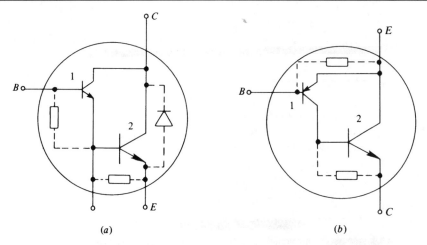

Fig. 2.15 High gain power bipolar transistors: (*a*) Darlington connection (npn, high current gain unit), (*b*) complementary (super α) connection (pnp, using an npn power device): 1 – signal or intermediate level device, 2 – main power bipolar transistor.

sometimes provided, with specified current–time extraction profiles, to reduce t_s and t_f even further.

2.3.1.2 Insulated gate bipolar transistor (IGBT)

The IGBT is intended to operate as a MOSFET from the control standpoint but with the advantageous features of a BJT at the main electrodes. Conductivity in the drain-drift region is controlled by the injection of minority carriers. A vertical cross-section of the structure is shown in Fig. 2.16. Compared with the MOSFET, Fig. 2.13(*a*), an additional layer is introduced at the drain end.

A heavily doped p^+-type structure has a lightly doped n-type region grown on to it by epitaxy. The emitter then found in this structure contains islands by windowing p-type material and subsequent n overdoping. Two silicon dioxide layers are then deposited. These have interconnected gates, with an interposing polysilicon layer which forms the gate of the overlaid signal MOSFET for each cell.

Fig. 2.16 shows that the emitter to collector current path is basically a pnp bipolar structure. Under the gate area a four-layer (three-junction) SCR-type parasitic structure exists – this is undesirable but unavoidable. In first-generation devices this parasitic element, Fig. 2.17, caused latch-up conditions so that the gate of the MOSFET could lose extinction control. The introduction of the further p^+ diffusion in the base body region in the fabrication of

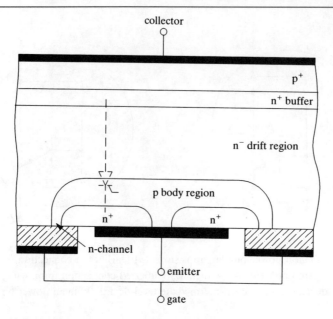

Fig. 2.16 Vertical cross-section of an IGBT.

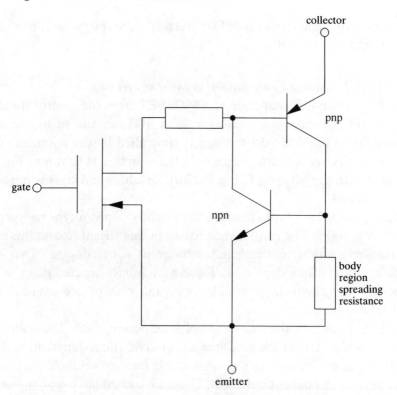

Fig. 2.17 IGBT equivalent circuit showing the constituent transistors of the parasitic SCR.

2.3 Compound devices

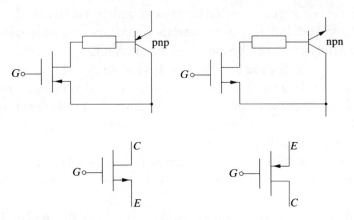

Fig. 2.18 Representation of IGBT function.

third-generation devices reduces the h_{FE} of the gate-emitter npn device. As a result, a slight increase of on-state dissipation occurs but the effects of the parasitic SCR device are almost totally suppressed.

In the n-IGBT compound device the signal n-channel MOSFET converts the pnp power bipolar transistor action into a positively controlled device. A representation of IGBT function is shown in Fig. 2.18. The emitter of the internal device is seen externally as the collector. The device therefore behaves as a voltage-controlled constant current source, of approximate rating 3–5 A/V, as shown in Fig. 2.19. Capacitance C_{GC} can be very significant in state transitions due to the 'Miller effect', namely the alteration of the apparent value of a component by virtue of it forming a feedback loop of an amplifier. The component may be parasitic or a separate circuit element. Then:

$$C_{in} = C_{GE} + C_{GC}K_m, \quad \text{where} \quad K_m = (1 + g_m R_L)$$

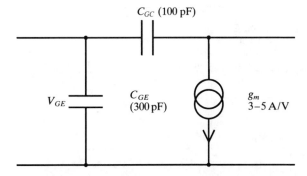

Fig. 2.19 Circuit representation of IGBT action.

An IGBT has an input capacitance approximately one-tenth that of an equivalent MOSFET but the G–C overlap capacitance is comparable and therefore the Miller effect can be more significant. Similar or identical control circuits are often used for the two devices. Like a MOSFET, there is a gate-emitter threshold voltage of 2–5 V which must be exceeded before significant current will flow. Because of device compounding, the saturation resistance has only a slight temperature coefficient which is positive at higher levels and negative at low levels of load current.

The basic action of IGBT turn-on is similar to that of a MOSFET, being a monotonic process that occurs rapidly, since no recombination process is involved.

Turn-off in an IGBT is a duplex process in which the power bipolar transistor now plays an dominant role. There is a distinct reduction in the rate of decay of current, Fig. 2.20, due to BJT action. The first stage, or MOSFET current reduction, is largely controlled by the rate at which charge is extracted from the gate capacity, i.e. it is under external control and occupies most of the fall.

The latter part or tail of the current–time characteristic is outside the control of external effects. This is controlled by charge recombination of the residual charge in the central plasma of the device. Although the BJT region is at or near the residual 10% of the trajectory, considerable switching dissipation occurs in it. Due allowance must be made for this dissipation, either by extension of the t_{off} time or by modification of the design.

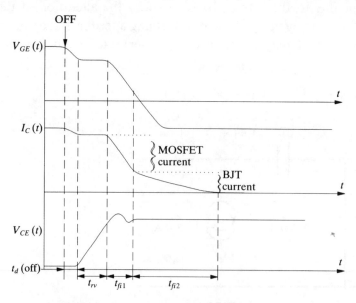

Fig. 2.20 Typical turn-off current and voltage waveform of an IGBT.

Table 2.1 *Some comparative properties of the IGBT and the MOSFET.*

IGBT	MOSFET
Minority/majority carrier conduction	Majority carrier conduction
Low forward drop	High $R_{DS(on)}$
Turn-off time 500 ns	Turn-off time < 100 ns. MHz operation possible
No internal diode	Reverse internal parasitic diode
4-layer device	3-layer device

An IGBT, despite being nonlinear and analogue in nature, is almost always used as a switch. Extinction is effected merely by allowing discharge through an external resistor R_{GE}. As a result of this, snubberless operation is possible as the dv/dt is then control electrode limited. Turn-off times of less than 10 µs may be achieved, the initial part of the fall being less than 0.2 µs.

When current levels exceed rated values the device behaves more like a MOSFET. Repetitive surges of overcurrent are permissible provided that thermal ratings are not exceeded. Reverse bias is not used in an IGBT so that RBSOAs do not exist.

Some manufacturers create a Zener diode across the gate–cathode, but others do not, so that care has to be taken to prevent the induction of electrostatic charges that may accidentally damage the gate. External back-to-back Zener diodes are often added across the gate–cathode terminals.

A comparison between certain properties of the IGBT and the MOSFET is given in Table 2.1. The use of the IGBT is reported to be more cost effective than the MOSFET in applications higher than 1 kW.

2.3.2 Cumulative feedback connected devices (thyristors)

The thyristor family is a class of devices which contain three or four semiconductor layers in which positive feedback is used to create trigger action. Semiconductor switches of the thyristor type were introduced in the 1960s and preceded power transistors by several years. Thyristor switches are still the best devices for operation at really high power (*PH*) levels.

With the fabrication referred to in Section 2.2.1.3 it was possible to create, with one diffusion process and an alloyed junction, a device that can handle hundreds of amperes of current from sources in excess of 1 kV rating. The increased complexity due to feedback has caused a relaxation in semiconductor design. The need for a base in order to achieve good values of h_{FE} is removed. Doping levels can be low resulting in the achievement of breakover voltages V_{FBO} greater than 3 kV, constituting a rugged device.

2.3.2.1 Basic thyristor theory

The basic four-layer device is shown in Fig. 2.21. Its basic operation is equivalent to the action of two bipolar transistors connected to create sufficient feedback to be regenerative, resulting in either the 'on' or 'off' state. For the device known as the silicon controlled rectifier (SCR) only the gate $G(p)$ is accessible to the external circuit. The following equations describe the currents in the two-transistor analogue circuit of Fig. 2.21(b).

At the nodes,

$$i_B(n) = i_C(n) + i_G(n) \tag{2.11}$$

$$i_B(p) = i_C(p) + i_G(p) \tag{2.12}$$

For the transistors,

$$i_B(p) = i_K(1 - \alpha_p m_p) - i_{CBO}(p) \tag{2.13}$$

$$i_B(n) = i_A(1 - \alpha_n m_n) - i_{CBO}(n) \tag{2.14}$$

For the overall device,

$$i_A = i_K - i_G(p) + i_G(n) \tag{2.15}$$

The base current gains α_p and α_n of the two bipolar transistors in Fig. 2.21(b) are current and voltage sensitive. Weighting factors m_p and m_n are associated with α_p, α_n to represent the voltage sensitivity, giving overall gain coefficients $\alpha_p m_p$ and $\alpha_n m_n$ respectively.

From (2.11)–(2.15) it is found that

$$i_A = \frac{\alpha_p m_p i_G(p) + (1 - \alpha_p m_p) i_G(n) + i_{CBO}(p) + i_{CBO}(n)}{1 - \alpha_n m_n - \alpha_p m_p} \tag{2.16}$$

$$i_K = \frac{(1 - \alpha_n m_n) i_G(p) + \alpha_n m_n i_G(n) + i_{CBO}(p) + i_{CBO}(n)}{1 - \alpha_n m_n - \alpha_p m_p} \tag{2.17}$$

In the absence of gate signal then, despite a positive anode voltage, only leakage current will flow as junction J_2 is reverse biased. If, in any way, the impedance of the junction falls and the resultant current passes through J_1 or J_3, further increases of current will occur due to bipolar and regenerative

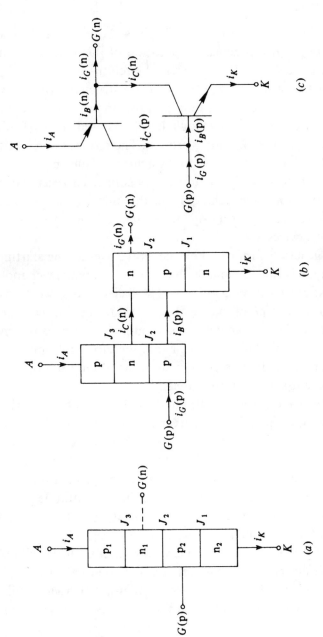

Fig. 2.21 Construction and two-transistor analogue of the thyristor and silicon controlled rectifier (SCR) (a) four-layer construction, (b) two-transistor analogue, (c) equivalent circuit.

action. If either or both of the gains $\alpha_p m_p$, $\alpha_n m_n$ rise such that the sum of the gains is unity, then

$$1 = \alpha_p m_p + \alpha_n m_n \tag{2.18}$$

Substituting (2.18) into (2.16) and (2.17) gives the result that i_A and i_K become infinitely large. The practical implication of this is that the anode–cathode current is limited only by the external circuitry.

The two bipolar transistors of Fig. 2.21 are constructed, practically, in a unicrystalline die of silicon. Each current gain requirement is low compared with the situation that exists when a single bipolar is used. This permits low doping levels to be used in the impurity process. The result of this fabrication form is that high breakdown voltages can be achieved. Thus, by constructing two, admittedly poor, low current gain devices connected regeneratively, a high power handling capacity is achieved. Furthermore, it can be controlled by exceedingly low power triggering signals, compared with even the highest gain power bipolar devices.

The thyristor is generally fabricated by successive overdoping starting with the anode. The anode is thus the one with the lowest doping level and hence J_3 withstands most of any reverse voltage. The bases of the two transistors need not, as with a single bipolar, be particularly thin. Thus, 'punch-through' device failures are much less likely and maximum current densities may be much higher than with bipolars. Thyristors are therefore not only generally more rugged than bipolars from a current surge standpoint but they possess much larger equivalent safe operating areas.

If a positive voltage $+v_{AK}$ exists at the anode with respect to the cathode, in the absence of gate signal, junction J_2 is reverse biased and only forward leakage current can flow. This is known as the forward blocking condition. If the impedance of this junction falls due to (say) the injection of some positive gate current, further leakage current will flow, passing through junctions J_1 and J_3.

If a negative voltage $-v_{AK}$ exists at the anode then the cathode potential is positive with respect to the anode, p–n junctions J_1 and J_3, Fig. 2.21, are reverse biased and the n–p junction J_2 is forward biased.

In the absence of gate signal, the reverse voltage inhibits forward diffusion current at J_1 and J_3 but permits the minority carrier thermal current to exist. This can be measured as reverse leakage current, of order up to a few milliamps.

With positive gate voltage, $v_{GK}(p)$, holes are injected into the p base section, Fig. 2.21.

The application of positive gate voltage to a reverse-biased thyristor should be avoided in order to prevent the failure of a junction by thermal runaway, as large increases of apparent leakage current occur.

2.3 Compound devices

A large increase of reverse-anode voltage can cause thyristor failure by punch-through of the reverse-biased junction J_3. In high voltage thyristors this is prevented by using a thick n-layer between the p-layers.

(A) Thyristor turn-on

To initiate conduction the anode voltage must be positive with respect to the cathode. Under this condition a thyristor may be caused to conduct by any of several techniques. Various members of the thyristor family of devices are associated with particular triggering methods, described below.
- (i) gate triggering (SCR, complementary SCR, SCS),
- (ii) forward breakover voltage (trigger diac, pnpn switch),
- (iii) irradiation methods (LASCR),
- (iv) dv/dt triggering,
- (v) temperature elevation, } usually fault conditions.

(i) Gate triggering

This is the most common method for controlling the point-on-wave at which conduction is initiated. The injection of additional carriers due to gate current increases the diffusion rate across junctions J_1 or J_3. When the gate current is sufficiently large, and provided v_{AK} exceeds its holding value of between 1 and 3 V, the polarity of the potential at junction J_2 reverses into the forward mode and avalanche conduction occurs. The change of state from forward leakage conduction (off) to full conduction (on) is depicted in the non-reversible static characteristics of Fig. 2.22. The anode current level at which ignition occurs is called the latching current and this remains largely constant as the forward voltage is increased. The latching current is always less than the minimum trigger current specified by the device data. It is usual to design and classify devices so that the minimum specified trigger current is considerably greater than the holding current. Triggering via the $i_G(p)$ electrode constitutes an SCR class of thyristor whereas triggering from the $i_G(n)$ gate electrode constitutes a complementary SCR class.

(ii) Forward breakover voltage

Increase of the anode forward voltage in the absence of gate signal causes the potential energy of the barrier at junction J_2 to increase. The carriers from electron–hole pair generation in J_2 are therefore accelerated at a greater rate. When the anode voltage (and therefore the junction reverse voltage) is sufficiently large avalanche breakdown of J_2 occurs. This condition should be avoided with SCRs but is the normal means of triggering in thyristor devices such as the pnpn switch and the trigger diac.

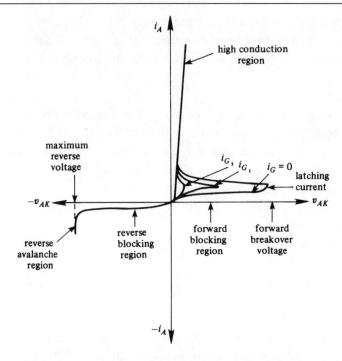

Fig. 2.22 Static voltampere characteristics of a thyristor device.

(iii) Irradiation methods

A thyristor device can be triggered on by irradiating any of the junctions but J_1 is usually chosen. Forms of radiation such as light, gamma or X-rays may be used but these can cause permanent damage in the crystal lattice. A suitable light source is provided by the xenon flash lamp. In the case of light, J_1 behaves as a photovoltaic cell causing a sophisticated carrier imbalance.

(iv) dv/dt triggering

A thyristor device may be triggered on by a rapidly rising anode voltage in the presence or absence of gate signal. This phenomenon often arises as a fault condition due to spurious 'spikes' of voltage emanating from other equipments – often other thyristor equipments. The so-called dv/dt effect occurs as a result of displacement current effect through J_2. The steeper the anode wavefront the more likely is the thyristor to trigger below its forward-blocking voltage rating as the capacitive reactance is less and more current flows. Typical dv/dt ratings to avoid spurious switch-on are in the range 10–20 V/μs. The response of a thyristor to a fast-rising voltage pulse can be affected by the previous conduction or blocking condition and, in particular, by how recently it was conducting forward current. Internally shorted gate devices can be constructed but are used infrequently on account of their poor $(di/dt)_{max}$ ratings.

(v) Temperature elevation

Any increase of the junction J_2 temperature of a thyristor device causes increase of the leakage current. Above 40 °C the leakage current theoretically doubles for every 7 °C rise of temperature. There is, also, an edge contamination effect. To avoid overheating it is customary to mount thyristor devices onto heat sink metal blocks – this is discussed in Chapter 1 above.

(B) SCR turn-off

In thyristor devices, due to the strong regenerative clamp action, once conduction is initiated the gate signal loses control. The control electrode level of hysteresis is infinite. The minimum current that the device will sustain in the absence of gate drive is called the holding current. To extinguish a conducting device the current must be reduced below the holding value long enough for turn-off to occur. Typical turn-off times lie in the range 10–300 µs.

In circuits with a.c. supply the usual method of SCR extinction is to permit the anode to follow excursions of the supply voltage. Therefore, for about half the time, the anode voltage is negative, which leads to natural commutation or switch-off. In circuits with d.c. supply where the supply polarity is time invariant, commutation of the SCRs has to be achieved artificially. This is often done by connecting an oppositely charged capacitor across the device and thereby establishing reverse-bias voltage.

Consider the mechanism of thyristor turn-off. When the excess minority-carrier concentration at junction J_3 reaches zero the junction becomes reverse biased and voltage builds up. After the forward current has decayed to zero, J_2 is forward biased but the junctions are still flooded with carriers. A short period of reverse current occurs, limited only by external circuit impedance, while the stored carriers reverse direction. Junction J_1 has a relatively low avalanche-breakdown voltage, compared with J_3 and J_2, because of high impurity concentrations on both sides. When the transient reverse current has decayed to zero the reverse voltage mostly appears across J_3. The carrier (hole) concentration near J_2 then reduced by recombination, occupying most of the turn-off time, to a value low enough to prevent junction breakdown on re-application of forward voltage. If the circuit contains significant inductance the transient reverse current can cause a large reverse-storage-voltage transient across the device. Furthermore, if a forward voltage is reapplied before the charge carriers have decayed the device may trigger spuriously.

(C) Ratings of thyristor devices

(i) The di/dt limitation

In some applications a thyristor switch is subjected to a very steep rise of current at and after switch-on. The 'vertical' regenerative triggering action occurs exceedingly quickly but the 'horizontal' migration of trigger charge in the base region takes place very slowly, i.e. between 10 and 50 m/s, as the translation takes place on drift field basis. The result is that only the region near the gate contact actually triggers. Therefore, unless limited by external means, the full load current passes through a small fraction of total die cathode surface near the gate. The thermal capacity of silicon is very small and it is a poor thermal conductor with the result that a hot spot is created. In this region the forward voltage across the device will then be lower than elsewhere as its temperature coefficient is negative. The charge associated with load current cannot spread readily. The result is damage to the crystalline structure of the die or cracks in it due to mechanical stresses as silicon is so brittle.

The electrical characteristics of the trigger signal have a significant bearing on this phenomenon. Flooding the gate almost instantly with a high level of charge (i.e. hard drive), even if it is reduced subsequently so as not to exceed the thermal rating of the gate structure, is much more satisfactory than applying charge to the gate gradually (i.e. soft drive).

This limitation of the time rate of current rise is common to all trigger controlled, silicon devices. It can be minimised by sophisticated fabrication techniques which increase the mutual gate–cathode line contact and so reduce the spreading resistance. Even so, there still remains an upper limit above which the device is likely to be damaged. This necessitates the designation of a parameter known as $(di/dt)_{max}$, which is paramount in high power devices. They will often withstand sustained overcurrent load or limited overheating but are likely to fail if this rating is exceeded. The load to be fed by the device must possess sufficient series inductance to limit the rate of rise of current – even when its initial and final magnitudes are low – to less than the $(di/dt)_{max}$ rating.

Developments over the last two decades have been mainly centred on improving the $(di/dt)_{max}$ rating of SCRs. Initially all gates were side gate oriented, many being point contact, with a corresponding high value of r_x.

This construction was often a cause of device failure. As the indexing of diffusion masks improved certain other devices emerged:

(a) centre gate,
(b) dual gate (consisting of two concentric gate rings – inside and outside an annular cathode – joined by a shorting strap),
(c) gate–cathode amplifying dual gate (master–slave device),
(d) gate–cathode amplifying dual gate (containing a second gate interdigitated with the main cathode).

The constructional feature known as interdigitation, now widely used in high power devices, consists of either of the forms below, or some combination of them:

- where many gate fingers radiate from the central amplifying region of the slice with arcing protrusions into the cathode recesses,
- interleaved comb-like structures of the gate and cathode.

(ii) Anode voltage ratings

Almost without exception devices are fabricated by commencing with low-dope p material which constitutes the anode. Junction J_3 is formed, as are the other junctions, by diffusing alternately n and p material.

It follows therefore that the highest resistivity silicon is on either side of J_3 and therefore it is this junction which 'holds off' reverse voltages, J_2 being forward biased. If junction J_3 reverse avalanches the device may well be damaged. Three different reverse voltage ratings may be specified by the manufacturer, who usually includes a safety margin: a peak reverse continuous value $V_{RC_{max}}$, a peak reverse repetitive rating $V_{RR_{max}}$ and a maximum reverse surge or transient rating $V_{RS_{max}}$ which may be 30% or more higher than $V_{RR_{max}}$. By stating the rating with open circuit gate the lowest possible value is specified.

The material on either side of J_2 has the same or a higher impurity level than the anode and therefore the junction J_2, which is reverse biased in the forward blocking state, normally has a lower value of 'hold off' forward voltage than that of J_3. The result is that if a device can 'block' a particular voltage in the forward direction it will certainly block that value and more in the reverse direction.

There are three different forward voltage ratings corresponding to the three reverse voltage ratings above. The forward blocking voltage V_{FC} is a continuous rating and represents the maximum value of forward anode voltage that the device will withstand, in the absence of gate current, without switching on. The V_{FC} rating varies inversely with temperature. Of higher value is the repetitive peak forward blocking voltage $V_{FR_{max}}$. The maximum forward voltage that the device will safely withstand at the given temperature is defined as the peak forward surge or transient rating $V_{FS_{max}}$, which is equal to or less than the corresponding reverse voltage $V_{RS_{max}}$. The application of a forward voltage greater than $V_{FS_{max}}$ is almost certain to damage the device. Between $V_{FS_{max}}$ and V_{FC} all devices of the type may switch on by breakover but will be undamaged.

(iii) The dv/dt limitation

If even a small-amplitude fast transient of forward voltage is applied to the device, it may well trigger as displacement current will flow through the depletion layer of J_2. To prevent this occurring manufacturers quote $(dv/dt)_{max}$ ratings for devices which must not be exceeded. The unwanted triggering of a thyristor can be avoided by the use of a snubber circuit. This is a series $R–C$ combination connected between the anode and the cathode. Snubber circuits are discussed in Chapter 3 below.

(iv) Anode current ratings

Semiconductor rectifiers, thyristors and transistors are basically designed and manufactured in type batches for current rating and subsequently selected and categorised on voltage ratings. The maximum mean forward current $I_{av_{max}}$ is the upper limit continuous current that may be permitted to flow in the forward direction under stated conditions of temperature, reverse voltage and current waveform. The thermal limitation in the immediate region of the die is the usual limiting factor. A thyristor device also has an r.m.s. current rating similar to the mean current ratings and applying concurrently with it. In a.c. applications that use SCRs in parallel or inverse parallel, it is important to note that the mean/r.m.s. current ratio changes with waveform. Also it is less than the mean/r.m.s. load current ratio.

SCRs with mean current ratings in excess of 3000 A and voltage ratings in excess of 6 kV are now (1995) available, Fig. 2.23. An outstanding feature of the silicon controlled rectifier (not shared by all other members of the thyristor family) is its ability to withstand surges of current whose absolute values may be orders of magnitude greater than the $I_{av_{max}}$ rating. The SCR will withstand temporary short circuit conditions where the current density through the die is very high. Provided it is not heated above a certain level the structure of the junctions will not be permanently disrupted.

The mechanisms of failure and their interactions are complex during excess surge conditions as different materials with differing mechanical thermal and electrical properties are involved. Thus, failure occurs at temperatures well below the one which gives rise to melting of the connecting solder or crystal-

2.3 Compound devices

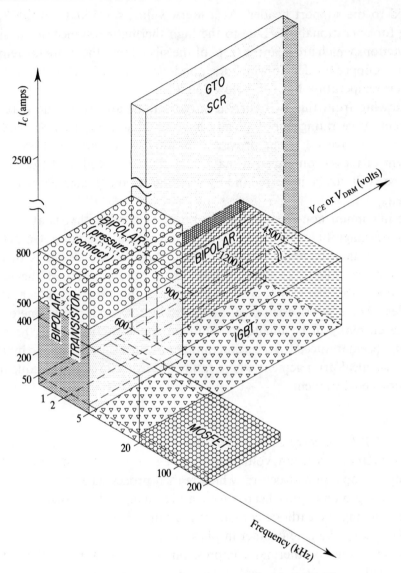

Fig. 2.23 Portfolio of power semiconductor devices (ref. TP 15).

line disruption. Suitable values for maximum ratings of surge current and encapsulation temperature can only be derived empirically. Manufacturers tend to be conservative in this area of rating but the following general guidelines apply. The highest temperature to which the device may be subjected is the inoperative storage temperature. This maximum rating may be as high as $T_{j\text{max}} = 250\,°\text{C}$. Mechanical stress is probably an important limiting factor as the die is at a uniform temperature but is soldered or thermal-pressure bonded to the support header. At a lower value, say 150 °C, is the $T_{case_{max}}$ rating for operational use. Due to the high thermal resistance of the silicon the junctions which are near the 'top' of the silicon and therefore far removed from the copper header, to which the anode/heat drain is soldered, will be at a higher temperature but still below 250 °C.

Following from this two current surge ratings are used. The maximum recurrent surge rating may typically have a value as high as one order of magnitude above $I_{av_{max}}$ but provided its duty cycle and the thermal time constant of the silicon are such that $T_{case_{max}}$ is not exceeded, the device performance will not be impaired in any way and its anticipated life will not be reduced.

The maximum non-recurrent surge rating is a value that is at least two orders of magnitude above $I_{av_{max}}$. The maximum operating temperature $T_{case_{max}}$ may then be temporarily exceeded. Provided the device is then disconnected and allowed to cool down before reapplication of the normal usage condition the device will perform the function for which it was intended. There is no assurance, however, that secondary parameters like leakage current will be unaltered. Sometimes the useful life of the device will only be guaranteed for a certain number of such surges, say one hundred. This is an ideal arrangement where faults are likely to arise infrequently in hazardous environments.

(D) Methods of SCR construction

Small (junction) area, low voltage devices, e.g. 1 A, 200 V, are produced by forming a simple pnp structure, which is then processed as follows:
   (a) fixing on a copper header with a high temperature solder,
   (b) alloying the cathode contact at the same time forming J_1,
   (c) welding the gate contact in place,
   (d) soldering (or thermal compression bonding) in place the cathode lead,
   (e) encapsulating.

The structure of a typical modern SCR is shown in Fig. 2.24.

As device areas (i.e. current ratings) increase, thermal bonding to the header reduces thermo-mechanical stress and permits a higher operating temperature. The diffusion formation of J_1 permits more precise fabrication and using planar techniques, which are possible because of this process, permits surface passivation.

The advantages of surface passivation – the deposition of a layer of glass over the surface – are twofold:

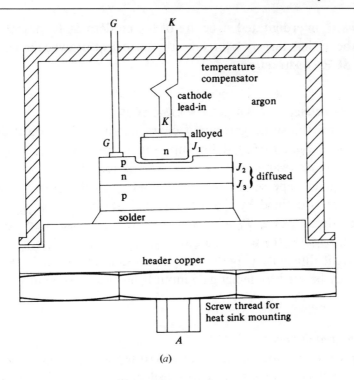

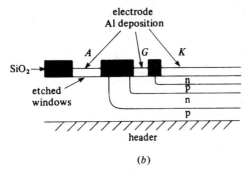

Fig. 2.24 SCR construction: (a) formation of a basic thyristor constructed on a pellet basis, (b) planar technique of construction incorporating a silicon dioxide passivation layer.

(a) edge contamination effects on leakage are reduced to insignificance as the junctions are fabricated after the silicon dioxide layer has been deposited by etching away 'windows' through which diffusion can take place,
(b) the encapsulation can be much simpler – plastic moulding techniques may be used rather than metal cans, which require glass–metal seals for lead-outs.

Switching devices and control electrode requirements

The fully diffused, interdigitated structure of the modern SCR, described in Section C(i) above, has largely overcome the historical drawbacks encountered in early SCR construction.

(E) Gate ratings and characteristics of SCRs

Figure 2.25 depicts the static gate input characteristics of a typical device where rejects have been excluded by imposing limits. The relatively low doping levels of SCRs results in considerable variation of the gate–cathode resistance for a given type of device. One boundary to the area of reliable firing in Fig. 2.25 is defined by the rectangular hyperbola $(i_G v_G)_{max}$ which defines the gate circuit power P_G max. When the gate is pulsed and the duty cycle is reduced, by (say) the use of narrower gate pulses, the $(i_G v_G)_{max}$ curve moves to the right thus enlarging the area of possible firing. An equivalent circuit which may be used to depict gate input behaviour is presented in Fig. 2.26.

(i) Static firing requirements

To drive a typical SCR gate circuit from a zero input impedance source at such a voltage level as to trigger the least sensitive device, i.e. 2 V, could well prove catastrophic for many of the devices of lower input impedance. A series

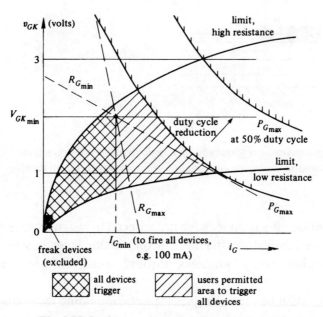

Fig. 2.25 Static gate characteristics of a typical SCR device.

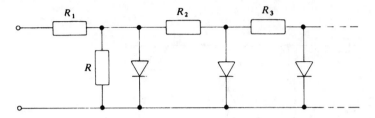

Fig. 2.26 Approximate equivalent circuit of the gate input characteristics.

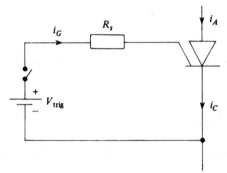

Fig. 2.27 Firing circuit incorporating external gate resistor.

resistance R_s, depicted in Fig. 2.27, must be included and the voltage increased in order that the lower impedance input devices are not flooded with charge and destroyed through gate burn-out. A range of suitable voltage sources and resistors then exist, shown by the limits $V_{GK_{min}}/R_{G_{min}}$ and $I_{G_{min}}/R_{G_{max}}$ in Fig. 2.25. Furthermore, these characteristics are for conditions when the device case is at 20 °C room temperature. The curves of all devices migrate downwards, i.e. V_G falls, if the case of the device is above 20 °C. Also the leakage through J_2 doubles every 7 °C above 40 °C and therefore the apparent sensitivity increases. The gate average power locus may also 'derate' with temperature, causing the $(i_G v_G)_{max}$ curve in Fig. 2.25 to move towards the origin.

It is necessary to provide a low resistance return path to earth in order that leakage current can readily flow and the device be desensitised. Also the reverse voltage rating of the gate (usually < 20 V) must not be exceeded either by the signal or by any interference. Fig. 2.28 shows a typical circuit which protects the device from such hazards. Diode D prevents the gate voltage from being more than (about) 1 V below the cathode voltage. The necessary value of R_s to ensure the triggering of all the devices of a given type, including the least sensitive, is not affected by D but is a function of R_G, as follows:

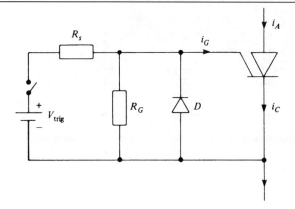

Fig. 2.28 Gate triggering arrangement incorporating gate voltage protection.

$$V_{trig} = \left(I_{G_{min}} + \frac{V_{G_{min}}}{R_G}\right) R_s + V_{G_{min}} \qquad (2.19)$$

With the typical values $I_{G_{min}} = 100\,\text{mA}$, $V_{G_{min}} = 2\,\text{V}$, $R_G = R_s = 30\,\Omega$, then

$$V_{trig} = \left(0.1 + \frac{2}{30}\right)30 + 2 = 7\,\text{V}$$

A device of limited current sensitivity but lower input resistance will receive more current but at lower voltage. A device of limited voltage sensitivity but higher input resistance will receive a greater voltage. Even if a pulse transformer coupling, which has an inherently low resistance path to cathode, is used for direct voltage isolation purposes it may be necessary to insert this type of network between it and the gate because of the low output impedance of the transformer when properly driven from a low-impedance source at its primary.

(ii) Pulse firing
The advantages of pulse firing have already been stated above, in connection with Fig. 2.25. If, however, the duty cycle is reduced below a certain level of time such that a pulse is only of a few microseconds duration, correspondingly more charge has to be supplied due to displacement current effects associated with the diodes of the equivalent circuit. Furthermore the maximum surge gate current rating may then be exceeded.

Where inductive loads have to be controlled, a single short pulse may be unsatisfactory as the current may not, by the time of pulse extinction, have risen to its sustaining value. Rather than increasing the length of the pulse, which would reduce the fundamental frequency of the spectral response, bursts of high frequency pulses should be generated in a train by (say) modulating an astable pulse circuit (e.g. at 20 kHz). Such waveforms can then be readily transmitted by dimensionally small, low rated pulse transformers with resultant economy.

2.3.2.2 Triac (bidirectional SCR)

The triac is a bidirectional single-gate SCR device that performs the circuit function of two SCRs connected in inverse-parallel, as shown in Fig. 2.29. Either of the electrodes T_1, T_2 can act as an anode and either as a cathode. The device can be triggered by either positive or negative voltages on the gate with respect to T_1. Operation is thus possible in any of the four quadrants of the anode voltage/current plane. The sensitivity of response to gating is different in the four quadrants because one SCR uses a p gate while the other uses an n gate, Fig. 2.29(c). The actual trigger current paths are shown in Fig. 2.30. Although the sensitivity of response to gate signal is non-uniform the sensitivities are sufficiently high as to make differences insignificant in use.

Now considerable difficulties arise over supplying the gate trigger signals of two separate SCRs used for full-wave rectification circuits or for a.c. voltage controller circuits. In each case the two gate signals must float with respect to each other and to earth, Fig. 2.29(b). With a triac only one trigger

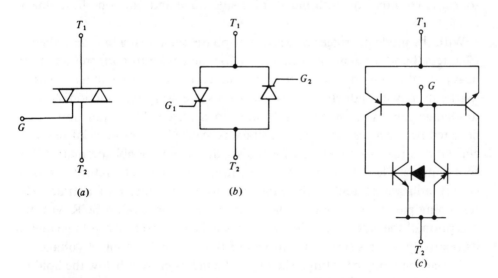

Fig. 2.29 The triac: (a) circuit symbol, (b) two-SCR equivalent, (c) transistor equivalent.

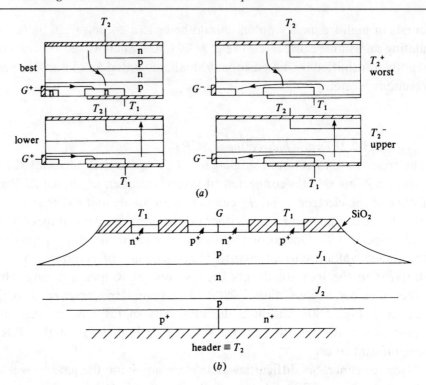

Fig. 2.30 Triggering of a triac: (*a*) trigger current paths, (*b*) device construction.

source is required for both halves of a sine wave and only one heat sink is needed.

With the mode of triggering known as phase-angle switching (described in Chapters 8 and 9 below) it is necessary for the device to go from a state of heavy conduction to one of high resistance at the zero voltage crossing (particularly with inductive loads). It is thus necessary for junction J_2 or J_3 to change rapidly from being saturated to depleted. If the junction is not depleted before a certain level of voltage is reached, the product of recovery current and reapplied voltage causes dissipation and possible permanent hot spot damage. The rate of rise of voltage rating $(dv/dt)_{max}$ therefore has to be strictly adhered to and is the most important parameter of a triac. The remaining parameters are essentially the same as those of an SCR with the exception of the gate, which does not avalanche with excessive voltage due to its relatively low resistance, regardless of the sign of the applied voltage.

A triac may only be extinguished by reducing its current below the holding current value for sufficiently long to permit recombination of the carriers to occur. Turn-off is usually achieved by reversal of the anode voltage. This book is concerned with the semiconductor switch as a circuit element. From a power flow standpoint it is more convenient and more general to consider the inverse-parallel SCR pair rather than the triac.

2.3.2.3 Gate turn-off thyristor (GTO)

The loss of control by the gate of SCRs and triacs, after triggering, can be highly inconvenient, except in a.c. circuits where reversal of the supply voltage causes inherent and desirable commutation in every cycle. In direct voltage supply applications special commutation circuits are needed which usually require a second device of similar rating to the one which needs commutating. Therefore, in d.c. traction and other similar applications, where direct voltage/current supply control is used, the adoption of a gate turn-off device would be logical and desirable. Silicon controlled switches of low power rating, with separate gate terminals for turn-on and turn-off, have been available for a number of years. These devices have now been superseded by the gate turn-off thyristor (GTO), which can be switched on or off from the same gate terminal. Switch-off can also be effected by anode voltage reversal, as with an SCR. High power devices are now available with ratings similar to those of the SCR, Fig. 2.23.

In considering the operation of a GTO it is reasonable to first consider why an SCR totally loses it gate control facility once it has been triggered. A high degree of regenerative feedback occurs due to the implicit tight coupling between the two transistor devices, Fig. 2.21. The upper pnp transistor J_2 and J_3 is the poorer of the two but, even so, this possesses a gain which is far too high to allow destruction of the clamping action of the positive feedback without device destruction due to A_{pc} being too low. Means must be introduced whereby the performance of one or both devices can be reduced by a predictable amount so that the total gain is much nearer to unity than to two in value. Moreover, values of less than unity must be achieved with practical voltage levels. The drastic step of shorting J_3 with carefully controlled resistive filaments was originally adopted. Alternatively, the material in the base of the upper transistor can be modified by gold doping. By either method the current which would normally flow through J_3 and cause positive feedback is routed through the equivalent of a low value resistance connected between $G(n)$ and the anode, lowering the level of hysteresis. The structure of a GTO is shown in Fig. 2.31. In more sophisticated, second-generation devices the interdigitation is so fine that indexing the shorting strips exactly under the cathode fingers is impossible to achieve consistently. Therefore, gold doping is adopted.

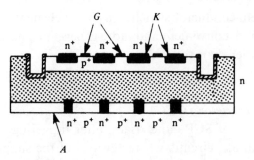

Fig. 2.31 Cross-sectional view of a GTO crystal.

With an SCR the $(di/dt)_{max}$ rating is a principal limiting parameter. With a second-generation GTO thyristor there is a high degree of structural interdigitation and the p layer is much smaller than in a conventional SCR (compare Fig. 2.31 with Fig. 2.30(a)). The result is that for a GTO the $(di/dt)_{max}$ rating is very high or may not even be quoted as a limitation at all on many commercial specifications.

A GTO thyristor has a low forward resistance but also a relatively low reverse resistance as J_1 constitutes the blocking junction and it has a low avalanche voltage. Reverse blocking has to be achieved by connecting a diode of similar rating in series or parallel with the GTO device – if this is not done device failure is likely to result. If reverse current flow is required the diode must be connected in inverse-parallel with the GTO thyristor so that a low impedance path is provided and the reverse voltage across the GTO is a forward voltage on the diode.

(A) GTO thyristor turn-on

The basic turn-on mechanism is very similar to that of an SCR of corresponding current rating. Internally there are considerable differences with regard to the shape and growth of the current plasma. Nevertheless, there is little externally observable difference, i.e. a sufficient positive current or voltage at the gate, with respect to the cathode, will trigger the device into a very low resistance 'on' state, after which the gate signal may be removed. If the gate current is insufficient then transistor action takes place. With large-area devices, negative gate or anode interference voltages are likely and a 'partial' off triggering may occur with possible damage to the device. This can be avoided by permitting, after a hard switch-on, a small residual positive gate voltage and current which is of no consequence from a dissipation standpoint but is of sufficient value to override the effects of any interference and maintain the existence of the full extent of the plasma.

(B) GTO thyristor turn-off

Because of the large line contact between gate and cathode, Fig. 2.31, the application of a negative voltage to the gate, with respect to the cathode, causes current (holes) from the plasma to flow to the gate from the p material. Hence, the external gate circuit must initially be able to cope quickly with considerable current flow, Fig. 2.32(a). The current flowing to the region of the cathode nearest the gate therefore reduces. Electrons cannot be emitted to the reverse-biased base and the plasma cannot be maintained. In less than a microsecond the whole of the region below the cathode becomes affected, the plasma being 'squeezed' down to small filaments which continuously reduce, Fig. 2.32(b), and switch-off occurs. During the 'off' state a slight negative voltage can usefully be left on the gate until a state change is required.

The switch-off mechanism has three distinct parts:
  (i)   a saturation time when the charge has to be extracted from the gate/base region, reducing the anode current to about 90%,
  (ii) a fall time when the plasma shrinks from 90% to 10%,
  (iii) a residual decay time when recombination takes place in areas remote from the fingers of the interdigitation and the current falls to zero.

All three regions contribute to commutation loss. Items (i) and (ii) can be drastically affected by external circuitry. Turn-off gains (i.e. ratio of anode current to gate current for switch-off) in excess of 10 can be realised. There is a critical level of anode current above which the application of negative gate pulses will cause device failure.

Like the SCR a GTO thyristor can also be extinguished by removal of the anode voltage or by the application of an appropriate negative anode voltage. This results in a significant reverse gate current. Both turn-on and turn-off occur abruptly, possibly in less than 1 µs, and therefore commutation losses can be small and junction hot-spot formation is unlikely. Because of this fast acting capability the GTO thyristor is useful in high frequency applications. When the external circuit contains inductance the fast turn-off of currents may result in large transient induced voltages which necessitate some form of protection. In gate turn-off power applications, turn-off still has to be preceded or accompanied by some adequate means of dissipating, diverting or storing the energy in the anode circuit before extinction can occur.

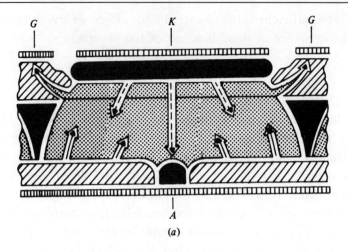

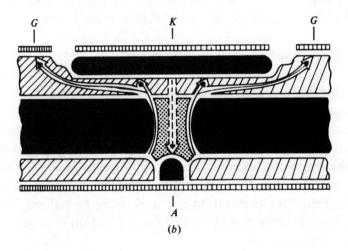

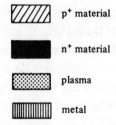

p⁺ material

n⁺ material

plasma

metal

Fig. 2.32 Conduction mechanism through a GTO crystal at turn-off: (*a*) during the 'on' state, (*b*) during the final stages.

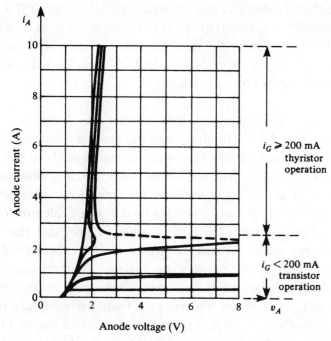

Fig. 2.33 Static voltampere characteristics of a GTO thyristor ($V_{PI} = 600\text{–}1000$ V, $I_{av} = 15$ A) (quoted from ref. 43).

(C) Static voltampere characteristics

Figure 2.33 shows typical anode current–voltage characteristics for a GTO thyristor. These differ considerably from those of an SCR, Fig. 2.22. For example, the latching current of a GTO thyristor is several amperes for large power devices, compared with 100–500 mA for SCRs of similar rating. If the gate current is less than the latching value the device behaves like a high voltage, low gain transistor and considerable anode current (up to 2 A) can flow. The associated incidental dissipation is then large.

(D) Limitations of power handling capability

The most significant parameter limiting power handling is the $(dv/dt)_{max}$ rate at which the voltage may be reapplied after turn-off by the gate, combined with the absolute value of the reapplied voltage. This parameter is considerably affected by the form of the negative signal applied to the gate, which affects the decay pattern of the plasma and the initiation of residual remote (away from the fingers) charge recombination. Good gate trigger sensitivity can be obtained, together with enhanced dv/dt capability, by second-generation techniques of a high interdigitation structure.

It is a disadvantage that the negative gate current required for turn-off is a significant fraction (typically one-tenth) of the anode current. In spite of this the use of GTO thyristors eliminates the need for the expensive SCR commutation circuits required in d.c. applications and in variable-frequency, force-commutated inverter drives for a.c. motors.

(E) Special trigger requirements for GTO devices

It was pointed out in Section 2.3.2.1(B) above that the great disadvantage of much SCR operation is the need for reliable commutation methods to extinguish the conducting devices. The need for commutation circuits adds significantly to the cost of SCR control in applications such as d.c. choppers and many a.c. inverter drives. This need is avoided by the use of gate turn-off devices such as the GTO thyristor. Turn-off is achieved provided that a suitable voltage waveform, which is negative with respect to the cathode, can be applied to the gate at the appropriate instant. Furthermore, in order to minimise switching losses when this reverse voltage is applied the current must rise quickly to a much higher value than the gate turn-on current I_{GT}. For example, for $I_A = 20\,\text{A}$, I_{GT_off} may well need to rise to $4\,\text{A}$. If external circuitry prevents this from occurring the decay of plasma filaments occurs slowly. The turn-off time then rises considerably with a result that the commutation losses also rise. It is absolutely essential therefore that a trigger circuit design ensures that the impedance of the gate turn-off current loop is as low as possible. The value of turn-off loss is also controlled by the external load and the way it affects the rise of reapplied voltage. Because of very high di_G/dt at turn-off, the designer must ensure minimal inductance in series with the gate when the 'off' trigger is applied. It may be necessary to use parallel shorting, solid dielectric capacitors across electrolytics on trigger supplies and short leads from signal source to gate. Turn-off loss is then minimised for a given load condition.

Although turn-off may be effected from quite low voltage sources, 1–2 V, it is usually derived from a 7–15 V bus. There is little virtue in exceeding 15 volts as the gate–cathode junction avalanches. If avalanching does occur damage does not arise because, due to the high doping levels used, breakdown occurs uniformly across the junction.

Turn-off is usually achieved via capacitance charging but an alternative resistive gate feed path should also exist, Fig. 2.34, to prevent spurious 'on' triggering during the 'off' state.

On account of load reactance the trigger current may fall to zero before the load current has risen to its latching value. The spread of the plasma in the device then becomes erratic and extinction may occur. The levels are not

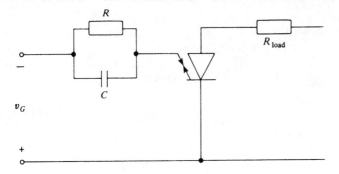

Fig. 2.34 Turn-off trigger circuit for a GTO thyristor.

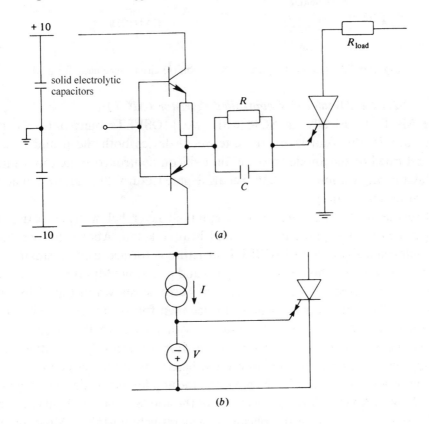

Fig. 2.35 Typical gate turn-off arrangement for a GTO thyristor: (*a*) circuit diagram, (*b*) equivalent circuit.

precise being dependent on junction temperature and geometry. It is therefore advantageous to have a gentle fall of gate current. A gate circuit for GTO thyristor operation is shown in Fig. 2.35.

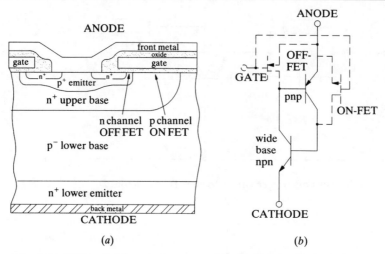

Fig. 2.36 MCT cell (*a*) typical cross-section (*b*) circuit schematic for p-MCT.

2.3.2.4 Metal–oxide controlled thyristor (MCT)

The MCT is basically an SCR with two MOSFETs built onto the gate structure. In this four-layer, three-terminal device both the p and n bases are returned to the anode 'surface' of a silicon controlled switch in a multicellular configuration, Fig. 2.36. Separate gate (return) and additional anode return terminals are provided.

The islands are diffused into an epitaxial layer below which exists the original cathode support material and heat sink tab. Above the two bases of each cell a signal-level MOSFET structure is formed using a single polycrystalline gate. By this means n-channel and p-channel MOSFET devices are automatically created above the relevant bases, as shown in Fig. 2.36.

A metallic layer over the whole surface then forms the anode. The application of negative or positive voltages, with respect to the anode, brings p-channel or n-channel devices into conduction, respectively. Negative anode voltage causes current to flow into the p base to turn the device on. Positive anode voltage short–circuits the n base to the anode causing device extinction, Fig. 2.36(*b*). An n^+ island at the centre of the anode of each cell aids the turn-off process by rendering the plasma degeneratively unstable. Alternative n-channel configured MCT devices are in the process of design and development.

The thyristor positive feedback effects, described in Section 2.3.2.1 above, occur during MCT triggering and extinction. Once either process is initiated the gate on each individual cell loses control. It is recommended that the 'on' and 'off' gate signals remain across the gate–anode during the static periods of the switching cycle.

It is essential that all the many (10^4+) cells change state simultaneously. This is particularly true at extinction, otherwise the load current merely redistributes to fewer cells. Due to hysteresis of the silicon controlled switch the fewer remaining active cells do not then extinguish even when the MOSFET channel resistance falls. Hot-spot failure results.

The gate trigger circuits must offer slew times faster than 200 ns, whereupon

$$t_{on} = t_{d_{on}} + t_r = 300 + 200 = 500 \text{ ns}$$
$$t_{off} = t_{d_{off}} + t_f = 700 + 1400 = 2100 \text{ ns}$$

Manufacturers' data sheets lay down stringent gating requirements which must be observed. The gate current required is not small because the gate input impedance is approximately a capacitance of 10 nF in parallel with a resistor of 10^9 ohms, being that presented by the MOSFETS. Miller amplification of the capacitance is insignificant.

With this device the gate leads must be short and the gate power supplies must be short-circuited at the trigger circuit by solid dielectric capacitors in order to minimise lead and source inductive effects. An external gate–cathode diode may be needed to prevent gate damage as there is no built-in Zener diode.

The switching times result in insignificant switching incidental dissipation but the on-state dissipation is very small. The maximum permissible current density during the on-state is the highest of any semiconductor switch and the dissipation is the lowest. A result of this is that the device encapsulation and associated heat transfer arrangements can be very small, particularly with low switching frequencies or soft loading applications. Voltages up to 20 V above the positive bus voltage are required to extinguish the device, which can be inconvenient as voltage level shifting is then required.

An inductive snubber must be used to limit di/dt effects. Voltage V_{DRM} suffers only 10% reduction due to dv/dt effects. With an MCT, reduction of the SOA rating requirements is needed where reactive components are required for switching aid purposes and then also serve as snubbers. In a.c. applications the MCT, like the IGBT, must be used with an inverse-parallel connected, fast recovery diode. The MCT is a new device. At the time of writing (1995) only pre-production samples of p-type devices, donated by Harris (Ref. TP16), were available for testing.

The authors' practical experience is that the MCT is a reliable and effective device. As alternative versions and triggering configurations become available the device will become widely adopted. It is particularly suited to high power, high voltage, non-alternating, low switching dissipation applications, particularly where equipment enclosure volume and weight considerations are decisive.

The triggering requirements of an MCT are easier than those of a GTO, particularly during turn-off, but more difficult than those of an IGBT. The on-state resistance is lower than that of a GTO and the absence of shorting strips reduces stress on the silicon. It seems likely that the MCT will supplant the GTO at higher power levels, in a corresponding manner to that in which the IGBT has supplanted it at lower power levels.

2.4 DEVICE SELECTION STRATEGY

As the variety of device types increases, the task of making the optimal choice of power switch for a specified circuit application becomes increasingly difficult.

The principal relevant factors are:
- voltage and current ratings
- switching frequency (slew rate)
- ruggedness under possible abuse
- ease of triggering
- availability and cost
- incidental dissipation
- need for aids and/or snubbers.

In the above list the first two items are the most important.

2.4.1 Voltage and current ratings

Device voltage and current ratings must be such as to satisfy the load impedance and supply bus voltage so that power can be delivered to the load.

2.4.2 Switching frequency (slew rate)

The maximum slew rate of the switching device dictates the fastest possible switching frequency. In turn, the switching frequency is related to the load impedance inertia and the maximum tolerable ripple. The lowest switching frequency gives the worst ripple while the highest frequency limits the maximum power delivery requirement. The interval between the switching frequency limits represents the range of power variation. Information about the switching frequencies of different classes of semiconductor power switches is included in Fig. 2.23 and Table 2.1.

In some switching applications it is necessary to use high slew rates. For example, there is the need to create pulse-width modulated square waves in inverters, with both high and low duty cycles. Where high slew rates are required, MOSFETs offer the best solution but problems may arise due to $(dv/dt)_{max}$ being exceeded.

2.4.3 Ruggedness against abuse

Abuse of power devices comes in the following forms:
  (a) occasional overvoltage, due to lightning or the proximity to other equipment, is inevitable, unavoidable and unpredictable, and
  (b) current surges due to accidental short circuiting of the output (i.e. reducing the load impedance to zero) will occur.

Attempts to minimise the effects of these hazards by circuit protection techniques are discussed in Section 3.2 of Chapter 3 below. SCRs are particularly rugged, being two decades more rugged than corresponding bipolars. MOSFETs are also rugged, being limited only by thermal considerations. On the other hand, bipolar devices, especially modern high-speed ones, are particularly poor in this respect.

In situations where repeated overcurrents are likely to occur, provided that these are not excessive (whereupon SMART devices can cut them out), MOS devices are less likely to suffer damage than BJT devices. This is because of the thin base sections of BJTs and the better the quality of BJT the poorer is the overcurrent performance. Moreover, a falling value of h_{FE} at collector current values, near the $I_{CE_{max}}$ rating, creates significant control electrode incidental dissipation. Excess voltage that causes the breakdown of semiconductor device junctions usually results in excess current surges.

Both single and compound devices possess phantom parasitic elements which, as a result of careful manufacture, are often rendered inert in normal applications. These phantom elements range from various forms of capacitance to loosely coupled BJT devices. When device failure occurs through abuse, this invariably happens because a phantom element of the device has been excited by excess electrical stimulation.

2.4.4 Ease of triggering

With higher current level BJT and Darlington devices, the base current, during both the on and off states, has to be accurately profiled which requires sophisticated triggering circuits. But the SOA of the base of a BJT is large and therefore damage is unlikely.

The GTO class of devices has probably the most challenging set of trigger circuit requirements. One feature is that the trigger circuit power supply has to float independently of the principal electrode power supply. For example, three-phase chopper systems require a minimum of four such trigger circuit floating power supplies. Such expenditure would only justify the adoption of GTO switches in installations of high power and high cost.

Trigger circuit requirements for an MCT system are much less stringent than for the GTO alternative but the switching action still needs to be accurately profiled.

MOSFET and IGBT devices have the simplest control circuit requirements. Both classes of devices require comparatively large voltage excursions (i.e. 10–12 V) to cause them to become properly saturated. This cannot be derived directly from logic 5 V level circuits so that amplification is required. Specially configured devices are available that can be fully saturated with logic level gate signals but the BV_{GS} and BV_{DS} values are presently limited to low levels.

2.4.5 Availability and cost

The cost of the switching devices must be held in relation to the cost of the total system design. In general it is better to use devices with ratings obtained in mainstream manufacturing batches. It is wise to avoid devices with unusual combinations of voltage and current ratings (e.g. high voltage with low current) as these may be difficult to duplicate and often represent freak extremes of production population batches.

Theoretically a BJT offers lower on-state voltage and dissipation than a MOSFET, because of its lower $R_{CE_{sat}}$. But, for $V_{DS} < 200$ V, the MOSFET has proved to be the better device. The falling price of MOS devices, per unit chip size, and the high production yields have resulted in very low values of $R_{DS_{on}}$ (e.g. for 25 A devices, $R_{DS_{on}} < 40$ mΩ). It therefore becomes economic to under-use MOS devices so that the BJT cannot compete in economic terms. At higher voltages the dopant concentration in the parasitic epitaxial BJT is lower, the $R_{DS_{on}}$ rises and the BJT device becomes dramatically less economic.

2.4.6 Incidental dissipation (ID)

The incidental dissipation of the switches affects the design and size of the controller enclosure. This usually has a greater bearing on cost than the choice of switching device, particularly with high capital cost installations.

If the design calls for the operation of (say) paralleled MOSFETs this will obviously affect the ID and therefore the necessary enclosure size and cost.

2.4.7 Need for aids and/or snubbers

The inclusion of a switching aid and/or snubber in a design also affects the heat transfer arrangements and the enclosure size because the aids or snubbers themselves dissipate heat energy. For high voltage systems, in particular, this may have a more significant effect than any other individual consideration.

2.5 REVIEW QUESTIONS AND PROBLEMS

Power transistors

2.1 With the aid of a vertical section diagram describe the construction of a triple diffused bipolar transistor pointing out its advantage over other methods of fabrication.

   A voltage is rapidly applied to the collector positively with respect to the emitter. Explain what happens in the device both with the base open and returned to the emitter via an external resistance. What are the external symptoms of the phenomenon? Under what situation is this condition dangerous and why?

   List two methods of obviating or minimising the problem, briefly explaining the basis upon which they operate.

2.2 If a bipolar device possesses an f_T of 2 MHz, why might it not operate satisfactorily in a 10 kHz chopper circuit?

2.3 Sketch a Bode plot depicting the forward current gain, $|h_{FE}|$, versus angular frequency of a bipolar transistor.

   Locate the transition frequency ω_T on the locus and explain why it is never measured directly. Say how it is measured.

   Discuss the significance of the plot with reference to the use of bipolar power devices in high frequency inverter circuits where square wave current amplification is required.

   By reference to the construction of a single diffused double sided bipolar device, explain why high values of ω_T and power handling capability cannot be achieved.

2.4 When a bipolar device is used in situations where the collector–emitter voltage is likely to alternate, special problems arise. Discuss how they occur and how they relate to the manufacturer's ratings of reverse safe operating area and reapplied $(dV_{CE}/dt)_{max}$.

Give methods of minimising the reverse dissipation and discuss their relative merits.

2.5 Describe the phenomena that occur if the collector of a bipolar device is
   (a) connected instantly to a positive bus with the base open circuit, and
   (b) thermally stabilised by a single resistor R_{BE} from base to earth, the bus then being subjected to switched alternating supply, as may occur in an inverter.

   Give three methods by which such undesirable combined operation can be prevented.

2.6 Draw the basic circuit of a simple power Darlington connection of a signal and a power bipolar transistor.

   With this device, it may be considered that it is only marginally saturated and some 'on' state dissipation saving is sacrificed in order that switching loss can be reduced. Explain this statement and refer in your discussion to the embedding of these devices in high frequency inverters.

2.7 What are the advantages of power bipolar transistors over FETs and SCRs for use in inverter circuits for power applications? What limits their usage?

2.8 With the aid of a vertical section diagram, describe the construction of a MOSFET.

   Explain the following:
   (a) how to turn on the device,
   (b) how to turn off the device.
   (c) why the device has no reverse-voltage blocking capability,
   (d) how the on-state resistance can be modulated.

2.9 In view of the answer to Example 2.8 Part (d), describe the modifications required to fabricate an insulated gate bipolar transistor (IGBT) from a MOSFET device. Illustrate your answer with a sectional diagram.

   In switched-mode applications, why does a bipolar device always have to be heavily driven with excessive base current? Hence, comment on the advantage of using IGBT devices as an alternative.

2.10 By recourse to an equivalent circuit including the parasitic elements, discuss the mechanisms of turn-on and turn-off of an IGBT. Explain why snubberless operation is achievable and indicate how external components connected to the gate circuit can control the turn-off time.

Thyristor family – SCRs

2.11 In normal fabrication the p gate sensitivity of an SCR type of thyristor is more than that of the n gate. Why is this the case?

2.12 Why does the gate input resistance of an SCR differ from that of a diode?

2.13 Why does an SCR have a $(di/dt)_{max}$ rating?

2.14 Interdigitation is used in both power bipolar and SCR devices but its effect within them is quite different and its introduction improves different parameters. State what these parameters are and how the improvement takes place in an SCR type of thyristor.

2.15 Explain why it is easier, on a quantity production basis, to manufacture high power SCRs than power transistors. Thus explain why SCRs have much higher current overload ratings than bipolars. State the difference between recurrent and non-recurrent surge ratings.

2.16 Draw the two bipolar transistor equivalent model of a silicon controlled switch and evolve a formula for the anode current identifying and establishing conditions for the various modes of triggering. What is the origin of the dV/dt mode of triggering?

Explain how the actual gate characteristic of a side gate SCR differs from that indicated by the model. By means of a more representative gate circuit, explain why such devices have a very poor $(dI/dt)_{max}$ specification. Briefly describe some of the techniques which are adopted in the manufacture of high current rating devices to minimise the effect.

2.17 The two bipolar transistor model of the thyristor family is only accurate for a relatively small-area, low dissipation device. The gate input characteristic is quite different for a large-area device. Draw an equivalent circuit which would more accurately represent the triggering behaviour of a side gate device controlling a maximum average current of, say, 10 A. Hence, explain why the inclusion of a gate–cathode connected capacitor would significantly improve the dV/dt rating of a small device but would fail to do so on a larger one.

2.18 Using a simple model of the gate input circuit of an SCR, explain why it is advantageous to use a 'hard' pulse drive rather than a steady current to trigger the device.

Multifilar wound transformers used in the 'starved' core mode are convenient for triggering SCRs in circuits where several devices operating at differing cathode voltages need to be triggered from the same source, particularly when the load is inductive. Explain why this is so, relating your discussion to the first part of the question.

2.19 For an SCR type of thyristor indicate on a sketch graph of the firing characteristics the following loci:

  (i) $I_{g_{max}}$
  (ii) $V_{g_{max}}$
  (iii) $P_{av_{max}}$
  (iv) P_{max}

under pulsed conditions with a 10% mark to space ratio.

Also indicate the area where (a) all devices of one type number will trigger and (b) none of the devices will trigger.

Explain why the device possesses a $(dI/dt)_{max}$ rating and comment on the likely subsequent external electrical symptoms if this rating is exceeded.

2.20 Whilst the trigger input characteristic of an SCR type of thyristor is basically that of a diode, its actual slope varies considerably from device to device within a single production batch. Explain why this is so and sketch the associated I_G–V_G characteristics for the upper and lower limit devices of the batch and indicate on your graph regions in which
   (a) all devices do not trigger,
   (b) some but not all devices trigger,
   (c) all devices trigger.

   Trace in the maximum gate dissipation locus and explain how it can be varied as the duty cycle of the trigger waveform is altered.

2.21 If a forward voltage is applied to the anode of an SCR type of thyristor which is below the breakover voltage it may well switch on, particularly if the device is in a relatively high ambient temperature or if the voltage is applied rapidly. Explain why this is so.

   A special device may be constructed which minimises these effects, particularly the latter. With the aid of a sectional diagram, describe its construction and why it is effective.

2.22 A batch of side gate type SCR thyristors has been fabricated and the gate input resistance characteristics measured.

   Draw a graph indicating typical upper and lower value loci that may be anticipated and explain why the variation exists.

   It is found that all devices trigger at levels below 100 mA or 3 V. Sketch these limits on your graph and indicate by shading the permissible triggering area and identify the other limiting factor.

   Differentiate between 'hard' and 'soft' drives and explain by recourse to an equivalent circuit why a hard low duty cycle drive is preferable to a soft one.

2.23 An SCR is to be triggered from a pulse transformer of zero output impedance capable of generating 10 V.

   Two resistors and a diode are available. Give a connection diagram of these and evolve a relationship between the two resistors which would ensure all devices were triggered and protected from reverse voltages. Choose two suitable values of resistance and state why you have selected them.

2.24 Draw the gate input characteristics of a batch of SCRs, indicating limit loci and user boundaries for all devices to be safely triggered. Explain why wide variations of input characteristics exist, and hence obtain an expression for the value of a series gate resistance which will protect the device when triggering from a specific direct voltage source.

   It is often found desirable to include also a further resistor and a diode, particularly if the trigger signal is capacitor fed. Explain the purpose of these components and draw a circuit which shows how these components should be connected. Evolve a new design formula for calculating the value of the series resistor if the second resistance is 50 Ω.

2.25 An SCR has a minimum trigger voltage of 1.75 V and a minimum trigger current of 100 mA. The gate–cathode resistor R_G has the value 20 Ω. If the trigger supply has a constant value of 6.2 V what is the smallest value of series gate resistance R_s to ensure turn-on of the least sensitive devices of this type?

2.26 When an SCR is used in a delayed angle firing application the device heat dissipation varies considerably with triggering angle despite the average current remaining the same. Explain, with the aid of a sketch graph of dissipation versus average current flow, why this is the case.

Thyristor family – triacs

2.27 A triac serves the same basic circuit purpose as an inverse-parallel connection of two silicon controlled rectifiers and yet the latter configuration is universally adopted at high power levels. Discuss this situation giving the reasons.

Describe, with the aid of a sectional diagram, the construction of a triac.

Draw an equivalent circuit of the device and explain why the trigger sensitivities in the quadrant modes of operation differ even though this is not apparent from the equivalent circuit.

2.28 (a) Draw the two bipolar device equivalent circuits of the thyristor family and cite the resultant expression for the anode current, giving the criterion for triggering.

List the various methods by which triggering can be evoked.

(b) Show how the equivalent circuit can be extended to represent the behaviour of a triac.

By means of diagrams, describe how the device has four different modes of triggering.

Also, explain why a $(dV_{T_1}/dT)_{comm}$ maximum rating is required.

Explain why this device is superior to an inverse-parallel pair of silicon controlled rectifiers in a.c., full wave, low power flow-control installations.

Thyristor family – gate turn-off thyristors (GTOs)

2.29 Draw the equivalent circuit model of a four-layer, three-junction semiconductor switching device.

Derive the equation for the anode current I_A, assuming that the current gains of the two constituent transistors are α_p and α_n, respectively.

Based on this derived formula, give the condition necessary for turn-on.

Draw the principal electrode static characteristics of a typical GTO, identifying two distinct areas of operation.

2.30 Two forms of fabrication exist for making GTOs. Describe, with the aid of a sectional diagram, the turn-off mechanism in the shorting strip type. Discuss the relative advantages and disadvantages of increased interdigitation and hence explain why only the other form can be used with fine geometry interdigitation.

Sketch typical static characteristics of a GTO device and point out where significant differences occur between the turn-on trigger requirements of this device and an SCR.

2.31 Draw a 'vertical' sectional diagram through the monolithic chip of a shorting strip type of GTO. By recourse to a simple two bipolar transistor equivalent circuit, explain why it is possible to achieve a greater power handling of direct voltage supplies than with a single bipolar device, a power Darlington or an insulated gate bipolar device.

Discuss what occurs if the anode/cathode voltage is reversed sufficiently to exceed the avalanche voltage of the gate–cathode junction.

Distinguish between the different trigger requirements of SCRs and GTOs.

2.32 What feature of the construction makes a GTO have its unique facility and what penalty is paid for it?

2.33 Draw the static characteristics of a typical GTO device. Significant differences exist between the turn-on characteristics of this device and those of an SCR. Identify them.

Explain why it is essential to turn the device 'off' from a near zero impedance voltage source, discussing typical 'off' trigger circuit precautions. Draw the basic composite trigger circuit if a 'hard on' drive is also adopted. What special precautions have to be taken in respect of the source impedance of the 'off' circuit?

2.34 Draw a circuit with two bipolar devices to represent the thyristor family of devices.

Use this diagram and an additional component to illustrate how the GTO device is realised, and explain the purpose of this component.

Draw the static characteristics of a GTO and identify two distinct regions of operation thereon. Then explain briefly why the device is more convenient than an SCR for control of power flow from direct voltage supplies.

Device selection strategy

2.35 Discuss the relative control and output electric characteristics of triacs, SCRs, power field effect transistors and bipolar transistors.

There are the following four applications in which power devices are required:

(a) a 10 W, 3–30 MHz transmitter output stage driven from a master oscillator via a buffer output stage incorporating a step-down transformer,

(b) a 1000 h.p. underground mine hoist operating an alternating current in adverse environment where short circuits frequently occur,
(c) a 300 W audio power amplifier output stage suitable for public address, and
(d) a 300 W sound-to-light convertor unit.

State which type of device you would incorporate in each design. Give the reasons why you have selected the device and also why you have rejected the remainder.

2.36 List the characteristics of an ideal electrically operated switch and write an essay on the comparative merits of power bipolar and field effect devices and SCRs, including triacs, when they are used in this capacity. Include in your discussion comments about the weak characteristics of each device and use your arguments to support your selection as a design engineer of the best devices in the following applications:
(a) high power alternating current supply traction systems in climatic conditions where lightning tends to cause frequent short circuits,
(b) low power theatre stage lighting with dimmer control,
(c) high frequency chopper stabilised direct voltage supplies,
(d) high efficiency direct voltage low level servo motor control and
(e) high level direct voltage traction.

2.37 Compare the relative advantages of triacs and SCRs and hence give two examples of application: one in which a triac is particularly suitable and the other where a silicon controlled rectifier type of thyristor is preferable. Give the reasons underlying your selection.

3
System realisation

3.1 INTRODUCTION

The final decision regarding the choice of switching devices and ancillary components must depend on the overall design for total system realisation. Any choice between device options is likely to have an interactive effect on other components. For example, the selection of a certain switching device for use in a particular circuit may or may not require switching aids and/or sophisticated protection. A relatively inexpensive electronic switching device may require ancillary circuitry that is prohibitive in cost, volume requirement, heat generation, excessive control circuit requirements, etc.

A consideration of some of the circuit features listed in the previous paragraph may be grouped under the general heading of 'circuit protection.' This, in turn, may be divided between the major classifications of 'preventive protection' and 'abuse protection.'

Preventive protection is required so that when a device is operational it is embedded in an environment which prevents certain secondary device ratings from being exceeded. A typical example of this is the use of snubber circuits, which is adopted to prevent spurious triggering due to transients.

Abuse protection is required to cope with fault conditions which cause the normal primary ratings of the device to be exceeded. Typical examples are overvoltage or current surges arising from short circuiting of the load. In terms of Fig. 2.2 above, the inside or circuit user shell then dilates (i.e. enlarges) to transgress the outer or absolute maximum limit shell of the particular device design. Under such fault conditions, the device environment must be changed as quickly as possible to remove the excesses of treatment and restore the original conditions.

System design is also contingent upon the integrity of the control electrode signals. The control circuit must be protected from contamination due to interference from the main circuit by conduction, electromagnetic coupling or by radiation. A faithful transference of the intended control signal to the normal electrode can be facilitated by the appropriate use of circuit layout, screening, interfacing and isolation.

3.2 PREVENTIVE PROTECTION CIRCUITRY

3.2.1 Voltage and current snubber circuits

3.2.1.1 Requirement for snubber circuits

Most semiconductor switching devices contain certain weak features in their construction. These weaknesses are not foreseeable in the conceptual device design stage and are usually not apparent under static conditions.

The rapid application or reapplication of excess voltage to a device, in the absence of a control electrode signal, can cause the device to trigger spuriously, or to turn on partially. This malfunction can damage not only the control circuit or the switches but also the source and/or load. When such rapid and excess surges are anticipated then voltage snubbers must be employed, as opposed to switching aids which do not respond to rapid surges.

Similarly, if the device current is allowed to rise too rapidly, damage can result due to current localisation and hot-spot generation at a device junction, resulting in device destruction. To prevent excessive rate of rise of current it is obligatory to use current snubbers.

3.2.1.2 Design of snubber circuits

In order to avoid damage the $(di/dt)_{max}$ rating of a device must not be exceeded. The easiest way to avoid this situation occurring is to include a current snubber in the form of inductance L in series with the load as depicted in Fig. 3.1(a). Fortunately, with many loads, there is inherently a considerable inductive component and separate protection may be unnecessary. The exponential rise of current after switch-on towards its final value $V_{bus}/R = I_{max}$ is depicted in Fig. 3.1(b) and satisfies the relations

$$i = I_{max}(1 - \varepsilon^{-Rt/L}) = \frac{V_{bus}}{R}(1 - \varepsilon^{-Rt/L}) \tag{3.1}$$

$$\frac{di}{dt} = I_{max}\frac{R}{L}\varepsilon^{Rt/L} \tag{3.2}$$

When $t = 0$,
$$\frac{di}{dt} = \left(\frac{di}{dt}\right)_{max} = \frac{RI_{max}}{L} = \frac{V_{bus}}{L} \tag{3.3}$$

For safe working the inductance L must be such that
$$L > \frac{V_{bus}}{\left(\dfrac{di}{dt}\right)_{max}} \tag{3.4}$$

If the circuit also contains a voltage snubber, (dashed in Fig. 3.1(a)), this can adversely affect the di/dt rating. When the SCR is off it has voltage V_{bus} across it. After switch-on the snubber capacitor, also charged to V_{bus}, discharges through the switch with a maximum (initial) current
$$I_{s_{max}} = \frac{V_{bus}}{R_s} \tag{3.5}$$

Strictly speaking the current rating is violated if the current jumps by only a small amount. Since this is an amplitude-sensitive phenomenon, however, small instantaneous changes can be tolerated.

The total SCR current at switch-on is therefore
$$\begin{aligned} I_{t=0} &\approx I_{max} + I_{s_{max}} \\ &= V_{bus}\left(\frac{1}{R} + \frac{1}{R_s}\right) \end{aligned} \tag{3.6}$$

Combining (3.3) and (3.6) gives a further expression for the minimum safe value of L to avoid di/dt breakdown

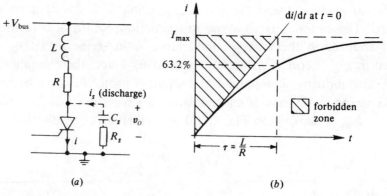

Fig. 3.1 Use of series inductance to limit the (di/dt) in an SCR circuit for d.c. application: (a) circuit diagram, (b) current versus time.

$$L > \frac{V_{bus}\left(\frac{R+R_s}{R_s}\right)}{\left(\frac{di}{dt}\right)_{max}} \qquad (3.7)$$

A comparison of (3.4) with (3.7) shows that if $R_s \gg R$ (which is true in many applications) the two expressions are identical so that the voltage snubber has little effect on the value of L in the current snubber.

In order to prevent accidental triggering by step changes of anode voltage, circuitry must be included to limit the rate of rise of voltage to a value which is less than its $(dv/dt)_{max}$ rating. The load reactance may be inductive and sufficiently large for the above condition to arise implicitly. Care must be taken, however. Measurements at low frequencies may be misleading as stray capacitance and resonance effects can often lead to a circuit which may appear to be inductive at power frequencies but is, in fact, capacitive at high frequencies, i.e. for sharply rising pulses.

The most satisfactory method is to include a capacitance C_s in parallel with and at close proximity to the device. This should form a time constant with the resistive part of the load, of such value as to limit the rate of voltage rise to less than that which would trigger the device spuriously.

(A) Resistive load

Suppose that the switch in Fig. 3.1(a) is off. After the sudden application of a step voltage V_1, the load is effectively the series R–L–C combination shown in Fig. 3.2(b).

Current i_s in Fig. 3.2 has an instantaneous value V_1/R and thereafter, if L is small, decreases exponentially

$$i_s = \frac{V_1}{R}\varepsilon^{-t/C_s R} \qquad (3.8)$$

Voltage v_o of Fig. 3.1 then becomes the charging voltage on the initially uncharged capacitor

$$v_o = \frac{1}{C_s}\int_0^t i_s\, dt = V_1(1 - \varepsilon^{-t/C_s R}) \qquad (3.9)$$

The time rate of change of v_o is

$$\frac{dv_o}{dt} = \frac{V_1}{C_s R}\varepsilon^{-t/C_s R} \qquad (3.10)$$

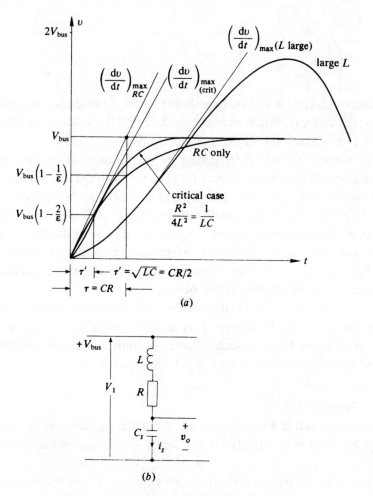

Fig. 3.2 Action of a snubber circuit: (*a*) response characteristics, (*b*) effective circuit.

which has a maximum value at $t = 0$. To prevent the $(dv/dt)_{max}$ rating of the switch from being exceeded it is seen that

$$C_{S_R} > \frac{V_1}{R\left(\dfrac{dv_o}{dt}\right)_{max}} \qquad (3.11)$$

where subscript R indicates a resistive load.

(B) Series inductance load

When the load contains significant series inductance L the action of the snubber corresponds to that of a series R–L–C circuit, Fig. 3.2(*b*), where R

3.2 Preventive protection circuitry

includes both the load and snubber resistances. Let a step voltage V_1 be applied to the previously unenergised circuit. The output voltage v_o is the voltage across the switch if the snubber resistor R_s is small compared with the capacitive impedance effect (which is usually the case), in Fig. 3.1(a).

The Laplace transform $v_o(s)$ to a step input $V_1(s)$ is given by

$$\frac{V_o}{V_1}(s) = \frac{1}{s}\frac{1}{LC_s}\frac{1}{s^2 + \frac{R}{L}s + \frac{1}{LC_s}}$$

$$= \frac{1}{s} - \frac{s + \frac{R}{2L}}{s^2 + \frac{R}{L}s + \frac{1}{LC_s}} - \frac{\frac{R}{2L}}{s^2 + \frac{R}{L}s + \frac{1}{LC_s}} \qquad (3.12)$$

The inverse Laplace transform of (3.12) gives

$$\frac{v_o}{V_1}(t) = 1 - \varepsilon^{-Rt/2L}\left(\cos\omega t + \frac{R}{2\omega L}\sin\omega t\right) \qquad (3.13)$$

where

$$\omega = \sqrt{\frac{1}{LC_s} - \frac{R^2}{4L^2}} \qquad (3.14)$$

Differentiating (3.13) and incorporating (3.14) gives the slope of the load voltage response,

$$\frac{d}{dt}\left(\frac{v_o}{V_1}\right) = \frac{R}{2L}\varepsilon^{-Rt/2L}\left(\cos\omega t + \frac{R}{2\omega L}\sin\omega t\right)$$

$$- \varepsilon^{-Rt/2L}\left(-\omega\sin\omega t + \frac{R}{2L}\cos\omega t\right)$$

$$= \varepsilon^{-Rt/2L}\left(\frac{\frac{R^2}{4L^2} + \omega^2}{\omega}\right)\sin\omega t$$

$$= \frac{\varepsilon^{-Rt/2L}\sin\omega t}{\omega LC_s}$$

or

$$\frac{dv_o}{dt} = \frac{V_1\varepsilon^{-Rt/2L}}{\omega LC_s}\sin\omega t \qquad (3.15)$$

The maximum value of this slope is obtained by further differentiation and equating to zero.

100   System realisation

$$\frac{d^2}{dt^2}\left(\frac{v_o}{V_1}\right) = \frac{1}{\omega LC_s}\left(\omega\varepsilon^{-Rt/2L}\cos\omega t - \frac{R}{2L}\varepsilon^{-Rt/2L}\sin\omega t\right)$$

$$= \frac{\varepsilon^{-Rt/2L}}{\omega L^2 C_s}\left(\omega L\cos\omega t - \frac{R}{2}\sin\omega t\right) \qquad (3.16)$$

$$= 0, \text{ when}$$

$$\tan\omega t = \frac{2\omega L}{R}, \text{ and}$$

$$\sin\omega t = \frac{\omega L}{|Z|} = \frac{\omega L}{\sqrt{\left(\frac{R}{2}\right)^2 + \omega^2 L^2}} \qquad (3.17)$$

Combining (3.14), (3.13) and (3.16) gives an expression for the maximum value of the slope of the output voltage, i.e. the maximum dv/dt across the device,

$$\left[\frac{d}{dt}\left(\frac{v_o}{V_1}\right)\right]_{\max} = \frac{\varepsilon^{-Rt/2L}}{\sqrt{LC_s}} \qquad (3.18)$$

Now the instant of time $t_{\max}$ at which the maximum slope has its maximum value is given by (3.16), from which

$$t_{\max} = \frac{1}{\omega}\tan^{-1}\left(\frac{2\omega L}{R}\right) \qquad (3.19)$$

Combining (3.15) and (3.16) gives

$$\left[\frac{d}{dt}\left(\frac{v_o}{V_1}\right)\right]_{\max} = \frac{\varepsilon^{-(R/2\omega L)\tan^{-1}(2\omega L/R)}}{\sqrt{LC_s}} \qquad (3.20)$$

It is seen in (3.20) that the introduction of load inductance reduces the necessary value of the snubber capacitance.

(i) Small L
When L is very small the behaviour reverts to the form described in Section 2.2.1.4(C) of Chapter 2. Output voltage $v_o(t)$ then has the single exponential form of (3.9) rather than the more complex form of (3.13) and the $(dv/dt)_{\max}$ rating is given by (3.11).

(ii) Critical damping value of L
The smallest value of L, which results in a tangent of zero slope at the origin at $\omega \to 0$, is given, from (3.14), by

$$L = \frac{R^2 C_s}{4} \tag{3.21}$$

The value of t to give maximum slope of $d/dt(v_o/V_1)$ is given in (3.16) when $\tan \omega t \to \omega t$ so that

$$t_{max} = \frac{2L}{R} = \frac{RC_s}{2} \tag{3.22}$$

Combining (3.18) and (3.22) gives

$$\left[\frac{d}{dt}\left(\frac{v_o}{V_1}\right)\right]_{max} = \frac{0.37}{\sqrt{LC_s}} \tag{3.23}$$

Substituting (3.21) into (3.23) and rearranging gives

$$C_{SL\text{crit}} = \frac{0.74 V_1}{R\left(\dfrac{dv_o}{dt}\right)_{max}} \tag{3.24}$$

It is seen that the value of snubber capacitance is reduced compared with the value for resistive load, in (3.11).

(iii) Large L

When L becomes very large $R/L \to 0$ and the load becomes resonant. The natural frequency of oscillation is obtained from (3.14),

$$\omega = \frac{1}{\sqrt{LC_s}} \tag{3.25}$$

With large L the voltage response $v_o(t)$ in (3.9) becomes a periodic function

$$v_o = V_1(1 - \cos \omega t) \tag{3.26}$$

as shown in Fig. 3.2(a).

The maximum slope of the dv_o/dt characteristic occurs at time t_{max} obtained from (3.16) and (3.25),

$$\tan \omega t_{max} = \frac{2}{R}\sqrt{\frac{L}{C_s}} \tag{3.27}$$

With large L, $\tan \omega t_{max}$ is always large and $\omega t_{max} \to \pi/2$ radians. Therefore, using (3.25),

$$t_{max} \simeq \frac{\pi}{2}\sqrt{LC_s} \tag{3.28}$$

From (3.26) the derivative dv_o/dt is

$$\frac{dv_o}{dt} = \omega V_1 \sin \omega t = \frac{V_1}{\sqrt{LC_s}} \sin \omega t \qquad (3.29)$$

This has its maximum value at $\omega t = \pi/2$ or $t = (\pi/2)\sqrt{LC_s}$, as seen in Fig. 3.2(a).

Therefore,

$$C_{s_L} = \frac{V_1^2}{L\left(\dfrac{dv_o}{dt}\right)^2_{\max}} \qquad (3.30)$$

With fixed $(dv_o/dt)_{\max}$ rating, the value of snubber capacitance C_s becomes small for very large L. Typical values are given in Example 3.1. As the maximum slope occurs at higher voltage with increasing L a conservative approach should be adopted in selecting C_s because $(dv/dt)_{\max}$ ratings are voltage level and step size sensitive. The variation is caused by the dependence of C_{dep} upon voltage.

3.2.1.3 Worked examples on snubber circuits

Example 3.1
An SCR has a $(dv/dt)_{\max}$ rating of 50 V/μs. It is to be used to energise a 10 Ω resistive load and it is known that step transients of 500 V occur.
(a) What is the minimal size of snubber required to avoid unintentional triggering?
(b) If the load is replaced by a 100 mH inductance – say an induction motor on low load – to what value can the capacitor be reduced to avoid spurious triggering?

Solution. Operation is represented by the circuit of Fig. 3.1(a). From (3.11),

(a) $\left.\dfrac{dv_o}{dt}\right|_{\max} > \dfrac{V_1}{C_s R_L}$

$$50 \times 10^6 > \frac{500}{C_s \times 10}$$

$$\therefore \ C_s = 1\,\mu\text{F}$$

(b) A load inductance of 100 mH represents a large inductance. From (3.30), the worst case condition is

3.2 Preventive protection circuitry

$$C_{SL} = \frac{V_1^2}{L\left(\dfrac{dv_o}{dt}\right)^2_{max}}$$

$$= \frac{(500)^2(10^{-6})^2}{100 \times 10^{-3}(50)^2} = 10^{-9}\,\text{F} = 1\,\text{nF}$$

Example 3.2

An SCR has a $di/dt|_{max}$ rating of 10 A/μs. It is to be operated from a 100 V supply.

(a) What is the minimum value of load inductance that will protect the device?

(b) If the recharge resistor of the snubber is 500 Ω and the load is 50 Ω what will be the new value of L?

Solution. (a) From (3.4) the inductance must satisfy the condition

$$L = \frac{V_{bus}}{\left(\dfrac{di}{dt}\right)_{max}} = \frac{100}{10} \times 10^{-6} = 10\,\mu\text{H}$$

(b) In the presence of a recharge resistor R_s of 500 Ω, while R is 50 Ω, it is seen from (3.7) that the minimum inductance L' becomes

$$L' = \frac{V_{bus}\left(\dfrac{R+R_s}{R_s}\right)}{\left(\dfrac{di}{dt}\right)_{max}} = \frac{100 \times \dfrac{550}{500}}{10 \times 10^6} = 11\,\mu\text{H}$$

Example 3.3

A d.c. chopper, Fig. 3.3, uses a MOSFET as a switch. The input voltage $V_{dc} = 30$ V and the chopper operates at a switching frequency $f_S = 40$ kHz, supplying a load current $I_L = 30$ A. The relevant switching times are $t_f = 30$ ns and $t_r = 80$ ns.

(a) For the condition of critical damping calculate values for C_S, L_S and R_S.

(b) Calculate the value of R_S if the maximum discharge current is limited to 10% of the load current.

(c) If the discharge time is one-third of the switching period, calculate the new value of R_S.

(d) What is the power loss due to the RC snubber circuit, neglecting the effect of inductor L_S on the voltage of snubber capacitor C_S, if $V_{sat} = 0$.

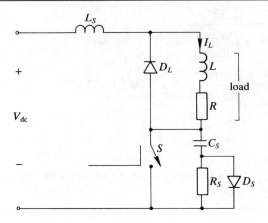

Fig. 3.3 D.C. chopper circuit in Example 3.3.

Solution. At turn-on,

$$\frac{di}{dt} = \frac{V_{dc}}{L_S} \qquad (i)$$

$$\frac{di}{dt} = \frac{I_L}{t_r} \qquad (ii)$$

At turn-off,

$$C_S \frac{dV}{dt} = I_L \qquad (iii)$$

$$\frac{dV}{dt} = \frac{V}{t_f} \qquad (iv)$$

(a) Equating (i) and (ii),

$$L_S = \frac{V_{dc} t_r}{I_L} = \frac{30 \times (80 \times 10^{-9})}{30} = 80 \, \text{nH}$$

From (iii) and (iv),

$$C_S = \frac{I_L t_f}{V_{dc}} = \frac{30 \times (30 \times 10^{-9})}{30} = 30 \, \text{nF}$$

For critical damping,

$$R_S = \sqrt{\frac{4L_S}{C_S}} = \sqrt{\frac{4 \times 80 \times 10^{-9}}{30 \times 10^{-9}}}$$

$$= \sqrt{\frac{32}{3}} = 3.27 \, \Omega$$

(b)
$$R_S = \frac{V_{dc}}{0.1 \times I_L} = \frac{30}{0.1 \times 30} = 10\,\Omega$$

(c)
$$\frac{1}{3}T = 4R_S C_S$$

$$\therefore R_S = \frac{T}{12C_S} = \frac{1}{12C_S f_s}$$

$$= \frac{1}{12 \times 30 \times 10^{-9} \times 40 \times 10^3}$$

$$= \frac{10^3}{1.2 \times 12} = 69.4\,\Omega$$

(d)
$$P_{loss} = \frac{1}{2} C_S V_{dc}^2 f_s$$

$$= \frac{1}{2} \times 30 \times 10^{-9} \times 30^2 \times 40 \times 10^3$$

$$= 0.54\,\text{W}$$

3.2.2 Ancillary environmental protection

In many power circuits two switches $S1$, $S2$ commutate with respect to one another, as shown in Fig. 3.4. Any overlap (i.e. simultaneous closure) between the two switches would short circuit the busbars during the state transition resulting in large (usually destructive) switching losses. Care has to be exercised to ensure that overlap does not occur, or that any effects resulting from it are minimised. There are two ways in which this problem can be treated: (a) surge limiting inductors, or (b) time cut strategies.

3.2.2.1 Current surge protection

A surge limiting inductor needs to be only a few microhenries of inductance, located in close proximity to the switch. For example, 5 cm of wire, arranged in the form of three or four turns, would invariably be adequate. This form of surge protection is also useful when the equipment is initially energised. Current surges frequently occur due to the inertia of rotating loads but this is often affected by a further much larger value per line component.

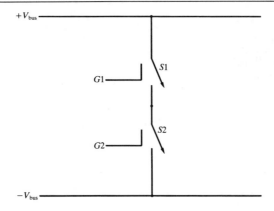

Fig. 3.4 Basic circuit of two semiconductor switches in series.

3.2.2.2 Time cut strategies

Time cut strategies are designed to prevent overlap between successive switchings by delaying the oncoming switch until the offgoing switch has completed its function. If sophisticated microprocessor or digital signal processing is available it is usually convenient to programme the necessary delays. For example, if the gate pulses of a power electronic switch are microprocessor-based then the delays can be built into the software. Mono-stable integrated circuits may be used where the delay time depends on the R and C circuit values.

Both current surge protection and time cut strategy are frequently used in conjunction with one another. The optimum delay time is often variable with pulse duty cycle. For example, variable delay occurs due to variation of the charge storage time in a Darlington transistor switched inverter. It is therefore prudent, during prototype development, to adopt excessive cut times which create distortion but ensure safe operation. Optimisation of a cut time can be left as a final activity after the basic operation of the system has been assured. Programmed delay is preferred rather than analogue type circuit delay. This optimisation task can be performed by monitoring variation of the no-load current. Excessive overlap protection causes reduction of the capability of the system or controller to deliver its full power. In inverters, for example, amplitude clipping causes waveform distortion as well as power delivery restriction and must be avoided.

3.2.2.3 Electromagnetic interference (EMI)

The switching or commutation of any switch generates voltage and current transients that are characterised by a whole spectrum of frequencies, ranging from the fundamental power circuit frequency up into the radio frequency band. Large values of electromagnetic radiation take place from the switching devices, heat sinks, connecting leads and output lines. For example, an individual diode or SCR in a power frequency (i.e. 50 Hz or 60 Hz) circuit radiates an annoying amount of electomagnetic interference in the range 200 kHz to 30 MHz. This interference can be easily detected on the AM band of any commercial radio receiver.

In addition to radiated interference there are components of interference due to conducted switching spikes and due to the electromagnetic coupling between adjacent wires or component devices. A common interference fault is the consequent spurious triggering of semiconductor switches, other than the one under intended commutation. Magnetic enclosure screening of the switching devices is essential.

All circuit lead wires should be twisted to reduce spurious pickup signals, including the 'in' and 'out' power leads of the main controller. All control gate lead wires should have a ferrite bead inserted to damp out any undesirable spurious signals. All control integrated circuit chips should have decoupling capacitors (0.1 µF typical) connected in close proximity between the supply and ground pins. The d.c. bus voltage rail should have electrolytic capacitors connected to maintain d.c. while short-circuiting any a.c. noise. Radio-frequency suppression filters should be fitted for all control circuits to minimise the effects of EMI.

3.3 ABUSE PROTECTION CIRCUITRY

The various approaches to abuse protection circuitry take two forms: (*a*) isolation of the supply from the load, and (*b*) removal of the input signal. An example of method (*a*) is the use of fuses for overcurrent protection.

3.3.1 Overcurrent protection

Overcurrent protection by the use of conventional fuses and circuit breakers tends to be inadequate in power semiconductor circuits. It is a feature of power semiconductor switches that overcurrent failure occurs very rapidly – even allegedly 'fast acting' fuses are too slow to protect the switches. For this reason fast acting fuses are usually adopted in a back-up capacity, to actuate when a persistent form of abuse occurs.

When load side faults of a short-circuit nature occur the quickest method of fault detection is to sense the excess current using a Hall plate detector. This fault signal is then used to stimulate the action of trigger circuitry so that

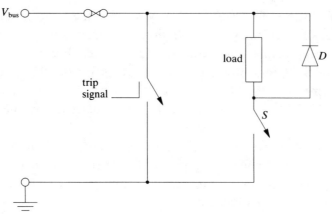

Fig. 3.5 Overvoltage protection employing a 'crowbar' circuit.

the conducting switches are extinguished. Since the fault detection process takes place at signal power level it can be effected even during a single cycle of the power switching sequence. Such protection circuits are commercially available in a single encapsulation and operable from a given logic output. Appropriate sensitivity can be realised by the proximity of the detector to one of the circuit electrode leads. An integrated gate drive IC will have a current sensing pin output to which a sensing detector can be connected. When the sensed current exceeds a certain pre-set value the relevant gate drive signal is disabled.

3.3.2 Overvoltage protection – crowbar

Semiconductor switches are better at withstanding overcurrent abuse rather than overvoltage abuse. This feature of operation is frequently employed, particularly in high power SCR circuits. The voltage excess is detected by separate sensing circuits and a separate 'crowbar' device is switched into its short-circuited state to cause fuses or circuit breakers to be actuated, Fig. 3.5. If this separate ancillary device has either (*a*) a higher voltage rating or is designed to avalanche, or (*b*) is a silicon controlled switch (SCS), none of the main circuit switches are endangered.

3.4 ISOLATION CIRCUITRY

At present (1995) only two-port, three-terminal semiconductor switches are readily available. It is often convenient to energise more than one switch from the same trigger circuit, either simultaneously or in some predetermined logic

3.4 Isolation circuitry

sequence. Regrettably, this is not possible as the reference main electrodes of the control electrodes of the various switches to be actuated are at different potentials. Direct voltage isolation is required. Also, some form of energy interchange process or boundary is needed. The two forms most frequently adopted are (a) pulse or isolation transformers, (b) opto-isolators, both of which are described in the following subsections.

3.4.1 Pulse isolation transformer

Pulse transformers are particularly useful in SCR and GTO thyristor trigger circuits. They remove the need for floating power supplies or opto-coupled gate devices by providing direct voltage isolation. Also, two or more devices can be triggered from the same source very economically by using multifilar secondary windings.

Either a full pulse may be faithfully transmitted or the transformer may be used in a 'starved core iron' mode to create a derivate processing of the input waveform so that essentially only the edges, which contain most of the high frequency content of the wave are transmitted. This latter condition is the one usually chosen as it has several advantages when pulse transformers are used in SCR trigger circuits.

Criteria for the two conditions are now evolved. A suitable pulse transformer coupling circuit is shown in Fig. 3.6. In the simplified equivalent circuit Fig. 3.6(c) it is seen that during the period when the bipolar transistor T_1 is saturated by a suitable base signal

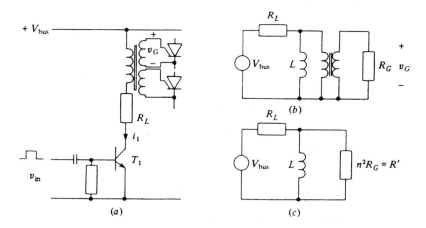

Fig. 3.6 Pulse transformer triggering: (a) circuit diagram, (b), (c) equivalent circuits.

$$V_{bus} = L\frac{di}{dt}\left(1 + \frac{R'}{R_L}\right) + i_L R_L \qquad (3.31)$$

which gives

$$i_L(t) = \frac{V_{bus}}{R_L}[1 - \varepsilon^{-tR_L R'/(R_L+R')L}] \qquad (3.32)$$

Resistor R_L serves as a current limiter. The transmitted voltage waveform is given by

$$\begin{aligned} e &= L\frac{di}{dt} \\ &= V_{bus}\frac{R'}{R_L + R'}\varepsilon^{-R_T t/L} \quad \text{where } R_T = R_L/R' \end{aligned} \qquad (3.33)$$

There are two operating conditions:
(a) When R_T is small or L is large such that say $L/R_T > 10T$, where T is the pulse width of the input to the base, then near faithful transmission of the pulse is achieved.
(b) When R_T is large or L is small such that say $L/R_T < T/10$, derivative action of the input pulse occurs, as shown in Fig. 3.7, i.e. the transformer takes the derivative of the input waveform and hence generates two 'spikes' rather than a single pulse. Such a shape is achieved by starving the core of iron which in itself is economical. The amplitude V_{bus} must be sufficient to trigger the least sensitive device of a batch. Therefore

$$V_G = V_{trig} = \frac{V_{bus}}{n}\frac{R'}{R_L + R'} \qquad (3.34)$$

or

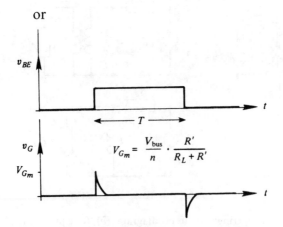

Fig. 3.7 Derivative output of a pulse transformer.

$$V_{bus} \geq V_{trig} n \left(1 + \frac{R_L}{R'}\right)$$

But $R' = n^2(V_{trig}/I_{trig}) = n^2 R_{trig}$ of the least sensitive device so that

$$V_{bus} \geq \frac{V_{trig}}{n}(n^2 + R_L/R_{trig}) \qquad (3.35)$$

In fact, under condition (b) a much larger V_{bus} may be adopted because the heating effect for a given initial amplitude is much less. The gate waveform is then the very type required to initially inject a large charge without significant heating of the gate. Such a 'hard' drive makes it possible to allow an SCR to be subjected to a high di/dt value immediately after triggering and is thus very desirable.

3.4.2 Opto-isolator

Opto-isolators employ an infrared light-emitting diode radiating onto a photo signal level Darlington pair device. The switching times of the phototransistors are very small, being typically $t_{on} = 2$–$5\,\mu s$ and $t_{off} \approx 300\,ns$, which imposes limits on high frequency applications. A simple form of opto-isolator has the two transistor devices in close proximity and uses lead-out wires. Slow and fast versions are available, the fast version being much more expensive. A second form of opto-isolator employs a fibre optic link between the two phototransistors. This can have significant advantages where radiation interference is severe and/or where the close proximity of the trigger circuits is not possible, for volumetric or other reasons.

Unlike the case of the pulse isolation transformer, secondary energisation is required with opto-isolators. This can often be derived from the main power sources. An example application of an opto-isolator driven IGBT

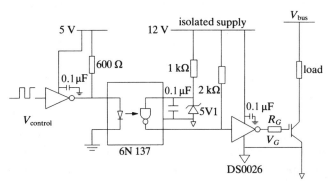

Fig. 3.8 Typical application of a 6N137 opto-isolator integrated circuit.

gate drive is given in Fig. 3.8. Note that the interbus capacitor (0.1 µF) is mounted close to the IC.

3.5 SYSTEM REALISATION STRATEGY

A high level of integration between control and power components is very often considered desirable in the quest to find a better circuit performance in terms of reliability and cost. Most of the design methods originate from block diagram design using discrete components. Control signals are often generated from analogue electronic components with demands deriving from a potentiometer, giving a continuous varying input voltage. With the advent of microprocessors, the control process can be performed by software, with the input interfaced via an analogue-to-digital (A/D) converter IC. For example, the generation of a pulse-width modulation (PWM) waveform for the control of inverters can be achieved using a microprocessor-based system with as few as five integrated circuits. The input signals are analogue voltage and frequency demands, read via input/output ports to the microprocessor via an A/D converter. The switching frequency is determined by the on-board crystal frequency and is chosen to match the maximum allowable switching frequency of the power switches. The control signals should also be generated in isolation from the power circuits for reliability reasons. The choice of control implementation again depends on the cost of the process involved in the power circuit. For example, a simple phase controlled triac is needed for a lamp dimmer application, while two separate on-board computers are needed to monitor the output voltage of a switched-mode power supply installed in a space shuttle.

With integration being the modern trend, the MOS gate driver has emerged as a standard component for driving power MOSFETs and IGBTs. The functions incorporated in a single chip are:
- undervoltage shutdown to ensure that the switch does not operate unsaturated,
- overvoltage shutdown to ensure that the switch does not have excessive voltage applied,
- simple current trip and current limiting to protect the switch against overcurrent.

Future trends are likely to be in the development of SMART power devices in which the driver and power semiconductor are manufactured on the same piece of silicon. The aim is to provide fast switches that can be directly driven by a low level logic voltage with added protection against overvoltages, overcurrent, thermal overload and capable of returning their status back to the controller.

Besides the development of new types of integrated gate drives and power semiconductors, integrated circuit technology has increased the performance and reliability of the control aspects. Microprocessors are now at the heart of most power electronics control technology, having the ability to handle complex control algorithms and to interface with the user. As tasks that have to be performed grow in number and size, one has to resort to multi-processor systems or ultra-high speed processors like the Digital Signal Processor (DSP). In order to decide upon which system to implement the control algorithms, the following features are noted:

- microprocessors are superior when large amounts of memory are required,
- microprocessors offer more versatile peripherals for interfacing purposes,
- development time for DSPs tends to be longer,
- DSP development tools are usually not as standardised as those for microprocessors, and tend to be expensive,
- DSPs and microprocessors use different software solutions for a given problem (e.g. a DSP calculates arithmetic functions whereas a microprocessor uses a look-up table).

An alternative to DSPs or microprocessors is to use an Application Specific Integrated Circuit (ASIC) to implement the control functions in a single chip. This approach reduces the numbers of chips, boards and connectors, and hence increases reliability, noise immunity, speed, temperature range, etc. It is also much more difficult to copy the design circuit as compared with discrete components, hence protecting the hardware/software manufacturers. A typical example of an ASIC application in power electronics is the digital pulse-width-modulator chip dedicated to the PWM three-phase voltage source inverter. The main drawback of an ASIC is the non-recurrent cost related to design and prototype development. Such investment is only justified for volume production, which is seldom the case in large power electronic equipments where production runs are under one thousand units. However, the tedious ASIC design effort has been reduced in recent years due to the fast reduction of the price/performance ratio of computer-aided design (CAD) workstations and the development of user-friendly ASIC design software packages (otherwise known as silicon compilers) at an affordable price. Using such packages, the power electronics designer can design the custom chip without special knowledge of the microelectronics. The functional schematics are drawn and tested via simulation and various input/output test vectors can be created for simulation and prototype testing purposes. Field Programmable Gate Array (FPGA) technology has vastly improved its hardware and software simulation capability, hence reducing the ASIC development time and cost of implementation.

3.6 PROTOTYPE REALISATION

3.6.1 Principles

Most of the useful power electronic circuit topologies have now been fully evaluated both practically and by design aids such as SPICE (ref. 14). Nevertheless, it is rarely possible to totally plagiarise an existing circuit and to extract and realise one from reference books without malfunction at the physical realisation stage. The information available is invariably incomplete so that construction, connection and energisation may result in failure. Rigid adherence to a strict regime of consolidation is vital.

A low power, low voltage, high value load resistance, snubber-free model should be constructed by starting with the processing circuits and using these as signal sources to trigger circuits. The connection of the multi-switch configuration should only be effected when those have been actuated individually. Loads and voltages of this circuit must be adequate to ensure that the switches function correctly, in principle. Usually 20–30 V and 100–200 Ω are sufficient to maintain the switching in the conducting state. Observation of the processing output should ensure that interaction malfunction does not occur.

When the switching aids and snubbers have been added, the applied voltages may then be increased gradually to the rated value, while watching the switching waveforms for evidence of malfunction, interference or other spurious transients. After the rated voltage is achieved, the load may be reduced to its rated power delivery value. If a motor or other form of complex load is to be driven, it is better if an electric circuit model is applied initially using resistance only with the reactive elements added later. Finally, the true load can be energised. In the case of multi-phase circuits only a single half-phase should be developed to the power level. This should then be extended to a half bridge before further extension is pursued.

3.6.2 Example – single-phase voltage control circuit

A single-phase voltage controller circuit is widely used for lamp dimming and motor control. The circuit is extensively discussed in Chapter 8 below. An example circuit layout is given at the bottom RH side of Fig. 3.9, showing a controller containing two SCRs. Let the target specification be 'to obtain proportional control of the load voltage'.

In engineering practice, even in a competitive industry, the first priority is the realisation of a reliable operational circuit. This is not likely to be unique. Simplification and refinement of the basic design are subsequent activities.

Several different approaches may be adopted in the present specification. A suggested schematic block diagram is shown in Fig. 3.9. Some further detail of how the various blocks can be realised is given in Fig. 3.10, which has been used by the authors as a student laboratory exercise.

It is advisable to realise the various circuit blocks sequentially, starting at the a.c. source and using each circuit block as a signal generator for realisation of the subsequent block function. Oscilloscope probes can be used to observe the associated waveforms and to ensure correct realisation and non-interaction of two separate signal currents. To ensure the correct transmission of signals to appropriate power switches the oscilloscope must be used in differential mode. Final calibration of the overall transfer function relating r.m.s. load voltage to the SCR gating signal requires the use of a true r.m.s. voltmeter.

3.7 DEVICE FAILURE – MECHANISMS AND SYMPTOMS

Device failure frequently occurs arising from unintentional abuse, which has to be removed when discovered. A common fault condition is the application of erroneous signals at the control electrode. This often takes the form of gate signals that exceed the device rating. Another common fault is the application of a random spurious signal that causes unintended gating with possible consequent damage to other devices connected in the same power circuit.

With BJT and Darlington devices, excess voltage can cause thermal runaway through overcurrent. Partial destruction is unlikely unless caught by protection circuits. Initially, as the resistance falls to near zero, the current rises causing temperature rise of the chip which then melts so that an open-circuit may result. Slight abuses can cause increases in leakage current through increased junction edge contamination.

SCR, triac and GTO devices, being compounds of BJT component devices, tend to have similar attributes but may be much more easily damaged due to abuses at the control electrode. This is particularly true of low current rated SCRs which do not have highly interdigitated gate–cathode structures and for which the trigger circuits must therefore be carefully designed.

MCT devices appear to be reliable but there is little experience of their use to date.

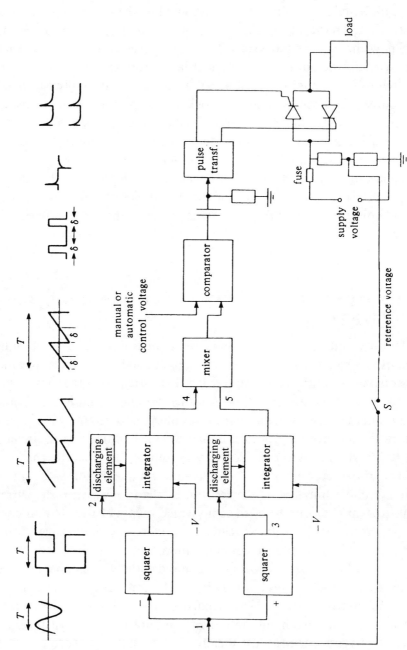

Fig. 3.9 Block diagram of firing-circuit for phase control ($0 < \alpha < 180°$).

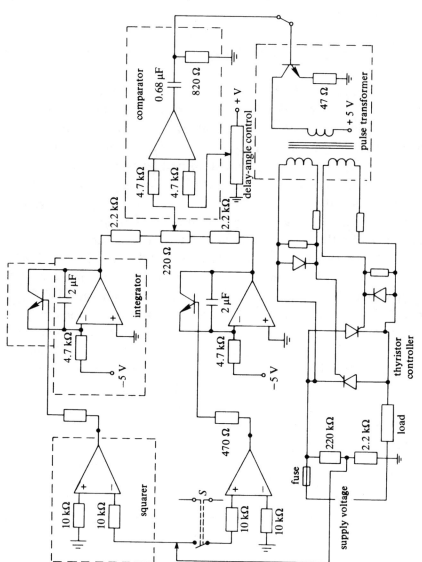

Fig. 3.10 Circuit diagram of firing-circuit for SCR phase control ($0 < \alpha < 180°$).

All IGBT devices contain a parasitic SCR which is responsible for latch-up failures. Despite this tendency, which is suppressed in later-generation versions, many such devices are in use. Third-generation IGBTs tend to be reliable but damage can occur due to very low gate capacitance which can be charged up to high voltages.

In a MOSFET, voltage surges excite the parasitic body BJT and many device breakdowns occur due to BJT type failures. A dv/dt snubber is often needed for turn-on.

Current surges in a MOSFET are limited only by thermal considerations. Unlike the BJT, the MOSFET does not inherently thermally 'run away'. Total destruction does not always result. Failure through abuse can occur gradually over a long period of time with the device superficially appearing to operate normally. If the critical temperature of parts of the chip is exceeded then those islands involved break down the glass of the gate. They then behave as bipolar devices in parallel with the remainder of the MOSFET, which is performing normally. The result is a fall of input resistance from its normal value of $10^9\,\Omega$ to a value even as low as $1\,\mathrm{k}\Omega$. This usually loads the output of the gate control circuits causing signal distortion. The island cells may or may not be destroyed, dependent upon the source resistance of the gate signal. The device may continue to operate, with a value of R_{DS} anywhere between its normal value and infinity dependent upon how many cells have been destroyed. As R_{DS} has a positive temperature coefficient, at each event of operation the chip temperature becomes progressively higher and so does the incidental dissipation. The result is a gradual deterioration, which may not be initially perceived, and reduced load control.

A MOSFET is vulnerable to stray induced voltages. Care should be exercised to ensure that $V_{GS_{max}}$, usually approximately 20 V, is not exceeded, with resulting overcurrent.

At very high power levels the greatest degree of ruggedness and reliability is provided by SCR devices. For slightly lower power levels, where they can be appropriately adopted, third-generation IGBT devices have been found to be extremely reliable. Moreover, they can readily be connected in parallel to increase a circuit current handling capacity.

3.8 REVIEW QUESTIONS AND PROBLEMS

3.1 Briefly explain what happens in an SCR device when the rate of rise of load current and the rate of rise of anode/cathode voltage are excessive.

By means of a circuit diagram, show how snubbers and their associated ancillary components are connected to an SCR driving a resistive load from a power supply.

Evolve formulae for the minimum value of the snubbers such that associated ratings are not exceeded and explain the purpose of the ancillary components.

3.2 An SCR has $(dI_A/dt)_{max} = 10$ A/μs and $(dV_{AK}/dt)_{max} = 40$ V/μs. It is to be used to energise a 10 Ω load from a 500 V bus.
   (i) Derive formulae for the minimal values of the snubber components.
   (ii) Evaluate the components required.
   (iii) Explain the purpose of each component in the snubber circuits.

3.3 A semiconductor switch is embedded in a high power circuit to width modulate the energy of a 1000 V supply to a load. The device has a $(dI/dt)_{max}$ rating of 30 A/μs. Calculate the minimum value of inductive snubber to be included in the circuit to prevent this rating being exceeded.

3.4 A situation exists in which a solid-state device, conducting heavily in its reverse mode, is subjected to an immediate supply voltage of 100 V in its forward direction. If the device has a $(dv/dt)_{max}$ rating of 10 V/μs and its load resistance is 10 Ω, evaluate the minimum size of snubber capacitor which is required to prevent the rating being exceeded. Explain what may occur if the snubber component is omitted.

3.5 In an SCR, briefly explain what happens if:
   (i) the load current rises instantly to a significant value equal to or less than its maximum,
   (ii) regardless of condition (i), a voltage of significant size but less than that of the bus is instantly applied to the anode.

   In both (i) and (ii) explain why the phenomena occur.

   With the aid of electrode waveforms, show how snubbers prevent these phenomena from occurring.

3.6 Explain why it is possible to manufacture an SCR with a much higher power handling capacity than a bipolar device.

   The rate of rise of current, dI/dt, and voltage, dV/dt, must be controlled in circuits so that manufacturers' ratings are not exceeded. Explain what may occur if these ratings are exceeded, and what the result is likely to be.

   Give circuit diagrams showing how components are included in circuits to limit operation in order that the ratings are not exceeded. Develop associated formulae.

3.7 An SCR circuit with a 25 Ω resistive load has a $(dv/dt)_{max}$ rating of 40 V/μs. The surge voltages that are expected to arise are steep transients of value 650 V. What size of snubber circuit capacitance is needed to prevent dv/dt triggering?

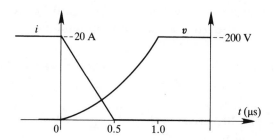

Fig. 3.11 Switching waveforms for problem 3.12.

3.8 If the SCR circuit of Problem 1.2 in Chapter 1 now includes a load inductance of 75 mH, calculate the revised value of snubber capacitance to prevent spurious triggering.

3.9 In the snubber circuit of Problem 3.7 what is the necessary value of capacitance if the critical value of series inductance is used? Calculate the critical inductance.

3.10 An SCR circuit has a load resistance of 15 Ω and operates from a 240 V rail supply. If the thyristor has a $(di/dt)_{max}$ rating of 12 A/μs, calculate the series inductance that must be included in the circuit to protect it.

3.11 Six GTOs are connected in a bridge configuration driven from a transformer to control the supply to a 1 Ω load. The devices are triggering sporadically and it is found that 200 V spikes are present on the supply lines from the transformer. The devices have a maximum voltage transient rating of 10 V/μs.

Calculate the minimum value of a capacitive snubber which would prevent this occurrence and show, by means of a circuit diagram, how it would be connected if a suitable resistor and diode were available to optimise performance.

If the resistive load were replaced by a motor of similar resistive load but with a large inductive reactance, how would this affect the value of capacitance?

3.12 What is the function of a turn-off snubber circuit?

A power switch has a turn-on time and turn-off time of 0.5 μs when switching a 20 A load on a 200 V d.c. rail. Design an *RCD* turn-off snubber circuit if the desired snubber action ceases 0.5 μs after the switch completes its switching action, Fig. 3.11. The maximum switching frequency its 2 kHz and the on-time duty cycle varies between 4% and 97%.

3.13 A snubber circuit for some power switches contains only the *R–C* series circuit with no diode. Explain why this is so.

4
Adjustable speed drives

4.1 BASIC ELEMENTS OF A DRIVE

A rotational mechanical load in which any one of a wide range of operating speeds may be required is often called an 'infinitely variable speed drive' or, more modestly, a 'variable speed drive' or 'adjustable speed drive'. Suitable operating characteristics to provide a given range of load torques T_L and speeds N might be provided by a pneumatic or hydraulic drive as well as by several forms of electrical variable speed drive. The basic components of such a drive, Fig. 4.1, are the motor or actuator that delivers torque to the load and controls its speed and the power controller that delivers power to the motor in a suitable form. In this book only those drives in which the drive motor is an electric motor are considered.

The output power developed by an electric motor is proportional to the product of the shaft torque and the shaft rotational speed. The value of the developed torque usually varies automatically to satisfy the demand of the load torque plus any torque associated with friction and windage. Increase of the shaft power due to increase of load torque is usually supplied by automatic increase of the supply current demanded by the motor. Any significant change in motor speed, however, must be obtained in a controlled manner by making some adjustment to the motor or to its electrical supply. The basic

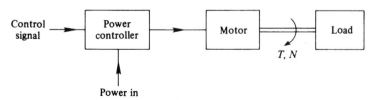

Fig. 4.1 Block diagram of a basic electrical adjustable speed drive.

variable to be controlled in an electric motor is its shaft speed and this contrasts with an electric generator where the basic variable to be controlled is the generated voltage at the machine terminals.

4.2 LOAD TORQUE–SPEED CHARACTERISTICS

Various industrial loads have forms of torque–speed characteristics that are typical of the application. Some of these are shown in Fig. 4.2. Good speed control requires that there should be a large geometrical angle (ideally 90°) between the drive characteristic T–N and the load characteristic T_L–N. This is to prevent excessive speed variation arising from small changes of load torque.

The load torque–speed characteristic is usually a general non-analytic function

$$T_L = f(N) \tag{4.1}$$

For the particular characteristics of Fig. 4.2 each example may be approximated to an analytic form:

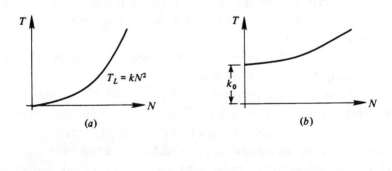

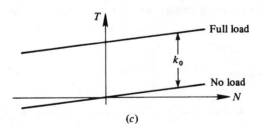

Fig. 4.2 Steady-state torque–speed characteristics of certain loads: (a) fan or pump, (b) compressor, (c) hoist or elevator.

$$T_L = kN^2 \qquad \text{for a fan or pump} \qquad (4.2)$$

$$T_L = k_0 + k_1 N + k_2 N^2 + \ldots \qquad \text{for a compressor} \qquad (4.3)$$

$$T_L = k_0 + k_1 N \qquad \text{for a hoist} \qquad (4.4)$$

4.3 STABILITY OF DRIVE OPERATIONS

4.3.1 Steady-state stability

In addition to the requirement that the drive line T and load line T_L intersect as nearly orthogonally as possible, there is a further condition on the nature of the intersection, illustrated in Fig. 4.3. Consider the dynamic stability of operation at each of the intersection points shown, in which an induction motor is used as the drive motor. What is the effect, on subsequent operation, of a small transient increase of speed? If immediately after the increase of speed, the load line T_L is greater than the drive line, T, it means that the drive is not delivering sufficient torque to maintain the increased speed. The result

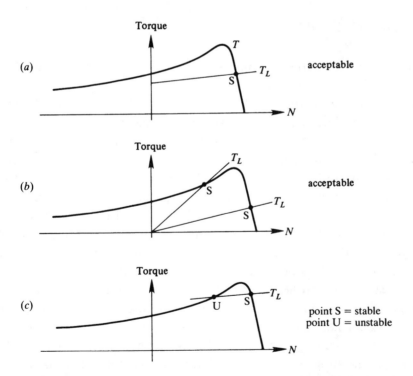

Fig. 4.3 Intersections of an induction motor drive line T with load lines T_L.

is that the speed falls back to the stable operating point, such as points S in Fig. 4.3. If, immediately after a transient increase of speed, the drive torque T is greater than the load torque T_L then the excess torque will accelerate the drive away from its former operating point. In Fig. 4.3(c), for example, the point U is dynamically unstable because the T line has a greater positive slope than the T_L line. In mathematical terms, the criterion of steady-state stability can be summarised as

$$\left[\frac{dT_L}{dN} - \frac{dT}{dN}\right] > 0 \tag{4.5}$$

Equation (4.5) and the discursive analysis that led to it are considerably limited in application. It applies only to small deviations from the steady state. It will not give a valid transient analysis.

The torque T developed by the drive has to supply the torque demand of the externally applied load T_L, overcome the friction and windage effects T_{FW} and to accelerate the inertial mass of the rotor during speed increases. If the polar moment of inertia of the load and drive machine is J and the friction consists of a static friction (stiction) term T_s plus a viscous friction term with coefficient B then

$$T = J\frac{dN}{dt} + T_s + BN + T_L \tag{4.6}$$

where $J\,dN/dt$ = inertial torque, $T_s + BN = T_{FW}$ = friction and windage torque and T_L = load torque.

In the steady state the speed N is constant, $dN/dt = 0$ and the inertial torque is zero. The friction and windage torque is negative for negative speeds, as shown by the no-load torque line T_{NL} in Fig. 4.2(c), in which T_s is zero. A friction and windage characteristic in which T_s is finite is given in the four quadrants of the torque–speed plane illustrated in Fig. 4.4 for the case of constant unidirectional load torque T_L. The necessary torque T developed by the motor undergoes a step reduction as the characteristic passes through zero speed into the positive torque/negative speed, braking region of quadrant II. If the developed torque T delivered by the motor is constant the torque applied to the load T_L has the form shown in Fig. 4.5.

The steady-state torque–speed characteristics of an ideal adjustable speed drive are shown in Fig. 4.6. Any preselected base (no-load) speed can be held almost constant as the load torque increases from (say) A to B. If very precise speed stability is required the drive lines in Fig. 4.6 should be vertical, representing zero speed regulation. Good stable drive performance requires an orthogonal or fairly orthogonal intersection between the drive

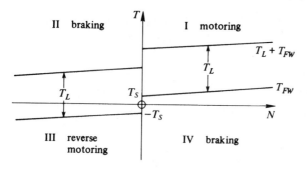

Fig. 4.4 Torque–speed characteristics with constant load.

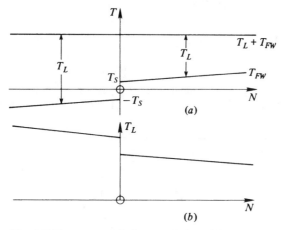

Fig. 4.5 Torque–speed characteristics with constant motor torque, T.

torque–speed characteristics and the load torque–speed characteristics. For the complete control of all types of load the drive torque–speed characteristics must extend into the reverse (braking) torque region – quadrant IV, and also into the reverse (overhauling) speed region – quadrants II, III of Fig. 4.6.

A crane or hoist application of the drive performance in the various quadrants of the torque–speed plane is given in Fig. 4.7. The tractive effort of the drive motor is represented by the arrows in the pulleys. Forward torque is represented as anticlockwise while reverse torque is represented as clockwise. The direction of the drive motor shaft and load is represented by the arrows alongside the hoist loads. In quadrant I, representing conventional motoring performance, the drive torque is moving the load upwards. Quadrant II represents the condition known as overhauling, where the forward torque of the drive is not sufficient to prevent the load reversing in direction. The drive motor is therefore exerting a torque to brake the tractive effort of the

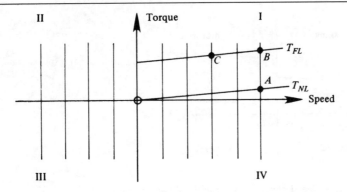

Fig. 4.6 Steady-state torque–speed characteristics of an ideal adjustable speed drive. T_{FL} = full load; T_{NL} = no-load.

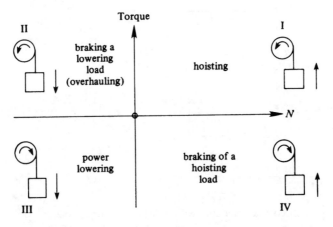

Fig. 4.7 Four-quadrant operation of a hoist or winch (the arrows in the pulleys represent the direction of the motor torque).

load which is causing it to rotate backwards. The condition of quadrant III is when the tractive effort of the drive has reversed so as to produce clockwise rotation of the pulley and downward motion of the load. Quadrant III therefore represents reverse direction motoring performance. In quadrant IV the reverse (clockwise) tractive effort of the drive is not sufficient to prevent the load moving upwards and driving the pulley and drive motor in the forward (anticlockwise) direction. Quadrant IV therefore represents the braking of a reverse torque drive.

There are many forms of variable speed drive in which only one direction of conventional motoring, i.e. quadrant I performance, is desired but the performance in one or more of the other quadrants may have to be considered as a fault condition.

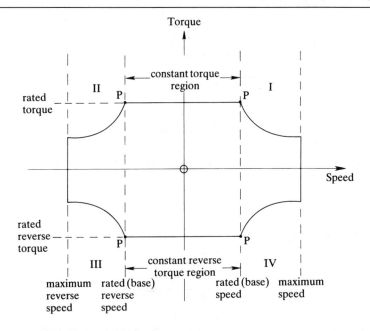

Fig. 4.8 Torque, speed and power boundaries.

For steady-state or continuous operation the boundaries of performance in the four quadrants are illustrated in Fig. 4.8. The most efficient points of operation are likely to be points P, where rated torque is delivered at rated (base) speed, which is the highest speed attainable at the rated motor flux. Between the forward and reverse rated speeds, including standstill, the torque level is limited by the motor current level, which is proportional to it.

At constant torque the power output is proportional to the speed and rated power occurs at points P. If overcurrent is to be avoided in the high-speed regions the torque must be reduced, indicated by the curved lines in Fig. 4.8. To maintain constant output power the curves in Fig. 4.8 should have the shape of rectangular hyperbolae in the torque–speed plane.

The maximum permitted speed is limited by mechanical considerations such as bearing friction and hoop stress on the rotating parts.

4.3.2 Transient stability

In addition to the steady-state drive performance of Fig. 4.6 a drive also has to satisfy design criteria with regard to its transient performance. The reaction of the drive to, for example, step changes of load torque or step changes of control signal, may be as important as the steady-state operation. Assume that the drive is operating, for example, at full speed with no-load repre-

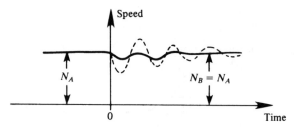

Fig. 4.9 Variation of speed with time after a step increase of load torque ($A \to B$ in Fig. 4.6).

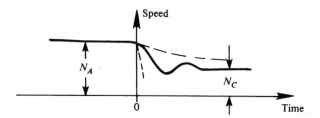

Fig. 4.10 Variation of speed with time after a step reduction of speed signal ($B \to C$ in Fig. 4.6).

sented by point A in Fig. 4.6. Let the load torque be now suddenly increased to the value B. The initial and final speeds for this load transition are shown in Fig. 4.9 and it is usually desired that the transition occur quickly with small overshoot and rapid settling time. The application of increased load torque will probably result in initial speed reduction below the final steady-state level and an acceptable speed response for many applications is the solid line in Fig. 4.9. A response corresponding to the dashed line in Fig. 4.9 would be too oscillatory and too underdamped for many drive applications.

Similarly, consider a step change of speed control signal at constant shaft load. Let the speed signal be suddenly reduced from a value corresponding to full speed, point A of Fig. 4.6, to a value corresponding to a low speed, point C. The speed cannot change instantaneously and an acceptable response in many applications is the form given in Fig. 4.10. A drive that is overdamped would result in a sluggish response indicated by the upper dashed line in Fig. 4.10. For a large speed reduction of this kind an underdamped response might result in an overshoot into the negative speed region that would be unacceptable in some applications.

The speed–time response of a drive to step changes of signal or step disturbances of any kind depends on the drive dynamics. In particular it depends on the physical parameters of the system, such as electrical resistances and inductances and mechanical friction and inertia. Proper system design can ensure that the transient response is satisfactory.

The transient torque–speed characteristics of a motor bear little resemblance to the steady-state torque–speed characteristics. For a three-phase induction motor, for example, the transient torque performance does not follow the smooth steady-state characteristic of Fig. 4.3, but is highly oscillatory. Transient torque performance is difficult to predict or to solve analytically.

4.4 PRINCIPAL FACTORS AFFECTING THE CHOICE OF DRIVE (REFERENCE TP1)

There is no order of merit or check list of essential requirements that is appropriate to every drive or even to most drives. The list of requirements that can arise is very diverse because of the wide range of drive applications. Also, the relative weighting of importance of various features will vary between particular applications. The following list of factors to be considered in the selection of a variable speed drive covers most requirements but is not exhaustive, and the order is not significant. Many of the following factors would also be applicable to mechanical and hydraulic drives as well as to electrical drives:

(*a*) rating and capital cost,
(*b*) speed range,
(*c*) efficiency,
(*d*) speed regulation,
(*e*) controllability,
(*f*) braking requirements,
(*g*) reliability,
(*h*) power-to-weight ratio,
(*i*) power factor,
(*j*) load factor and duty cycle,
(*k*) availability of supply,
(*l*) effect of supply variation,
(*m*) loading of the supply,
(*n*) environment,
(*o*) running costs.

The above principal factors affecting the choice of drive are now each briefly discussed with particular reference to features arising due to power electronics control techniques.

4.4.1 Rating and capital cost

The capital cost varies roughly with the rating of the drive and is mostly due to the cost of the drive motor and its control gear. In many thyristor controlled motor systems, especially induction motor drives, the cost of the thyristors and associated control circuits and protection is greater than the cost of the motor. A cage induction motor is about one-third to one-half the cost of a d.c. motor of the same rating but the thyristor and associated equipment is more expensive for induction motor systems. Other items in the above list of principal factors such as speed range and power/weight ratio affect the drive rating and therefore the first cost.

Assessment of the operating features and capital costs of certain forms of thyristor drives is made in Tables 4.1–4.3. It is emphasised that these tables represent the viewpoints of certain individual authors or organisations. Also the various items in a table may have been subject to uneven price revisions since the original date of publication.

4.4.2 Speed range

A drive with a wide speed range is more difficult to realise than one with restricted speed range and involves consideration of motor efficiency, power factor and speed regulation. Adjustable speed drives of wide speed range may require thyristors or transistors of higher rating to accommodate the higher currents associated with low-speed operation. Stall and creep speed operation of adjustable speed drives present additional control and protection problems and can severely reduce the power/weight ratio of the drive motor. The capital cost of the drive varies roughly with the required speed range. In some applications a range of discrete operating speeds may be acceptable rather than a continuously variable speed and this is likely to be much more economic. If only two or three different speeds are needed this will profoundly affect the choice of drive and drive motor and also the control apparatus.

4.4.3 Efficiency

The efficiency of a drive is the ratio of output power to input power and low efficiency has two serious disadvantages:
    (*a*) wasted energy has to be paid for at the same cost as useful energy,
    (*b*) the wasted energy may cause excessive heating of the drive components, especially the motor.

Table 4.1 *Typical variable speed drive characteristics (ref. TP4).*

	Commutator motor drives			Induction or continuous motor drives					
	Ward–Leonard drive	d.c. thyristor drive	a.c. commutator motor (stator-fed)	Secondary resistance control	Primary voltage control	Invertor drives	Cyclo convertor drives	Static slip-energy recovery	Eddy-current coupling
Output range (kW)	0.2–1000+	0.2–1000+	1.5–1000+	37–1000+	0.75–300	0.1–1000+	0.1–1000+	10–1000+	0.75–100
Typical speed range	8:1	20:1	3:1	2:1	4:1	20:1	6:1	4:1	17:1
Typical speed regulation (%)	5–0.1	5–0.1	5–0.5	poor	5–1	5–0.05	5–0.05	5–0.5	5–0.5
Constant power load	●●	●●●	●●●	●●●	●●●	●●●	●●●	●●●	●●●
Constant torque load	●●●	●●●	●●●	●●●	●●●	●●●	●●●	●●●	●●●
Varying torque load	●●●	●●	●●	●●●	●●	●●	●●●	●●●	●●●
High starting torque	●●●	●●●	●	●	●●	●	●●●	●	●●
Frequent stops and starts	●●●	●●●	●	—	●●	●	●●●	—	●
Frequent forward and reverse	●●●	●●●	—	—	—	●●●	●●●	—	●
Rapid acceleration and deceleration	●●●	●●●	●●●	—	—	●●●	●●●	—	—
High efficiency	75–85%	85–90%	90–95%	—	—	80–90%	80–90%	90–95%	—
2–5% accuracy of speed variation	●●	●●	●●	●	●●	●●	●●	●●	●●
Better than 1% accuracy of speed variation	●●●	●●●	●	—	●	●●●	●●●	●●	●
Atmosphere contaminated by gas, dust etc.	●	●	●	●●	●	●●●	●●●	●●	●
Easy maintenance	●	●	●	●●	●●	●●●	●●●	●●	●●●

Key: ●●● most suitable; ●● suitable; ● possible; — unsuitable.

Table 4.2 Cost comparisons for typical motor controllers (ref. TP4).

Drive system	Non-regenerative schemes				Regenerative schemes			
	7.5 kW	75 kW	750 kW	7500 kW	7.5 kW	75 kW	750 kW	7500 kW
Fully controlled d.c. thyristor bridge	1.0	1.0	1.0	1.0	1.0	1.0	1.0	1.0
Static slip-energy recovery scheme	—	1.3	1.2	1.3	—	1.1	1.0	1.0
Invertor drive	1.8	3.2	2.6	3.0	2.3	2.5	2.3	2.5
Cycloconverter drive	—	4.3	2.0	2.1	—	2.1	1.3	1.2

Table 4.3 Comparisons of variable speed drives (ref. TP9). 37 kW, (50 h.p.) 1500 r.p.m. (max) three-phase

Method	d.c.	Variable frequency	Kramer (SER)	Hydraulic		Switched reluctance
Electric motor	Shunt wound	Energy efficient induction	Wound rotor slip-ring	a.c. induction		Switched reluctance
Minimum speed as fraction of full speed	1%	10% (a)	80% (b)	1%		1%
Overall efficiency at full speed	90%	83%	91% (c)	83%		92%
Overall efficiency at half speed	80%	74%	87%	66%		89%
Power factor at full speed	0.79	0.79	0.59	0.88		(e)
Power factor at half speed	0.44	0.44	0.48	0.77		(f)
Size of control cabinet	Medium	Large	Medium	Small		Large
Can be exploited in hazardous zone	No	Yes	No	Yes		Yes
Maintenance cost	High	Low	Low	Medium		Low
Disturbance to a.c. supply	Maximum	Mean	Mean	Minimum		Mean

(a) Below 10% full-speed rotational instability occurs which could be avoided by additional precautions.
(b) Kramer is not limited to 80% full-speed. See text for relationship between speed range and cost.
(c) Higher efficiencies are attainable in high powers.
(d) Efficiency data were available only at 70% full-speed.
(e), (f) Power factors are similar to those originating from an a.c. induction motor, controlled in a similar way.

One ultimate limitation on the use of an electric motor is the permissible temperature rise. Excessive temperature may necessitate the installation of expensive forced cooling or even derating of the drive motor to avoid overheating of the winding insulation. Continuous high temperature working may cause deformation of metal parts or even bearing failure.

The efficiency of electric motors at their rated speed is about 60% for 1 horsepower machines and increases with rating to over 90% for very large drives of several thousand horsepower. The operation of a motor at a speed well below its rated speed is usually inefficient and continuous low-speed operation often causes temperature rise problems. For small electric motors (i.e. less than 1 h.p.) the low efficiency is accompanied by a low power/weight ratio. Note that 1 h.p. = 0.746 kW.

4.4.4 Speed regulation

Speed regulation is the fractional reduction of speed due to the application of load torque. In many drives some degree of speed regulation, say 5%, is acceptable and even desirable. Zero speed regulation can be realised by the use of a frequency controlled synchronous motor or by the use of an induction motor in a closed loop control system with tachometric negative feedback. A rigorous specification for speed regulation may have to be 'traded off' in an acceptance of loss of efficiency or loss of power/weight ratio.

4.4.5 Controllability

It was discussed in Section 4.3 above that an adjustable speed drive must have acceptable performance with regard to both steady-state operation and transient response. Steady-state operation is concerned with the accuracy of control and (say) how closely the shaft speed of the drive motor follows slow excursions of the drive control signal. In control systems terminology the accuracy of steady-state operation is called the 'servo performance' of the drive for both open-loop and closed-loop operation. One important consideration in the choice of drive is the ease or otherwise with which the servo loop can be closed since so many applications require close speed holding or very low speed regulation.

A further consideration is the speed of response of the system to rapid changes of control signal or load. It was pointed out in Section 4.3 above that the physical parameters of the motor and drive system are significant in determining the transient response. An important 'figure of merit' affecting the transient response of a motor is, for example, the torque/inertia ratio. In

4.4 Principal factors affecting the choice of drive (reference TP1)   135

considering the transient response of a thyristor controlled electric motor it is important to note that the speed of response of thyristors and power transistors and their associated electronic gating circuits is virtually instantaneous. This lack of delay in response is an advantage in most control applications and leads to better steady-state accuracy.

4.4.6 Braking requirements

In an electric drive it is sometimes acceptable to use gravity to brake the system by simply switching off the electrical supply and letting the drive coast to rest as its stored rotational energy is expended in overcoming friction and electric circuit losses. There are many applications however in which rapid deceleration is required and some form of applied electrical braking must be used.

If the stored rotational energy can be returned to the supply by suitable electrical connections the operational mode is described as 'regenerative' braking. In this mode of operation the electrical machine is temporarily acting as a generator by converting its stored mechanical energy into electrical energy.

A common form of electrical braking is known as 'dynamic' braking in which the drive motor again temporarily produces decelerating torque by generator action. With dynamic braking however the electrical energy converted from stored mechanical energy is not returned to the supply but is dissipated as heat in braking resistors external to the machine or, sometimes, in the drive machine itself. Dynamic braking usually involves additional control gear and sometimes d.c. or low frequency a.c. auxiliary power supplies. Both dynamic and regenerative braking are feasible for thyristor controlled motors but, in general, the paths of the external braking currents must be separate from the paths used by motoring currents. This involves the expense of additional hardware plus extra protection and interlocking between the motoring and braking paths.

On many adjustable speed drives it is required by law in the UK to use not only electrical braking but to incorporate some form of mechanical friction brake.

4.4.7 Reliability

The squirrel-cage induction motor and the permanent magnet reluctance motor are likely to be more reliable and require less maintenance than any form of motor, such as d.c. motors and a.c. commutator motors, that incorporate sliding electrical contacts. In addition the cage induction motor is acknowledged for the robustness that is derived from its simple construction. Consideration of the reliability of thyristor controlled drives also involves consideration of the reliability of the thyristors and their associated control gear. Industrial experience of large thyristor drives is encouraging from the viewpoint of reliability.

4.4.8 Power-to-weight ratio

In a few drive applications such as aircraft and missile systems the essential requirement is for lightness of weight, irrespective of other considerations. In general, however, the requirement of low power/weight or, sometimes, low power/size will involve a choice between different forms of electric motor. Thyristor controlled motor systems often have a power/weight and sometimes a power/size advantage over systems that involve a group of rotating machines such as, for example, the two- or three-machine Ward–Leonard set.

4.4.9 Power factor

Alternating current machine adjustable speed drives usually operate at a lagging power factor. The presence of a rectifier and/or an inverter or some form of thyristor chopper generally reduces the drive power factor still further. The low power factor of an induction motor operating from a conventional three-phase supply can be easily compensated by parallel terminal capacitance at the supply point. For thyristor controlled drives the use of terminal capacitance is not always helpful and alternative methods of power factor compensation may require to be investigated. The use of terminal capacitance or other power factor correction methods is likely to affect also the extent of supply current distortion and improvement in both these respects may justify the economic outlay.

4.4.10 Load factor and duty cycle

The temperature rise of an electric motor is a function of the duty cycle as well as the operating current level. Current overload can often be accepted if the duty cycle requires only intermittent working at the overload rate. Load factor becomes important for large drives because a discontinuous burden on the supply can cause voltage dips on the supply lines, possibly to the inconvenience of other customers.

With thyristor drives overload capacity is critical since thyristors have only a very restricted overload capacity. Moreover the protection of thyristors against overcurrent may require the use of high rupture capacity fuses that are very expensive.

4.4.11 Availability of supply

In the UK a three-phase, four-wire, 50 Hz supply is readily available in most locations. This can be used directly or rectified to provide a d.c. source if required. With a 50 Hz supply the maximum speed of an induction motor is 3000 r.p.m. and higher speeds must be realised by some form of frequency conversion. The standard frequency in aircraft electrical systems is 400 Hz. Although the availability of supply is a factor in the choice of a drive it need not prescribe the form of drive motor. An induction motor can be used as the final drive motor in a system for which only a d.c. supply is available just as a d.c. motor may be the best choice of drive motor in many cases where only a three-phase a.c. supply is available.

4.4.12 Effect of supply variation

The fidelity of the electricity supply in the UK is prescribed so that only voltage variations of $\pm 5\%$ and frequency variations of $\pm 1\%$ are permitted. In general purpose drives variations of voltage or frequency within these limits may be inconvenient but are not likely to be seriously disruptive or damaging. There are a few special forms of drives, however, where supply changes may have serious consequences on (say) the accuracy of the drive motor position or speed. Where this is likely to occur some form of stabilised supply or servomechanism compensation of supply changes must be incorporated.

4.4.13 Loading of the supply

Except for the cases of very large drives the public electricity supply is likely to be an 'infinite bus' or source of fixed voltage, frequency and sinusoidal waveform. Where a drive is of such a rating as to constitute the major load on a supply line some special arrangement between the supply authority and the customer may be needed, especially at switch-on.

The use of thyristor controlled drives usually involves the flow of harmonic currents in the supply lines. If these are of sufficient magnitude they can cause harmonic voltage drops in the supply system impedances that create harmonic voltages at the supply point of the offending customer and also at other supply points in the system. In the UK the waveform fidelity is governed by Electricity Council Engineering Recommendation G5/3, 1976. This gives permitted limits of harmonic current level in various converter and a.c. regulator systems and also prescribes harmonic voltage distortion limits at the various voltage levels of the supply system from 132 kV to the domestic supply level of 415 V (line). Corresponding guidelines in the USA are provided in the publication ANSI/IEEE Standard 519–1981 and in 'IEEE Recommended Practices and Requirements for Harmonic Control in Electric Power Systems', IEEE–PES and IEEE Static Converter Committee of IAS, 1993.

4.4.14 Environment

Adjustable speed drives quite often have to be sited in adverse, sometimes hostile, industrial environments. If the atmosphere is corrosive or explosive the motor will need to be totally enclosed. It is common to find a specification that calls for a motor that is splash-proof or drip-proof or even for one that is flame-proof and capable of heavy spraying with water. Again, the need may be for a motor that will operate immersed in water or in a high-vacuum chamber. In totally enclosed motors the cooling may take place via a closed air circuit which thereby requires a motor that is bulky and expensive.

The presence of high ambient temperature also creates special problems and if special insulating materials are needed or some degree of derating is necessary this clearly adds to the cost.

4.4.15 Running costs

The running costs for electrical energy are obviously related to the size, efficiency and power factor of the drive. Maintenance costs have to be accepted for any drive and the initial choice of drive should bear in mind the type of maintenance that will be needed. Thyristor controlled drives involve the use of elaborate and expensive electronic control gear for which the maintenance costs can be high. The increasing use of modular construction of electronic and microprocessor controllers, however, often means that routine servicing can be carried out by technicians who are not necessarily experts in electronic engineering.

4.5 TYPES OF ELECRIC MOTOR USED IN DRIVES

The commonest forms of electric motor used in adjustable speed drives are:
(a) d.c. motors,
(b) a.c. synchronous motors,
(c) a.c. induction motors.

4.5.1 D.c. motors

A direct current (d.c.) motor has two basic components, the field windings (invariably mounted on the frame or stator) and the armature winding (invariably mounted on the rotor). These two sets of windings have their axes mounted in electrical space quadrature (see Fig. 5.1) and both are supplied with power from direct current electrical sources. A commutator on the rotor is connected to the rotating armature conductors. This acts as a mechanical frequency changer or rectifier to maintain unidirectional armature circuit current through the brushes at all speeds.

For the study of speed variation the motor can be represented by equations representing its terminal properties of electrical input and mechanical output. These equations are developed in Section 5.1 of Chapter 5 below. The d.c. motor drive supplied from the d.c. source, via a chopper circuit, is described in Chapter 5. The d.c. motor drive supplied with rectified voltage and current, from either a single-phase or a three-phase a.c. source, is described in Chapter 6.

4.5.2 Synchronous motors

The many different types of synchronous motor may be classified into a number of groupings:
(1) wound-field motors,
(2) permanent magnet motors,
(3) synchronous reluctance motors,
(4) self-controlled (brushless) motors,
(5) stepping (stepper) motors,
(6) switched reluctance motors.

All types of synchronous motor have in common that their speed of rotation is precisely related to the frequency of the a.c. supply. For a range of discrete constant frequencies the torque–speed characteristics are represented by the vertical lines in Fig. 4.6. With wound-field, permanent magnet, synchronous reluctance and brushless motors there is a distributed, polyphase armature winding on the stator. Stepper motors and switched reluctance motors are energised by pulse trains applied to appropriate stator windings.

Because of the absence of a commutator, synchronous motors are not limited to maximum speed and have lower weight, volume and inertia compared with d.c. motors of the same rating. Most types of synchronous motors have a rotor structure that is salient in form.

4.5.2.1 Wound-field synchronous motors

When the magnetic excitation is provided by an external d.c. source, the excitation windings are wound onto the rotor and supplied via brushes and slip-rings. The speed of rotation is constant, with constant supply frequency, and is given by

$$N_1 = \frac{60 f_1}{p} \tag{4.7}$$

where N_1 is the motor speed in r.p.m., f_1 is the supply frequency in hertz and p is the number of pairs of poles for which the stator winding is wound. Operation can occur at the various discrete speeds specified by the integer values p. Torque is developed by the interaction of the stator and rotor magnetic fields. There is no synchronous starting torque and the motor is run-up from standstill by induction action.

High-speed and high power motors usually employ a cylindrical or round-rotor structure, Fig. 4.11(a), with a uniform air-gap. The motor equations are such that operation can conveniently be represented in terms of a per-phase equivalent circuit, using sinusoidal a.c. circuit theory (see, for example, references, 35, 40, 41). With a salient-pole structure, Fig. 4.11(b), the air-gap is non-uniform so that the motor develops a reluctance torque in addition to the synchronous torque of the two interacting fields. Alternating current generators (or alternators) in hydro-electric systems are usually synchronous machines with salient-rotor structure and many pairs of poles.

The characteristic feature of wound-field synchronous machines that makes them unique is that the field current can be varied independently. For the same a.c. supply voltage and frequency, with the same load torque, at the fixed value of synchronous speed, the field current can be varied over a wide range. For low values of field current the motor operates at a lagging power factor. For high values of field current the motor operates at a leading power factor. Synchronous motors are often intentionally overexcited so that their leading currents will power factor compensate the lagging currents

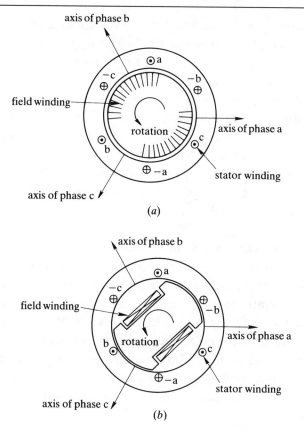

Fig. 4.11 Representation of an idealised wound-field synchronous machine: (a) cylindrical (round) rotor, (b) salient-pole rotor.

drawn by parallel-connected induction motors. At any load torque there is some value of field current that will result in unity power factor operation.

4.5.2.2 Permanent magnet synchronous motors

The rotor excitation field flux in a synchronous motor can be obtained by replacing the electromagnet field poles of Fig. 4.11 by permanent magnets. Magnetically hard materials such as ceramic ferrites or alloys of iron, nickel and cobalt may be used. The modern trend is to use samarium–cobalt–rare-earth materials because their high remanence and high coercive force permit reductions of magnet size and weight, but they are expensive. Ferrite materials are much cheaper but their low remanence requires larger magnets.

An inherent problem with permanent magnet motors is the inability to adjust the field current (and field flux). The internal e.m.f. remains proportional to speed, even in the overspeed (constant horsepower) range. With conventional fixed frequency operation this does not matter since the speed is constant. In adjustable speed (i.e. adjustable frequency) applications, calling for overspeed, high terminal voltages may be required, necessitating increased rating of the drive inverter.

A permanent magnet motor eliminates the need for a d.c. supply and avoids excitation winding copper losses. Because the excitation flux is constant, however, the motor cannot be used for power factor control. Also, the constant excitation flux tends to inhibit the induction action starting torque, which is lower than for a wound-field motor. Because of their high power/weight ratio permanent magnet synchronous generators are used in high-speed, 400 Hz aerospace applications.

4.5.2.3 Synchronous reluctance motors

A reluctance motor is a synchronous motor with salient rotor structure, but no field windings or permanent magnet excitation at all. The air-gap flux is provided entirely from the stator side a.c. armature windings which results in a low lagging power factor, typically 0.65–0.75 at full-load. Reluctance torque is developed as the salient-rotor poles align themselves with the synchronously rotating stator magnetic field. Like wound-field and permanent magnet machines, the synchronous reluctance motor has no synchronous starting torque and runs up from standstill by induction motor action in an auxiliary starting winding.

Early versions of the reluctance motor were limited to fractional horsepower ratings because the cheap, robust and reliable construction was counterbalanced by the disadvantages of low efficiency, low power factor and high starting current.

Subsequent design modifications involved the introduction of a segmented rotor construction to effect a flux barrier in each pole. This has increased the pull-out torque, the power factor and the efficiency. The simple construction of the reluctance motor makes it particularly useful in applications where several motors are required to rotate in close synchronism.

4.5.2.4 Self-controlled (brushless) synchronous motors

Use of the term 'brushless motor' is imprecise and even ambiguous in much electrical engineering literature. For example, permanent magnet synchronous motors and synchronous reluctance motors do not contain brushes at all and are therefore brushless motors.

So-called, 'self-controlled' synchronous motors are used in adjustable speed drives which require a range of operating frequencies. The self-control of a synchronous motor involves adjustment of the armature supply fre-

4.5 Types of elecric motor used in drives 143

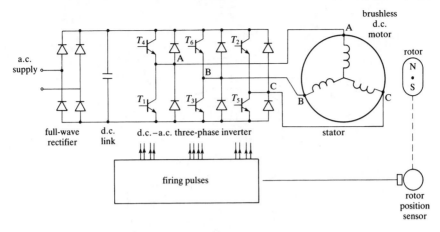

Fig. 4.12 Three-phase, full-wave brushless d.c. motor drive (ref. 31).

quency, proportionally as the rotor speed is varied, so that the armature field always moves at the same speed as the rotor. The rotating magnetic fields of the stator (armature) and the rotor (excitation system) are then always in synchronous motion producing a steady torque at all operating speeds. This is analogous to the d.c. motor in which the armature and excitation fields are synchronous but stationary for all operating speeds. Synchronous motor self-control requires the very accurate measurement of rotor speed and position and the very precise adjustment of the stator frequency. Rotor position sensing is done by an encoder which forms part of a control loop delivering firing pulses to the electronic switches of an adjustable frequency inverter feeding the armature windings.

The combination of an a.c. synchronous motor with permanent magnet rotor, three-phase inverter and rotor position sensor, Fig. 4.12, is often called a 'brushless d.c. motor.' The rotor position sensor and the inverter perform the function of the brushes and commutator of a d.c. motor. Commercial brushless d.c. motors have low torque ripple and are used in high-performance servo drives. If an a.c. supply is used, as in Fig. 4.12, to supply a cycloconverter or a d.c. link inverter, some references refer to the system as a 'brushless a.c. motor.'

4.5.2.5 Stepping (stepper) motors

The various forms of stepper motor are doubly salient in structure. There are an unequal number of pole projections on the stator and the rotor to ensure self-starting and to permit bidirectional rotation. Typically the stator and rotor pole numbers differ by two. A typical form of variable reluctance

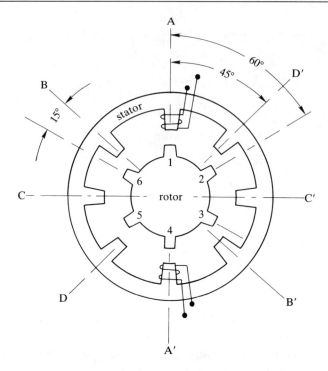

Fig. 4.13 Single-stack, four-phase, eight/six-pole variable reluctance stepper motor (only the phase A winding is shown) (ref. 31).

stepper motor is shown in Fig. 4.13, having eight stator poles and six rotor teeth. The rotor contains no windings.

Magnetisation of the stator poles is obtained by energising the stator coils sequentially. Reluctance torque is developed as the rotor teeth attempt to align with the magnetised stator poles. Each current pulse moves the rotor through a fixed angle, which facilitates angular position control without the need for shaft encoding and feedback circuitry. With the pole numbers of Fig. 4.13 an angular step-length of 15° can be realised. Much smaller step-lengths are possible by appropriate sequencing of successive pole energisation.

Stepper motors are ideally suited to applications requiring precise position control such as computer peripherals or numerically controlled machine tools. Variable reluctance stepper motors have high torque/inertia ratio, resulting in fast response with high acceleration, making them ideal as servo motors. They are usually built in the fractional horsepower or small integral horsepower ranges.

A stepper motor responds to unidirectional pulses of current in its field windings. This feature fits in well with modern digital electronics since a pulse of current is translated into a precise increment of angular shaft rotation. The electronic pulse source has to be engineered to match the electrical and mechanical requirements of the motor and its load.

4.5.2.6 Switched reluctance motors

The switched reluctance drive (SRD) is a development of great promise that is still little used. The motor is a high power development of the single-stack, variable reluctance stepper motor described in the preceding section. A doubly salient structure generally has a different number of poles on the rotor and the stator. There are no rotor windings and short stator end windings which facilitate easy cooling. The concentrated stator windings are energised by a pulse train resulting in a continuous rotational speed that is synchronous to the pulse frequency. The energising pulses are provided by a power electronics switching package that is custom designed to match each particular motor. Very high speeds can be realised and the high torque/inertia ratio makes it ideal as a servomotor. The motor has zero open-circuit voltage and contributes zero current to an external fault. Its efficiency is comparable with that of the equivalent induction motor. Closed-loop control via rotor positional feedback permits very precise position control.

A switched reluctance motor torque is due entirely to reluctance variation developed by the saliencies of one or two pairs of stator and rotor poles. This is in contrast with the synchronous reluctance motor, with which it has several important properties in common, where the entire air-gap surface contributes to the energy conversion process. Both the switched reluctance drive and the synchronous reluctance motor have in common that they require only unipolar current energisation since the rotor is not polarised. This reduces the number of switching devices required in the electronic controller package. Sometimes bipolar drive circuits are used. It is necessary to recover energy from the excited windings at current switch-off, which can be realised by the use of bifilar windings on the stator poles.

The SRD can provide four-quadrant operation that is comparable in performance, efficiency and cost with more established forms of induction motor or synchronous motor drives up to 200 kW. Careful hardware design and complex algorithms are needed to address the problems of torque pulsations and acoustic noise. The very high level of controllability of a switched reluctance motor makes it a natural competitor to the brushless d.c. motor drive and the vector controlled induction motor.

4.5.3 Induction motors

The three-phase induction motor is the work-horse of modern industry. The growth of modern electronics and the trend to microprocessor control of machines and systems has not significantly affected the position of the induction motor as the drive element that requires, most usually, to be controlled.

A significant feature of a.c. induction motor control is the emergence of field-oriented (vector) control as a commercial reality in high performance drives. So-called 'vector' control involves control of the spatial orientation of the air-gap flux and the secondary magneto-motive force (m.m.f.). The purpose is to decouple that portion of the stator current involved in producing air-gap flux from the portion involved in the direct production of torque, thereby providing independent control of torque and flux. The technique is usually implemented by a PWM inverter incorporating a current control loop and could be termed a current regulated induction motor drive.

The principle of indirect field-oriented control has achieved wide acclaim and acceptance for high performance servo-type drives. Its impact in the high volume, low performance market has been negligible. This is mainly due to the necessity of the very expensive pulse encoder required for precise speed measurement. Alternatives to the speed encoder method are the subject of much current investigation and typify the evolutionary nature of drives research.

It is conceivable that the well established situation of the induction motor might eventually be eroded by the switched reluctance drive in applications where wide speed variation, with precise control, is required. For constant speed operation, however, and for applications requiring only a restricted

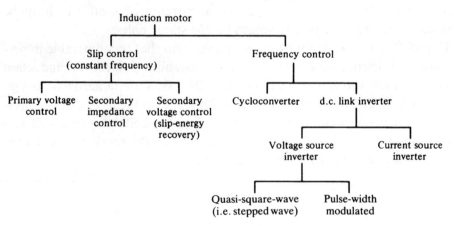

Table 4.4 *Methods of induction motor speed control.*

range of speed the induction motor appears to hold an unchallengeable monopoly.

Table 4.4 shows the main methods of induction motor speed control. Primary voltage control and secondary resistance control are covered in Chapter 9. Slip-energy recovery is covered in Chapter 10. Various methods of variable frequency inverter control are discussed in Chapters 11 and 12. The topic of cycloconverter frequency changing is not covered in this present text due to limitations of space.

4.6 DIFFERENT OPTIONS FOR AN ADJUSTABLE SPEED DRIVE INCORPORATING AN ELECTRIC MOTOR

Many forms of adjustable speed drive, incorporating the electric motor, will give a good approximation to the steady-state motoring operation (quadrant I) of Fig. 4.6. Some of these are listed below, in an arbitrary order:

(a) shunt or separately excited d.c. motor with armature voltage control,
(b) Ward–Leonard system (armature voltage control of a separately excited d.c. motor),
(c) three-phase induction motor with primary voltage or secondary resistance control, on closed-loop,
(d) three-phase induction motor incorporating pole changing techniques,
(e) three-phase induction motor in a slip-energy recovery scheme,
(f) three-phase a.c. commutator motor (e.g. Schrage motor),
(g) three-phase induction motor or synchronous motor fed by a voltage source, adjustable frequency inverter,
(h) three-phase induction motor fed by a current source, adjustable frequency inverter, with or without vector control,
(i) three-phase induction motor fed by a cycloconverter frequency controller,
(j) single-phase or three-phase stepper motor,
(k) switched reluctance motor.

4.7 A.C. MOTOR DRIVES OR D.C. MOTOR DRIVES?

Before the large-scale introduction of solid-state semiconductor switches in the 1970s d.c. motor drives were almost universally used for wide speed range, adjustable speed drives. A few applications utilised wound rotor induction motors with secondary resistance control but a.c. motors (both

Table 4.5 *Relative features of a.c. drives and d.c. drives.*

d.c. drives	a.c. drives
Advantages	
Well-established technology	Motor reliability
Simple and inexpensive power converter	Motor cost (size/weight)
Simple control	Environmentally insensitive
No zero speed problem	Good a.c. line power factor (PWM control)
Wide speed range	Lower first cost in some applications
Fast response	Cost strongly dependent on (developing) solid-state technology
Disadvantages	
Motor cost (size/weight)	Complex power converter
Environmentally sensitive	Complex control
Motor maintenance	Zero speed control (on open loop)
Cost not strongly dependent on solid-state technology	Developing technology
Poor power factor (for a.c. rectified supplies)	

synchronous and induction) were mainly regarded as constant speed machines. Conditions have changed in the last few years with the widespread use of power semiconductors that are able to handle ratings up to 6000 V and 3500 A without device paralleling. The application of these devices is now widespread in d.c. drives, almost totally replacing mechanical control methods due to improved performance and cost. The adoption of a.c. motors for adjustable speed drives has been slower due to the necessity for developing new technology in inverter design and control methods.

The relative advantages and disadvantages of a.c. drives and d.c. drives are summarised in Table 4.5. For d.c. drives the main component of cost is the expensive motor whereas the power converter and control system are relatively cheap and part of a mature and well established design technology. With a.c. drives, on the other hand, the motor (e.g. cage induction or synchronous reluctance) is cheap but the power converter and control system are complex and expensive. It is significant that the relative prices of a.c. and d.c. motors are not likely to change greatly. The very great competitive development work on solid-state devices, systems and control techniques all over the world, however, is likely to result in improvements of performance and relative cost that will favour a.c. drives more than d.c. drives. Improved control algorithms and more reliable hardware are making a.c. drives preferable in the lower ratings, with extensive use of power transistors, and gradually increasing the share of the market in medium and large drives.

4.8 TRENDS IN THE DESIGN AND APPLICATION OF A.C. ADJUSTABLE SPEED DRIVES

4.8.1 Trends in motor technology and motor control

The application of adjustable speed drives forms only a small fraction (<10%) of the world market for electric motors. Most motor applications call for constant speed operation with satisfactory start-up. An increased use of converters and inverters has led to improvements in standard induction motor design. These have addressed such issues as good starting torque, well damped response to load transients, rapid protection against current overload or loss of supply and adequate speed regulation. Removal of the starting/running design compromise in induction motors results in lower secondary resistance and higher full-load speed in which thyristor voltage control can provide soft start-up. No dramatic change is likely in induction motor design until adjustable frequency inverter drives take a major share of the market.

The use of high energy permanent magnetic materials such as neodymium–iron–boron and the samarium–cobalt–rare-earth ferrites has resulted in improved permanent magnets used in small synchronous reluctance motors. Market penetration of these specialised motors may have been limited by the excessive cost of the magnets in spite of the fact they offer higher efficiency and higher power density than equivalent induction motors.

In synchronous machine applications the use of square-wave excitation is an alternative to the more traditional sinusoidal flux distribution. The now foreseeable possibility of using high permanent magnet excitation in large synchronous machines is likely to have dramatic effects on motor and generator design.

4.8.2 Trends in power switches and power converters

Bipolar junction transistors (BJTs) have now largely replaced thyristors in inverter applications below ratings of 100 kW. Elimination of the need for auxiliary commutation circuits has resulted in cost reductions. Increased switching frequencies and new switching strategies, for current- as well as voltage-control, have been realised in the design of pulse-width modulated

(PWM) inverters. The use of faster switching speeds and higher frequency switching results in decreased harmonic distortion and decrease in the size of the filter components.

In the 1980s a significant technological advance took place due to the introduction of high power gate turn-off thyristors (GTOs), described in Chapter 2 above. The pace of evolutionary change in power electronics is well illustrated by the fact that before the GTO can fully supplant the conventional thyristor, it will itself be supplanted.

Design of the next generation of power converters is likely to be dominated by the use of the MOS controlled thyristor (MCT) and, especially, by the insulated gate bipolar transistor (IGBT). These gate on/off devices, described in Chapters 1–3 above, offer advantages in terms of low forward voltage drop and sufficiently high ratings to satisfy most of the present drives market. The ultimate aim is for a switch that possesses bidirectional voltage blocking capability, bidirectional current conduction capability and bidirectional turn-off. This will constitute a four-quadrant gate turn-off device that is completely comprehensive.

4.9 PROBLEMS

Steady-state stability

4.1 If friction effects are ignored the equilibrium state of a drive can be represented by the equation $T = J dN/dt + T_L$. For a small deviation from the equilibrium the equation becomes

$$(T + \Delta T) = J \frac{d}{dt}(N + \Delta N) + (T_L + \Delta T_L)$$

Show that solution of the incremental equation above leads to the criterion of equation (4.5).

4.2 Some possible intersections of motor lines and load lines, for steady-state torque speed performance, are shown in Fig. 4.14. Use the criterion (4.5) to determine whether each intersection represents likely stable (S) or unstable (U) operation.

4.3 A motor with polar moment of inertia J develops a torque defined by the relationship $T = aN + b$. This motor drives a load defined by the torque–speed relationship $T_L = cN^2 + d$. If the coefficients a, b, c, d are all positive constants, of such values that start-up from rest is possible, determine the equilibrium speeds and specify if each equilibrium speed represents stable or unstable steady-state operation.

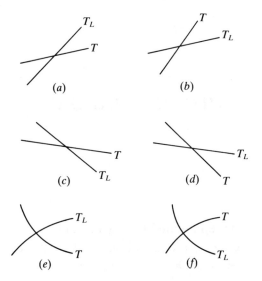

Fig. 4.14 Intersections between motor (drive) lines T and load lines T_L in quadrant I of the torque–speed plane.

4.4 The friction and windage loss in a certain motor is characterised by a friction torque characteristic $T_{FW} = KN$. Sketch (i) the charcteristics of motor torque T versus speed N if the load torque T_L is constant and (ii) the characteristics of load torque T_L versus speed if the motor torque T is constant. In each case assume that T_{FW} at full speed is equal to about 25% of the load torque.

5

D.c. motor control using a d.c. chopper

In this textbook the d.c. motor is represented largely in terms of its terminal properties. Those readers interested in the morphology of d.c. machines are referred to the many excellent texts, such as reference 35.

5.1 BASIC EQUATIONS OF MOTOR OPERATION

The injection of direct current through the motor field windings, Fig. 5.1, establishes an excitation current which sets up a field of flux in the motor air-gap. In terms of instantaneous variables,

$$e_f = i_f R_f + L_f \frac{di_f}{dt} \tag{5.1}$$

$$\phi_f = f(i_f) \tag{5.2}$$

If the motor operates on the linear part of the magnetisation characteristic of its mutual flux path, Fig. 5.2, there is a linear relationship between the steady-state field flux Φ_f and the steady-state value I_f of the field current

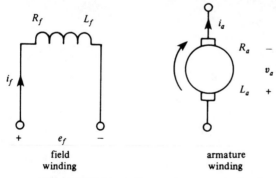

Fig. 5.1 Field winding and armature of a d.c. machine

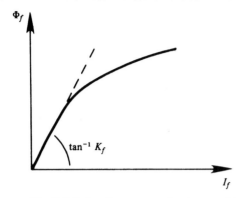

Fig. 5.2 Saturation curve of a d.c. machine.

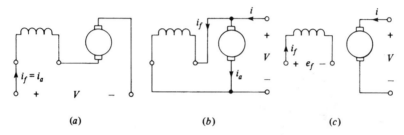

Fig. 5.3 Field and armature interconnections: (*a*) series machine, (*b*) shunt machine, (*c*) separately-excited machine.

$$\Phi_f = K_f I_f \tag{5.3}$$

Under steady-state conditions there is no time variation of the field current. Further, no e.m.f. is induced in the field windings due to armature circuit effects so that, in the steady state,

$$E_f = R_f I_f = \frac{R_f}{K_f} \Phi_f \tag{5.4}$$

The armature circuit and the field circuits may be interconnected in the three basic ways shown in Fig. 5.3. Each connection results in particular motor performance and each is suited to particular load applications. In the series connection, Fig. 5.3(*a*), the armature and field currents are identical and the motor output (i.e. torque–speed) characteristic is of the hyperbolic form shown in Fig. 5.4. The shunt connection, Fig. 5.3(*b*), and the separately excited connection, Fig. 5.3(*c*), each deliver an output characteristic of the form in Fig. 5.5.

When the armature conductors carry current, forces are exerted on them due to the interaction of this current with the steady air-gap flux Φ (which

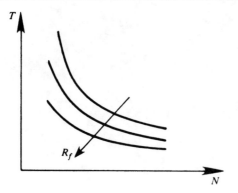

Fig. 5.4 Torque–speed characteristics of a d.c. series motor.

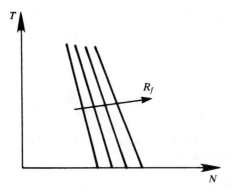

Fig. 5.5 Torgue–speed characteristics of a d.c. shunt motor.

consists, very largely, of the field current component Φ_f). The resulting instantaneous torque $T(t)$ developed by the motor is given by

$$T(t) = K_T \phi i_a \tag{5.5}$$

In terms of steady-state, time average values the torque is given by

$$T = K_T \Phi I_a \tag{5.6}$$

Rotation of the armature conductors in the flux field causes an e.m.f. to be induced in the armature circuit of such polarity as to oppose the flow of armature current. This induced e.m.f. is usually known as the reverse e.m.f. or back e.m.f. In instantaneous variables the armature back e.m.f. e_b is given by

$$e_b = K_E \phi n \tag{5.7}$$

where n is the instantaneous speed. Taking time average values, for steady-state operation, results in a speed equation

$$E_b = K_E \Phi N \tag{5.8}$$

In the SI system of units the constants K_T and K_E are identical and have the dimensions newton metres per weber ampere or volt seconds per weber radian. Combining (5.6) and (5.8) then gives alternative expressions for the internal power P developed by the motor

$$P = TN = \frac{K_T}{K_E} E_b I_a = E_b I_a \text{ (in SI units)} \tag{5.9}$$

Each of the motor equations (5.1)–(5.9) applies to all three motor connections in Fig. 5.3.

The armature circuit instantaneous voltage equations are

$$v_a = i_a R_a + L_a \frac{di_a}{dt} + e_b \tag{5.10(a)}$$

(for the shunt and separately excited connections)

$$v_a = i_a(R_a + R_f) + (L_a + L_f)\frac{di_a}{dt} + e_b \tag{5.10(b)}$$

(for the series motor).

For steady-state operation the inductive effects are usually negligibly small and equations (5.10) reduce to

$$V = I_a R_a + E_b \text{ (shunt, sep. exc. motors)} \tag{5.11(a)}$$

$$V = I_a(R_a + R_f) + E_b \text{ (series motor)} \tag{5.11(b)}$$

Terminal voltage V is usually fixed, while I_a and E_b represent time average values. The difference between the magnitudes of E_b and V is usually only a few per cent, even at full load.

Combining (5.6), (5.8) and (5.11(a)) gives useful expressions for the speed control of shunt and separately excited motors.

$$N = \frac{V - I_a R_a}{K_E \Phi} = \frac{V}{K_E \Phi} - \frac{R_a}{K_T^2 \Phi^2} T \tag{5.12}$$

If (5.12) is combined with (5.3) and (5.4) then, neglecting saturation and since $E_f = V$, the torque–speed characteristic of a shunt motor is given by

$$T = V^2 \frac{K_T K_f}{R_a R_f} \left(1 - \frac{K_E K_f}{R_f} N\right) \tag{5.13}$$

Variation of the torque–speed characteristics, with R_f as the parameter, is shown in Fig. 5.5.

For a separately excited d.c. motor the above equations give

$$T = \frac{K_T K_f E_f}{R_a R_f}\left(V - \frac{K_E K_f E_f}{R_f} N\right) \tag{5.14}$$

when $E_f = V$, (5.14) reduces to (5.13).

For a d.c. series motor the flux varies with the motor armature current. Neglecting magnetic saturation the motor equations give an expression for average motor speed

$$N = \frac{V - I_a(R_a + R_f)}{K_E \Phi} = \frac{V}{K_E K_f I_a} - \frac{R_a + R_f}{K_E K_f} \tag{5.15}$$

Combining (5.15) with (5.4) and (5.6) gives

$$T = \frac{K_T K_f^2 V^2}{[K_E K_f N + (R_a + R_f)]^2} \tag{5.16}$$

If the motor operates in deep saturation, due to excessive field current, (5.3) is no longer valid and the torque–speed equations (5.13), (5.14) and (5.16) require modification.

The dynamic equation for the developed torque T of an adjustable speed drive is

$$T = K_i J \frac{dN}{dt} + T_s + BN + T_L \tag{5.17}$$

where K_i is constant, $J\,dN/dt$ = inertial torque, $T_s + BN = T_{FW}$ = friction and windage torque and T_L = load torque. Some further discussion of this is given in Section 4.2, of Chapter 4 above.

Control of the speed of a d.c. motor is usually obtained by
 (i) variation of the field circuit resistance R_f (which changes the flux Φ), or
 (ii) variation of the armature voltage V (which changes the armature current I_a).

The resistance method wastes significant energy, particularly in some traction applications.

Adjustment of the armature voltage by means of a solid-state power converter is an efficient and economic method of d.c. motor speed control. Two basic methods may be used, illustrated in Fig. 5.6, to obtain a variable direct voltage. If the source is a fixed direct voltage, from (say) a battery, it is possible to use an a.c. link converter, Fig. 5.6(a), consisting of an inverter

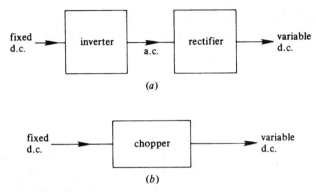

(a)

(b)

Fig. 5.6 Methods of obtaining an adjustable armature voltage: (a) a.c. link converter, (b) d.c. chopper.

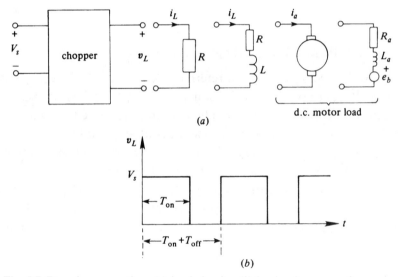

Fig. 5.7 D.c. chopper action: (a) load circuits, (b) load voltage waveform.

and a rectifier or to obtain direct conversion by means of a d.c. chopper, Fig. 5.6(b). If the source is an alternating current supply it is necessary to use a controlled rectifier, as discussed in Chapter 6 below.

5.2 D.C. CHOPPER DRIVES

The action of a d.c. chopper is to apply a train of unidirectional voltage pulses, Fig. 5.7(b), to the load, Fig. 5.7(a). By regulating the mark–space ratio or duty cycle of the conduction pulses the average load voltage can be controlled. In Fig. 5.7(b) it can be seen by inspection that the average output voltage is given, in terms of the switch times T_{on} and T_{off}, by

$$\text{Average of } v_L(t) = V_{L_{av}} = \frac{T_{on} V_s}{T_{on} + T_{off}} = \gamma V_s \tag{5.18}$$

where

$$\gamma = T_{on}/(T_{on} + T_{off})$$

The usual application requirement is for a fixed value of $V_{L_{av}}$ and therefore fixed values of T_{on} and T_{off}. If T_{on} is varied, with the overall period $T_{on} + T_{off}$ constant, the resultant voltage wave represents a form of pulse-width modulation. If T_{on} is constant but T_{off} is varied, the resultant voltage wave then represents a form of frequency modulation.

5.2.1 Basic (class A) chopper circuit

The basic chopper circuit, often referred to as a class A chopper, is shown in Fig. 5.8. It consists of a semiconductor switch S (often an SCR) and an uncontrolled rectifier (diode) D. The commutation circuit essential for controlled extinction when a thyristor switch is used can take many forms but is not described here. The interested reader is referred to the many existing texts on the subject, some of which are given in the bibliography.

The load voltage waveform of Fig. 5.7(b) is true for all passive impedance loads and also for motor loads when the current is continuous. This form of chopper connection is sometimes called a 'buck' converter because the output

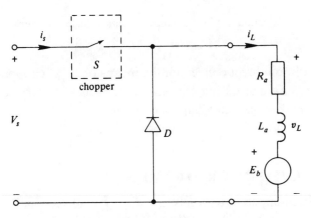

Fig. 5.8 Basic (class A) semiconductor chopper circuit (commutation circuit not shown).

voltage cannot exceed the input voltage level V_s. A chopper in which the output voltage can exceed the input voltage is sometimes called a 'boost' converter, briefly described in Section 5.2 below.

When switch S conducts, in the circuit of Fig. 5.8, the supply voltage V_s is applied to the load. While switch S is off the load voltage is held at zero by the action of the diode D if load current continues to flow (which will occur if the load contains inductance). For all loads the current is unidirectional and the polarity of the load voltage is non-reversible. Operation takes place only in the positive voltage, positive current quadrant of the load voltage/load current plane so that the circuit is referred to as a one-quadrant chopper.

With a constant thyristor firing-angle the load voltage waveform is fixed. The load current waveform, at the same firing-angle, depends on the nature of the load and the magnitude of its impedance.

With resistive loads diode D has no effect and the load current waveform $i_L(t)$ is identical to the load voltage waveform $v_L(t)$. With inductive and motor loads, typical waveforms are shown in Fig. 5.9. For high values of duty cycle γ the load current fluctuates in magnitude but is likely to be

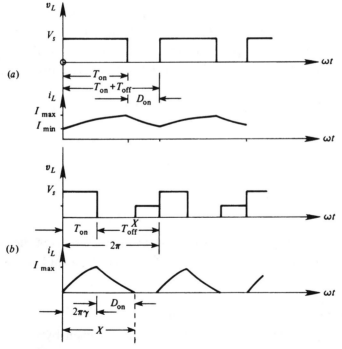

Fig. 5.9 Load current and voltage waveforms with inductive load: (a) continuous load current, (b) discontinuous load current.

continuous. For low values of γ, especially with low inductance, the load current may fall to zero during the off periods of the switch. For separately excited and shunt motors back e.m.f. E_b is invariant with armature current. For series motors E_b is dependent on motor current but may be considered constant for continuous current operation.

5.2.1.1 Analytical properties of the load voltage waveform

Let the repetition periodicity $T_{on} + T_{off}$, Fig. 5.9, be designated as 2π radians to facilitate harmonic analysis,

$$T_{on} + T_{off} = 2\pi \tag{5.19}$$

Since the independent variable is chosen as ωt the periodic time of the overall (on + off) cycle is

$$\text{periodic time} = \frac{2\pi}{\omega} \tag{5.20}$$

The frequency of the chopper operation is the inverse of the periodic time,

$$\text{chopping frequency} = f_c = \frac{\omega}{2\pi} \tag{5.21}$$

Typical chopping frequencies are usually in the range $100 < \omega/2\pi < 1000$ Hz for thyristor choppers and up to 10 kHz for transistor choppers. In low power applications MOSFET switches can be used at frequencies in excess of 200 kHz.

The on period of the chopper T_{on} is, from (5.18),

$$T_{on} = \gamma(T_{on} + T_{off}) = 2\pi\gamma \tag{5.22}$$

The on-time of the switch in seconds is therefore

$$\text{switch on-time} = \frac{2\pi\gamma}{\omega} \tag{5.23}$$

The terms T_{on}, T_{off} now serve the double purpose of identifying the conduction state of switch S in Fig. 5.8 and also defining its period of conduction or extinction in radians.

(i) Continuous armature current operation
The output voltage waveform $v_L(\omega t)$ in Fig. 5.9(a) is given by

$$v_L(\omega t) = V_s|_0^{2\pi\gamma} + 0|_{2\pi\gamma}^{2\pi} \tag{5.24}$$

The time average value of the periodic function $v_L(\omega t)$ is given by

$$V_{L_{av}} = \frac{1}{2\pi} \int_0^{2\pi} v_L(\omega t) \, d\omega t$$

$$= \frac{1}{2\pi} \int_0^{2\pi\gamma} V_s \, d\omega t \quad (5.25)$$

$$= \frac{V_s}{2\pi} [\omega t]_0^{2\pi\gamma}$$

$$\therefore V_{L_{av}} = \gamma V_s$$

which confirms the result of (5.18). Because $V_{L_{av}} \leq V_s$ the circuit of Fig. 5.8 is a 'buck' converter.

The r.m.s value of the load voltage waveform is given by

$$V_L = \sqrt{\frac{1}{2\pi} \int_0^{2\pi} v_L^2(\omega t) \, d\omega t}$$

$$= \sqrt{\frac{V_s^2}{2\pi} [\omega t]_0^{2\pi\gamma}} \quad (5.26)$$

$$\therefore V_{L_{rms}} = \sqrt{\gamma} \, V_s$$

The ripple factor, defining the ratio of the a.c. components to the average value, is given by

$$\text{ripple factor} = \frac{\sqrt{V_{L_{rms}}^2 - V_{L_{av}}^2}}{V_{L_{av}}}$$

$$= \frac{V_s \sqrt{\gamma - \gamma^2}}{V_s \gamma} \quad (5.27)$$

$$\therefore RF = \sqrt{\frac{1-\gamma}{\gamma}}$$

For full conduction $\gamma = 1$ and $RF = 0$.

The Fourier coefficients for the nth harmonic of the load voltage waveform are given by

$$a_n = \frac{1}{\pi}\int_0^{2\pi} v_L(\omega t)\cos n\omega t\, d\omega t$$

$$= \frac{V_s}{\pi}\int_0^{2\pi\gamma}\cos n\omega t\, d\omega t \qquad (5.28)$$

$$= \frac{V_s}{n\pi}\sin 2\pi n\gamma$$

$$b_n = \frac{1}{\pi}\int_0^{2\pi} V_L(\omega t)\sin n\omega t\, d\omega t$$

$$= \frac{V_s}{\pi}\int_0^{2\pi\gamma}\sin n\omega t\, d\omega t \qquad (5.29)$$

$$= \frac{V_s}{n\pi}(1 - \cos 2\pi n\gamma)$$

The peak amplitude $\hat{V}_{L_n}$ and phase-angle ψ_n of the nth load voltage harmonic are therefore given by

$$\hat{V}_{L_n} = c_n = \sqrt{a_n^2 + b_n^2}$$

$$= \frac{V_s}{n\pi}\sqrt{[\sin^2 2\pi n\gamma + (1 - \cos 2\pi n\gamma)^2]} \qquad (5.30(a))$$

$$= \frac{2V_s}{n\pi}\sin n\pi\gamma \qquad (5.30(b))$$

$$\psi_n = \tan^{-1}\left(\frac{a_1}{b_1}\right) = \tan^{-1}\left(\frac{\sin 2\pi n\gamma}{1 - \cos 2\pi n\gamma}\right) \qquad (5.31(a))$$

$$= \frac{\pi}{2} - n\pi\gamma \qquad (5.31(b))$$

(ii) Discontinuous armature current operation

If the load current becomes zero during part of the cycle, as in Fig. 5.9(b), this will occur when switch S and diode D are both off. The load is then isolated from the supply. Any e.m.f. E_b present across the motor brushes will then be registered across the load terminals. For the interval $X \le \omega t \le 2\pi$ in Fig. 5.9(b) it is seen that there is a load voltage $v_L(\omega t) = E_b$.

If E_b is constant then

$$v_L(\omega t) = V_s|_0^{2\pi\gamma} + E_b|_X^{2\pi} \qquad (5.32)$$

The average value of this load voltage is

$$V_{L_{av}} = \frac{1}{2\pi} \int_0^{2\pi} v_L(\omega t)\, d\omega t$$

$$= \frac{1}{2\pi} \left(\int_0^{2\pi\gamma} V_s\, d\omega t + \int_X^{2\pi} E_b\, d\omega t \right) \quad (5.33)$$

$$= \gamma V_s + \left(1 - \frac{X}{2\pi}\right) E_b$$

Diode extinction angle X can be obtained from equation (5.50) below. For full conduction $X = 2\pi$ and (5.33) reduces to (5.25). The r.m.s. load voltage is

$$V_{L_{rms}} = \sqrt{\frac{1}{2\pi} \int_0^{2\pi\gamma} V_s^2\, d\omega t + \int_X^{2\pi} E_b^2\, d\omega t}$$

$$= \sqrt{\gamma V_s^2 + E_b^2 \left(1 - \frac{X}{2\pi}\right)} \quad (5.34)$$

With discontinuous current the Fourier components of the load voltage waveform are obtained by substituting (5.32) into the defining integrals of (5.28), (5.29) to give

$$a_n = \frac{V_s}{n\pi} \sin 2\pi n\gamma - \frac{E_b}{n\pi} \sin nX \quad (5.35)$$

$$b_n = \frac{V_s}{n\pi} (1 - \cos 2\pi n\gamma) - \frac{E_b}{n\pi} (1 - \cos nX) \quad (5.36)$$

The fundamental ($n = 1$) component of the load voltage, for example, is given by

$$v_{L_1}(\omega t) = \hat{V}_{L_1} \sin(\omega t + \psi_1) \quad (5.37)$$

where

$$\hat{V}_{L_1} = c_1$$
$$= \frac{1}{\pi} \sqrt{(V_s \sin 2\pi\gamma - E_b \sin X)^2 + [V_s(1 - \cos 2\pi\gamma) - E_b(1 - \cos X)]^2} \quad (5.38)$$

$$\psi_1 = \tan^{-1} \left[\frac{V_s \sin 2\pi\gamma - E_b \sin X}{V_s(1 - \cos 2\pi\gamma) - E_b(1 - \cos X)} \right] \quad (5.39)$$

For continuous current operation $X = 2\pi$ and (5.38), (5.39), reduce to (5.30), (5.31) respectively, with $n = 1$.

5.2.1.2 Analytical properties of the load current waveform

With a motor load, having the equivalent circuit representation of Fig. 5.7(a), the current equation is, in general,

$$v_L(\omega t) = R_a i_L + L_a \frac{di_L}{dt} + E_b$$

or

$$\frac{di_L}{dt} + \frac{R_a}{L_a} i_L = \frac{v_L - E_b}{L_a} \tag{5.40}$$

(i) Continuous armature current operation

With switch S on (and D off)

At $\omega t = 0^+$, in Fig. 5.9, $v_L = V_s$ and $i_L(\omega t) = I_{\min}$. Solution of the first-order linear differential equation (5.40) gives the result, for $0 < \omega t < 2\pi\gamma$,

$$i_L(\omega t) = \frac{V_s - E_b}{R_a}(1 - \varepsilon^{-\omega t/\omega\tau}) + I_{\min}\varepsilon^{-\omega t/\omega\tau} \tag{5.41}$$

where $\omega\tau = \omega L_a/R_a$.

When $\omega t = 2\pi\gamma$, in Fig. 5.9, $v_L = V_s$ and $i_L(\omega t) = I_{\max}$. Substituting into (5.41) gives

$$I_{\max} = \frac{V_s - E_b}{R_a}(1 - \varepsilon^{-2\pi\gamma/\omega\tau}) + I_{\min}\varepsilon^{-2\pi\gamma/\omega\tau} \tag{5.42}$$

Equation (5.42) is not time dependent and remains true after S switches off.

With switch S off (and D on)

With $\omega t = 2\pi\gamma^+$, in Fig. 5.9, $v_L = 0$ and $i_L(\omega t) = I_{\max}$. In the interval $2\pi\gamma < \omega t \leq 2\pi$ the load current is given by

$$i_L(\omega t) = \frac{-E_b}{R_a}[1 - \varepsilon^{-(\omega t - 2\pi\gamma)/\omega\tau}] + I_{\max}\varepsilon^{-(\omega t - 2\pi\gamma)/\omega\tau} \tag{5.43}$$

But at $\omega t = 2\pi$, $i_L(\omega t) = I_{\min}$. Equation (5.43) may then be rewritten

$$I_{\min} = \frac{-E_b}{R_a}[1 - \varepsilon^{-2\pi(1-\gamma)/\omega\tau}] + I_{\max}\varepsilon^{-2\pi(1-\gamma)/\omega\tau} \tag{5.44}$$

The simultaneous solution of (5.42) and (5.43) yields

$$I_{\max} = \frac{V_s}{R_a}\left(\frac{1 - \varepsilon^{-2\pi\gamma/\omega\tau}}{1 - \varepsilon^{-2\pi/\omega\tau}}\right) - \frac{E_b}{R_a}$$

$$I_{\min} = \frac{V_s}{R_a}\left(\frac{\varepsilon^{2\pi\gamma/\omega\tau} - 1}{\varepsilon^{2\pi/\omega\tau} - 1}\right) - \frac{E_b}{R_a} \tag{5.45}$$

5.2 D.c. chopper drives

For full conduction of switch S in Fig. 5.8, $\gamma = 1$ so that $I_{max} = I_{min} = (V_s - E_b)/R$. The use of (5.45) permits solution of (5.41) and (5.43) to determine a value for $i_L(\omega t)$ at any instant of the cycle during continuous conduction. In many chopper applications the maximum current value is limited by device ratings. Performance of the type in Fig. 5.9(a) is then known as current limit control (CLC).

(ii) Discontinuous armature current operation

Under certain circuit conditions the load current $i_L(\omega t)$ may fall to zero resulting in the discontinuous pulse pattern of Fig. 5.9(b). This waveform consists of two parts that may be classified according to the conduction of switch S, as in section (i) above.

The maximum current is now, in general, different from the value I_{max} obtained with continuous operation and it occurs for a different value of T_{on}. The minimum current for discontinuous operation is, by definition, zero.

With switch S on (and D off)
In the interval $0 < \omega t < 2\pi\gamma$ of Fig. 5.9(b) the current equation (5.40) applies, with the limits that $v_L = V_s$ and $i_L(\omega t) = 0$ at $\omega t = 0+$. This gives

$$i_L(\omega t) = \frac{V_s - E_b}{R_a}(1 - \varepsilon^{-\omega t/\omega \tau}) \tag{5.46}$$

Maximum $i_L(\omega t)$ occurs at $\omega t = 2\pi\gamma$ so that

$$I_{max_d} = \frac{V_s - E_b}{R_a}(1 - \varepsilon^{-2\pi\gamma/\omega\tau}) \tag{5.47}$$

The terminology I_{max_d} defines the maximum current value for discontinuous operation.

With switch S off (and D on)
When S switches off in Fig. 5.8 the load voltage $v_L(\omega t)$ falls to zero due to the conduction through diode D. The circuit differential equation (5.40) is then modified to

$$\frac{di_L}{dt} + \frac{R_a}{L_a} i_L = \frac{-E_b}{R_a} \tag{5.48}$$

This has the solution (5.43) except that the maximum current is now given by (5.47) to result in

$$i_L(\omega t) = \frac{-E_b}{R_a}\left[1 - \varepsilon^{-(\omega t - 2\pi\gamma)/\omega\tau}\right]$$
$$+ \frac{V_s - E_b}{R_a}\left(1 - \varepsilon^{-2\pi\gamma/\omega\tau}\right)\varepsilon^{-(\omega t - 2\pi\gamma)/\omega\tau} \quad (5.49)$$

Let current extinction occur at $\omega t = X$ in Fig. 5.9(b), where X is the current extinction angle. Putting $i_L(X) = 0$ and $\omega t = X$ into (5.49) gives an explicit expression for X,

$$X = \omega\tau \ln\left[\varepsilon^{2\pi\gamma/\omega\tau}\left\{1 + \frac{V_s - E_b}{E_b}\left(1 - \varepsilon^{-2\pi\gamma/\omega\tau}\right)\right\}\right] \quad (5.50)$$

Equation (5.49) therefore defines the current in the region $2\pi\gamma < \omega t < X$ where X is found from (5.50). Equation (5.50) is indeterminate if $E_b = 0$.

Let switch S in Fig. 5.8 have a conduction period of particular value T'_{on} radians that represents the boundary between continuous and discontinuous operation. Then, from (5.22),

$$\gamma' = \frac{T'_{\text{on}}}{2\pi} = \frac{T'_{\text{on}}}{T'_{\text{on}} + T_{\text{off}}} \quad (5.51)$$

At this particular value of conduction the minimum current $I_{\min}$ defined by (5.45) just falls to zero so that, rearranging,

$$\frac{E_b}{V_s} = \frac{\varepsilon^{2\pi\gamma'/\omega\tau} - 1}{\varepsilon^{2\pi/\omega\tau} - 1} \quad (5.52)$$

The relationship between E_b/V_s and γ' is shown in Fig. 5.10 with the factor $2\pi/\omega\tau$ as parameter. If a circuit operates with a specified value of γ, defined by (5.18), then the criteria for continuous or discontinuous operation are

$$\begin{aligned}\gamma > \gamma', &\text{ continuous current} \\ \gamma < \gamma', &\text{ discontinuous current}\end{aligned} \quad (5.53)$$

where γ' is defined by (5.52).

Since the ratio $2\pi/\omega$ is the periodic time of the overall cycle, the parameter $2\pi/\omega\tau$ of the 'state of conduction' curves, Fig. 5.10, is

$$\frac{2\pi}{\omega\tau} = \frac{\text{period of the overall (on + off) cycle}}{\text{time constant of the load impedance}} \quad (5.54)$$

If the circuit is passive, $E_b = 0$ and (5.52) can only be satisfied by $\gamma' = 0$. In other words, there is no finite value of T_{on} that will result in discontinuous operation. Although the current may become small it is finite and operation is therefore continuous.

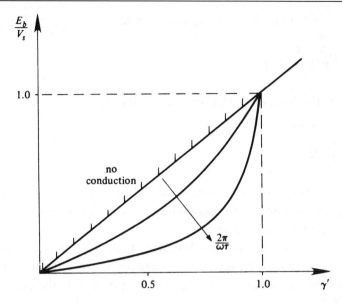

Fig. 5.10 Current continuity criteria.

The motor armature inductance L_a is obviously an important parameter in determining current continuity. For small motors of laboratory size, L_a is of the order 10–50 mH. For larger motors the inductance is smaller, being typically 2–10 mH.

5.2.1.3 Average current, r.m.s. current and power transfer

The average load current is given by the basic equation

$$I_{L_{av}} = \frac{1}{2\pi} \int_0^{2\pi} i_L(\omega t) \, d\omega t \tag{5.55}$$

For continuous current operation equations (5.41), (5.44) are substituted into (5.55) for the intervals $0 \leq \omega t \leq 2\pi\gamma$ and $2\pi\gamma \leq \omega t \leq 2\pi$ respectively. For discontinuous current operation equations (5.46), (5.49) are substituted into (5.55) for the intervals $0 \leq \omega t \leq 2\pi\gamma$ and $2\pi\gamma \leq \omega t \leq X$ respectively. In either case it is found that the average load current is given by

$$I_{L_{av}} = \frac{V_{L_{av}} - E_b}{R} \tag{5.56}$$

For continuous conduction, from (5.25) and (5.56),

$$I_{L_{av}} = \frac{1}{R}(\gamma V_s - E_b) \tag{5.57}$$

For discontinuous conduction, from (5.33) and (5.56),

$$I_{L_{av}} = \frac{1}{R}\left(\gamma V_s - \frac{X}{2\pi} E_b\right) \tag{5.58}$$

The r.m.s. value of the load current is defined in the classical way by

$$I_{L_{av}} = \sqrt{\frac{1}{2\pi}\int_0^{2\pi} i_L^2(\omega t)\, d\omega t} \tag{5.59}$$

Calculation of the r.m.s. current involves the substitution of (5.41) and (5.43) or (5.46) and (5.49) into (5.59), depending on whether the current is continuous or discontinuous, respectively.

These calculations are long and tedious because of the mathematical nature of the instantaneous current equations. A measure of the difficulty in a formal calculation of the r.m.s. load current can be pictured from the current waveforms in Fig. 5.9.

An approximation to the r.m.s. load current can be found by considering only the first few terms of the harmonic series

$$I_{L_{rms}}^2 = I_{L_{av}}^2 + I_{L_1}^2 + I_{L_2}^2 + \ldots \tag{5.60}$$

The various harmonic load voltages are given by (5.30), (5.31) for continuous current operation and (5.35), (5.36) for discontinuous current operation. For the nth harmonic current the impedance offered by the load is

$$Z_{L_n} = \sqrt{R^2 + (n\omega L)^2} \tag{5.61}$$

With $n = 1$, for example, the r.m.s. value of the fundamental component of current for continuous current operation is obtained by combining (5.61) with (5.30).

$$\begin{aligned}
I_{L_1} &= \frac{V_{L_1}}{Z_{L_1}} = \frac{V_s}{\sqrt{2}\pi}\sqrt{\frac{\sin^2 2\pi\gamma + (1 - \cos 2\pi\gamma)^2}{R^2 + \omega^2 L^2}} \\
&= \frac{V_s}{\sqrt{2}\pi}\frac{2\sin\pi\gamma}{\sqrt{R^2 + \omega^2 L^2}}
\end{aligned} \tag{5.62}$$

Numerical applications of this technique are given in Examples 5.2 and 5.4 below.

The average power transferred to the load may be expressed by the basic relationship

5.2 D.c. chopper drives

$$P_L = \frac{1}{2\pi} \int_0^{2\pi} v_L i_L \, d\omega t \tag{5.63}$$

In the equivalent circuit of Fig. 5.8 the load power can be written, in a form more convenient for calculations, as

$$P_L = I_{L_{rms}}^2 R_a + E_b I_{L_{av}} \tag{5.64}$$

The term $E_b I_{L_{av}}$ in (5.64) represents the components of power transferred from the motor to the mechanical load plus the friction and windage. Motor iron losses are not shown explicitly in any of the equations.

Since the input voltage V_s is constant, average power is only transferred from the supply to the chopper by the combination of V_s with the zero frequency (i.e. time average) component of the input current $I_{s_{av}}$

$$P_{in} = V_s I_{s_{av}} \tag{5.65}$$

Instantaneous supply current $i_s(\omega t)$ flows only while switch S is conducting. In Fig. 5.9,

$$i_s = i_L\big|_0^{2\pi\gamma} + 0\big|_{2\pi\gamma}^{2\pi} \tag{5.66}$$

The average value of $i_s(\omega t)$ is defined by

$$I_{s_{av}} = \frac{1}{2\pi} \int_0^{2\pi\gamma} i_L(\omega t) \, d\omega t \tag{5.67}$$

For the case of continuous conduction, substituting (5.41) and (5.44) into (5.67) gives

$$I_{s_{av}} = \gamma \frac{V_s - E_b}{R_a} - \frac{\omega \tau}{2\pi} \frac{V_s}{R_a} \frac{(\varepsilon^{2\pi/\omega\tau} - \varepsilon^{2\pi\gamma/\omega\tau})}{(\varepsilon^{2\pi/\omega\tau} - 1)} \left(1 - \varepsilon^{-2\pi\gamma/\omega\tau}\right) \tag{5.68}$$

For discontinuous conduction, substituting (5.46) into (5.67) gives

$$I_{s_{av}}(\text{discont}) = \frac{V_s - E_b}{R_a} \left[\gamma - \frac{\omega\tau}{2\pi}\left(1 - \varepsilon^{-2\pi\gamma/\omega\tau}\right)\right] \tag{5.69}$$

The difference between the input power and the chopper output power is the switching losses in the switching device and diode

$$P_{in} - P_L = \text{switching losses} \tag{5.70}$$

In many cases the switching losses are negligible so that the chopper input and output powers may be considered equal.

The diode current is given by that portion of the $i_L(\omega t)$ curve in Fig. 5.9(a) between the limits $2\pi\gamma \leq \omega t \leq 2\pi$. Its average value is therefore defined as

$$I_{D_{av}} = \frac{1}{2\pi} \int_{2\pi\gamma}^{2\pi} i_L(\omega t)\,\mathrm{d}\omega t, \text{ continuous current}$$

or (5.71)

$$I_{D_{av}} = \frac{1}{2\pi} \int_{2\pi\gamma}^{X} i_L(\omega t)\,\mathrm{d}\omega t, \text{ discontinuous current}$$

Values for $I_{D_{av}}$ can be calculated by substituting (5.41) and (5.43) or (5.46) and (5.49) into (5.71).

Power transfer from the supply to the motor, in Fig.5.8, will only occur when the switch conducts. In the periods of load current conduction but switch extinction, labelled D_{on} in Fig. 5.9, the power dissipated is obtained by reduction of the energy stored in inductor L_a and in the rotating mass of the motor and its mechanical load. If the motor inertia is low there will be a speed oscillation following the pattern of the current oscillation but lagging it in time because of the time constant of the mechanical system. In many chopper applications any speed oscillation is negligibly small and it is acceptable to use average values of speed in performance calculations.

5.2.2 Class A transistor chopper

The basic class A chopper circuit of Fig. 5.9 can be used with a power transistor as the controlled switch, as shown in Fig. 5.11. The R_s–C_s combination across the transistor is a snubber circuit of the kind described in Chapter 3. If the series inductor L_s is lossless, the diode has conduction resistance R_D and the transistor has conduction resistance R_T, the circuit equations are

$$v_L = L\frac{\mathrm{d}i_L}{\mathrm{d}t} + R i_L = (i_s - i_L) R_D \tag{5.72}$$

$$v_T = i_T R_T = v_c + (i_s - i_T) R_s \tag{5.73}$$

$$C_s \frac{\mathrm{d}V_c}{\mathrm{d}t} = i_{s_n} = \frac{i_s R_T - v_c}{R_s + R_T} \tag{5.74}$$

$$L_s \frac{\mathrm{d}i_s}{\mathrm{d}t} = V_s - v_T - v_L$$

$$= V_s - \frac{R_T}{R_s + R_T}(i_s R_s + v_c) = (i_s - i_L) R_D \tag{5.75}$$

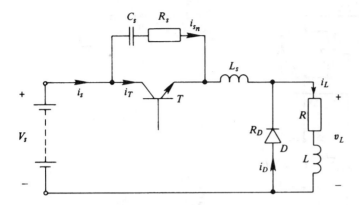

Fig. 5.11 Basic (class A) power transistor chopper circuit (base circuit not shown).

The technology described and the power circuit equations developed in Section 5.2.1 above are also applicable here. An appropriate numerical example is given in Example 5.6 below.

5.2.3 Class B chopper circuits (two-quadrant operation)

If the supply voltage V_s is not reversible the regeneration of load current can be realised in the circuit of Fig. 5.12(a). With thyristor T_1 switched on the supply voltage is clamped across the separately excited motor and positive or 'motoring' current flows, Fig. 5.12(b), causing diode D_1 to be reverse biased. When T_1 is switched off the potential of point p drops from V_s to zero. The load current cannot change instantaneously and a return path is provided via diode D_1, Fig. 5.12(c). When this current has been driven down to zero by the back e.m.f. e_b thyristor T_2 switches on to provide a path for negative armature current, Fig. 5.12(d). Thyristor T_2 is then switched off and the instantaneous negative armature current is transferred through diode D_2 to the supply, Fig. 5.12(e), and constitutes a regenerative current pulse. The opposition of supply voltage V_s reduces the negative current to zero and thyristor T_1 is switched on to restart the cycle of events. The circuit of Fig. 5.12 therefore operates in the two positive voltage quadrants of the load voltage/load current plane, if $e_b > V_s$.

An alternative form of two-quadrant chopper is shown in Fig. 5.13. With thyristors T_1 and T_2 switched on the motor current is positive and load voltage $v_L = +V_s$. When the thyristors are switched off a path for the positive motor current is provided via diodes D_1 and D_2. The load voltage is now $v_L = -V_s$ and the supply current has reversed. The average value of the load

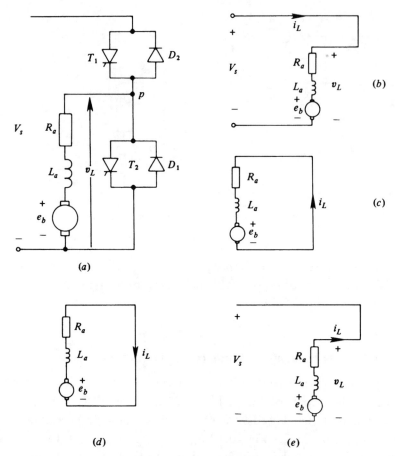

Fig. 5.12 Class B, thyristor/diode chopper operation: (*a*) circuit, (*b*) motoring current mode, (*c*) circulating armature current, (*d*) reverse circulating armature current, (*e*) regenerative current mode.

voltage is determined by the mark–space ratio or duty cycle γ of the thyristor switching, as illustrated in Fig. 5.14. With $\gamma > \frac{1}{2}$ the average load voltage $V_{L_{av}}$ is positive whereas for $\gamma < \frac{1}{2}$ it is negative. Operation therefore takes place between the two positive current quadrants of the voltage/load current plane. The motor power in the circuit of Fig. 5.13 can be made to regenerate into the supply, via D_1 and D_2, if the motor back e.m.f. is reversed (by reversal of the separately excited field current).

If the four switch branches of Fig. 5.13 are each replaced by an SCR in parallel with a reverse-connected diode it is possible to obtain four-quadrant operation in the output voltage output current plane. In all the above cases, whether for one-, two- or four-quadrant operation, the output voltage level is

5.2 D.c. chopper drives

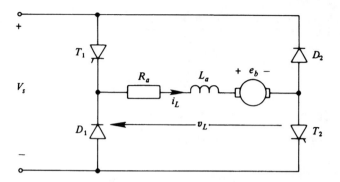

Fig. 5.13 Alternative two-quadrant, class B thyristor/diode chopper circuit.

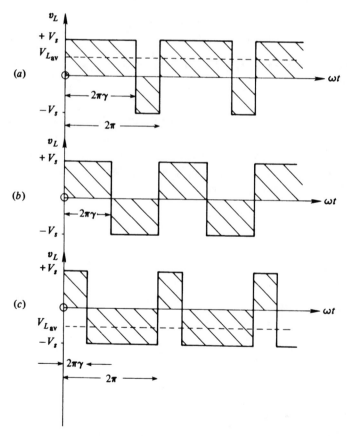

Fig. 5.14 Load voltage waveforms for the class B chopper circuit of Fig. 5.13: (a) $\gamma > 1/2$, (b) $\gamma = 1/2$, (c) $\gamma < 1/2$.

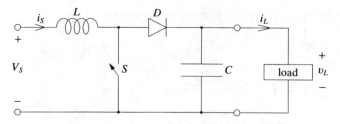

Fig. 5.15 Schematic circuit of a d.c. chopper boost converter.

equal to or less than the supply voltage so that the chopper action results in buck converter (step-down) operation.

Chopper operation in which the output voltage level can exceed the input voltage can be realised by the boost converter (step-up) action of the circuit of Fig. 5.15. Some of the textbooks in the list of references describe this circuit in detail.

5.3 WORKED EXAMPLES

Example 5.1
A class A chopper circuit is supplied with power from an ideal battery of terminal voltage 100 V. The load voltage waveform consists of rectangular pulses of duration 1 ms in an overall cycle time of 2.5 ms. Calculate the average and r.m.s. values of the supply voltage, the r.m.s. value of its fundamental component and the ripple factor RF.

Solution. The specified waveform has the shape shown in Fig. 5.7(b), where $V_s = 100$ V.

From (5.18),

$$\gamma = \frac{T_{on}}{T_{on} + T_{off}} = \frac{1}{2.5} = 0.4$$

From (5.25),

$$V_{s_{av}} = \gamma V_s = \frac{1}{2.5} \times 100 = 40 \text{ V}$$

From (5.26),

$$V_{s_{rms}} = \sqrt{\gamma} V_s = \frac{100}{\sqrt{2.5}} = 63.25 \text{ V}$$

From (5.27),

$$RF = \sqrt{\frac{1-\gamma}{\gamma}} = \sqrt{\frac{1-0.4}{0.4}} = 1.225$$

The magnitude of the fundamental component is given by (5.30), with $n = 1$. For $\gamma = 0.4$ this becomes

$$\hat{V}_{L_1} = \frac{2 \times 100}{\pi} \sin 0.4\pi$$

$$V_{L_1} = \frac{2 \times 100}{\sqrt{2}\pi} \times 0.951$$

$$= 42.8 \text{ V}$$

Example 5.2
In a class A chopper circuit an ideal battery of terminal voltage 100 V supplies a series load of resistance 0.5 Ω and inductance 1 mH. The SCR is switched on for 1 ms in an overall period of 3 ms. Calculate the average values of the load voltage and current and the power taken from the battery.

Solution. The ratio of SCR on-time to total period time is specified as

$$\gamma = \frac{1}{3}$$

The periodic time is specified as 3 ms so that, from (5.20),

$$\frac{2\pi}{\omega} = \frac{3}{1000}$$

Angular frequency ω is therefore

$$\omega = \frac{2000\pi}{3} = 2094.4 \text{ rad/s}$$

The time constant τ of the load impedance is

$$\tau = \frac{L}{R} = \frac{1}{1000 \times 0.5} = 2 \text{ ms}$$

The design parameter defining the state of conduction is therefore

$$\frac{2\pi}{\omega T} = \frac{3}{1000} \times \frac{1000}{2} = 1.5$$

For a passive load $E_b = 0$ and the boundary between continuous and discontinuous operation, equation (5.52), occurs with $\gamma' = 0$. The actual value of γ is specified as $\gamma = 1/3$. Since $\gamma > \gamma'$, in (5.53), operation is continuous.

The minimum and maximum values of the continuous load current are not needed to solve the present problem but are included for interest.

From (5.45),

$$I_{max} = \frac{100}{0.5}\left(\frac{1 - \varepsilon^{-0.5}}{1 - \varepsilon^{-1.5}}\right)$$

$$= \frac{100}{0.5}\left(\frac{1 - 0.6065}{1 - 0.223}\right) = 101.3\,\text{A}$$

$$I_{min} = \frac{100}{0.5}\left(\frac{\varepsilon^{0.5} - 1}{\varepsilon^{1.5} - 1}\right)$$

$$= \frac{100}{0.5}\left(\frac{0.6487}{3.482}\right) = 37.3\,\text{A}$$

The average load voltage, (5.18), is

$$V_{L_{av}} = \gamma V_s = \frac{100}{3} = 33.33\,\text{V}$$

The average load current, (5.57), is

$$I_{L_{av}} = \gamma \frac{V_s}{R} = \frac{100}{3 \times 0.5} = 66.7\,\text{A}$$

The average supply current is given by (5.68), in which

$$\omega T = \frac{\omega L}{R} = \frac{2000\pi}{3} \times \frac{1}{1000 \times 0.5} = 4.19\,\text{rad}$$

$$\varepsilon^{2\pi/\omega T} = \varepsilon^{1.5} = 4.482$$

$$\varepsilon^{2\pi\gamma/\omega T} = \varepsilon^{0.5} = 1.65$$

$$\varepsilon^{-2\pi\gamma/\omega T} = \varepsilon^{-0.5} = 0.6065$$

5.3 Worked examples   177

In (5.68),

$$I_{S_{av}} = \frac{100}{0.5}\left[\frac{1}{3} - \frac{4.19}{2\pi} \times \frac{(4.482 - 1.65)}{(4.482 - 1)}(1 - 0.6065)\right]$$

$$= \frac{100}{0.5}\left(\frac{1}{3} - \frac{4.19}{2\pi} \times \frac{2.832 \times 0.3935}{3.482}\right)$$

$$= \frac{100}{0.5}(0.333 - 0.213)$$

$$= 24 \text{ A}$$

This value is seen to be, as expected, significantly lower than the average load current. The input power is, from (5.65)

$$P_{in} = V_s I_{S_{av}}$$
$$= 100 \times 24$$
$$= 2400 \text{ W}$$

The impedance of the load to currents of fundamental frequency (5.61) is

$$V_{L_1} = \sqrt{R^2 + \omega^2 L^2}$$

$$= \sqrt{(0.5)^2 + \left(\frac{2094.4}{1000}\right)^2}$$

$$= \sqrt{0.25 + 4.385} = 2.153 \, \Omega$$

The fundamental component of the load voltage has the peak value, (5.30),

$$\hat{V}_{L_1} = \frac{2V_s}{\pi} \sin \pi\gamma$$

$$= \frac{200}{\pi} \sin 60°$$

$$= \frac{200}{\pi} \times \frac{\sqrt{3}}{2} = 55.13 \text{ V}$$

The r.m.s. value of the fundamental current is therefore

$$I_{L_1} = \frac{\hat{V}_{L_1}}{\sqrt{2}Z_{L_1}} = \frac{55 \times 13}{\sqrt{2} \times 2.153} = 18.1 \text{ A}$$

Taking only the first two terms of the harmonic series in (5.60) gives an approximate value for the r.m.s. load current

$$I_{L_{rms}} \simeq \sqrt{I_{L_{av}}^2 + I_{L_1}^2}$$
$$= \sqrt{(66.7)^2 + (18.1)^2}$$
$$= 69.1\,\text{A}$$

The load power is therefore

$$P_L = I_{L_{rms}}^2 R$$
$$= (69.1)^2 0.5$$
$$= 2388.2\,\text{W}$$

In this example the circuit switching losses $P_{in} - P_L$ are negligibly small.

Example 5.3
A separately excited d.c. motor with $R_a = 0.3\,\Omega$ and $L_a = 15\,\text{mH}$ is to be d.c. chopper speed controlled over a range 0–2000 r.p.m. The d.c. supply is 220 V. The load torque is constant and requires an average armature current of 25 A. Calculate the range of mark–space ratio required if the motor design constant $K_E\Phi$ has a value of 0.001 67 Vs per revolution.

Solution. In the steady state the armature inductance has no effect. The required motor terminal voltages are given by (5.11(a)).

At $N = 0$, $E_b = 0$ so that

$$V = I_a R_a = 25 \times 0.3 = 7.5\,\text{V}$$

At $N = 2000$ r.p.m., from (5.8),

$$K_E\Phi = 0.001\,67\,\text{V/r.p.s.}$$
$$= 60 \times 0.001\,67\,\text{V/r.p.m.}$$
$$\therefore E_b = 60 \times 0.001\,67 \times 2000$$
$$= 200.4\,\text{V}$$

In (5.11(a)),

$$V = 25 \times 0.3 + 200.4 = 207.9\,\text{V}$$

To give motor voltages of the above values requires the following values of γ:

$$\gamma_0 = \frac{7.5}{220} = 0.0341$$

$$\gamma_{2000} = \frac{207.4}{220} = 0.943$$

If the chopper was to be switched fully on, then the full supply voltage of 220 V would be applied to the motor

$$E_b = 220 - 25 \times 0.3 = 212.5\,\text{V}$$

The speed would then be

$$N = \frac{E_b}{K_E \Phi} = \frac{212.5}{60 \times 0.00167} = 2121\,\text{r.p.m.}$$

which might be acceptable in the particular application since it saves SCR switching losses.

Example 5.4
The class A chopper and 100 V battery of Example 5.2 are applied to a separately excited d.c. motor with $R_a = 0.2\,\Omega$ and $L_a = 1\,\text{mH}$. At the lowest speed of operation the back e.m.f. E_b is found to have a value of 10 V, with the SCR switched on for 1 ms in each overall period of 3 ms. Calculate the average values of the load current and voltage. Also calculate, approximately, the switching losses in the semiconductor diode and SCR and the efficiency of the drive.

Solution.

$$\gamma = \frac{T_{on}}{T_{on} + T_{off}} = \frac{1}{3} = 0.333$$

Since the periodic time is 3 ms the frequency of SCR switch-on is 1000/3 or 333.3 Hz and $\omega = 2\pi \times 333.3 = 2094.4\,\text{rad/s}$.

Now,

$$\tau = \frac{L}{R_a} = \frac{1}{1000 \times 0.2} = 5\,\text{ms}$$

Therefore,

$$\frac{2\pi}{\omega \tau} = \frac{2\pi \times 1000}{2\pi \times 333.3 \times 5} = 0.6$$

Since $E_b = 10\,\text{V}$ then

$$\frac{E_b}{V_s} = \frac{10}{100} = 0.1$$

We must now find the critical value γ' defining the boundary between continuous and discontinuous conduction, using (5.52),

$$0.1 = \frac{\varepsilon^{0.6\gamma'} - 1}{\varepsilon^{0.6} - 1}$$

from which

$$\gamma' = 0.132$$

Since $\gamma > \gamma'$, then from (5.53), the current is continuous.

The average load voltage, (5.18), is

$$V_{L_{av}} = \gamma V_s = \frac{100}{3} = 33.33\,\text{V}$$

The average load current, for continuous current operation, is found from (5.57),

$$I_{L_{av}} = \frac{1}{R_a}(\gamma V_s - E_b)$$
$$= \frac{1}{0.2}(33.33 - 10) = 116.67\,\text{A}$$

The output power is given by (5.64), which requires calculation of the r.m.s. load current.

The harmonic impedances of the first three load current a.c. harmonics are

$$Z_{L_1} = \sqrt{(0.2)^2 + \left(\frac{2094.4}{1000}\right)^2} = 2.104\,\Omega$$

$$Z_{L_2} = \sqrt{(0.2)^2 + (2 \times 2.094)^2} = 4.193\,\Omega$$

$$Z_{L_3} = \sqrt{(0.2)^2 + (3 \times 2.094)^2} = 6.282\,\Omega$$

Equation (5.30(a)) gives the corresponding peak load voltages, where $\gamma = 1/3$,

$$V_{L_1} = \frac{100}{\pi}\sqrt{0.75 + 2.25} = 55.12 \text{ V}$$

$$V_{L_2} = \frac{100}{2\pi}\sqrt{0.75 + 2.25} = 27.57 \text{ V}$$

$$V_{L_3} = 0$$

The r.m.s. values of the first two a.c. harmonic currents are

$$I_{L_1} = \frac{V_{L_1}}{\sqrt{2}Z_{L_1}} = \frac{55.13}{\sqrt{2} \times 2.1} = 18.56 \text{ A}$$

$$I_{L_2} = \frac{V_{L_2}}{\sqrt{2}Z_{L_2}} = \frac{27.57}{\sqrt{2} \times 4.193} = 4.65 \text{ A}$$

Including the average value, the r.m.s. load current (5.60) is

$$I_{L_{\text{rms}}} = \sqrt{(116.67)^2 + (18.56)^2 + (4.65)^2}$$
$$= \sqrt{13978} = 118.23 \text{ A}$$

The load power is therefore, from (5.64),

$$P_L = I_{L_{\text{rms}}}^2 R_a + E_b I_{L_{\text{av}}}$$
$$= (118.23)^2 0.2 + 10 \times 116.67$$
$$= 2795.7 + 1166.7$$
$$= 3962.4 \text{ W}$$

The average battery current is given by (5.68) where

$$\omega T = \frac{2\pi}{0.6} = 10.472$$

$$\varepsilon^{2\pi/\omega T} = \varepsilon^{0.6} = 1.822$$

$$\varepsilon^{2\pi\gamma/\omega T} = \varepsilon^{0.2} = 1.221$$

$$\varepsilon^{-2\pi\gamma/\omega T} = \frac{1}{1.221} = 0.819$$

Then

$$I_{S_{av}} = \frac{1}{3} \times \frac{100-10}{0.2} - \frac{10.472}{2\pi} \times \frac{100}{0.2} \frac{(1.822-1.221)}{(1.822-1)} (1-0.819)$$

$$= 150 - 833 \times \frac{0.6}{0.822} \times 0.181$$

$$= 150 - 110 = 40\,\text{A}$$

The input power is given by (5.65),

$$P_{in} = V_s I_{S_{av}}$$
$$= 100 \times 40 = 4000\,\text{W}$$

The difference between P_{in} and P_L is the chopper switching losses,

$$\text{switching losses} = P_{in} - P_L$$
$$= 4000 - 3959.4 = 40.6\,\text{W}$$

Note that at this low speed only $E_b I_{L_{av}} = 1166.7\,\text{W}$ is transferred to the mechanical load. If the motor iron losses are neglected the efficiency of the drive is

$$\eta = \frac{E_b I_{L_{av}}}{P_{in}} = \frac{1166.7}{4000} = 29.2\%$$

Example 5.5
In a chopper-controlled, separately excited motor drive $R_a = 0.2\,\Omega$ and $L_a = 1\,\text{mH}$. The switch is switched on for a period of 2 ms in each overall control period of 3 ms and the average speed now results in an average back e.m.f. of 85 V. If the d.c. supply is 100 V, calculate the average values of the load current and voltage and the drive efficiency.

Solution.

$$\gamma = \frac{T_{on}}{T_{on} + T_{off}} = \frac{2}{3} = 0.667$$

The periodic time is 3 ms so that, from (5.20),

$$\omega = \frac{2\pi}{3} \times 1000 = 2094.4\,\text{rad/s}$$

The armature circuit time constant is the same as in the corresponding Example 5.4:

$$\tau = \frac{L_a}{R_a} = \frac{1}{1000 \times 0.2} = 5\,\text{ms}$$

Therefore,

$$\frac{2\pi}{\omega\tau} = \frac{2\pi \times 3 \times 1000}{2\pi \times 1000 \times 5} = 0.6$$

The critical value γ' of the duty cycle is found from (5.52)

$$\frac{E_b}{V_s} = \frac{85}{100} = \frac{\varepsilon^{0.6\gamma'} - 1}{\varepsilon^{0.6} - 1}$$

which gives $\gamma' = 0.53/0.6 = 0.883$. Since $\gamma < \gamma'$, from (5.53), the armature current is now discontinuous.

The extinction angle of the current is defined by (5.50) where

$$\omega\tau = 2094.4 \times \frac{5}{1000} = 10.472\,\text{rad}$$
$$\varepsilon^{2\pi\gamma/\omega\tau} = \varepsilon^{0.4} = 1.492$$
$$\varepsilon^{-2\pi\gamma/\omega\tau} = \varepsilon^{-0.4} = 0.67$$
$$\varepsilon^{2\pi/\omega\tau} = \varepsilon^{0.6} = 1.822$$
$$\frac{V_s - E_b}{E_b} = \frac{100 - 85}{85} = 0.176$$

Therefore

$$\begin{aligned}X &= 10.472\ln\{1.492[1 + 0.176(1 - 0.67)]\}\\ &= 10.472\ln 1.578\\ &= 4.78\,\text{rad}\\ &= \frac{4.78}{2\pi} \times 360 = 274°\end{aligned}$$

The load current waveform is shown in Fig. 5.16.

For discontinuous operation the maximum current I_{max_d} is given by (5.47). In this case

$$\begin{aligned}I_{\text{max}_d} &= \frac{100 - 85}{0.2}(1 - \varepsilon^{-0.4})\\ &= 75 \times 0.33 = 24.75\,\text{A}\end{aligned}$$

184 D.c. motor control using a d.c. chopper

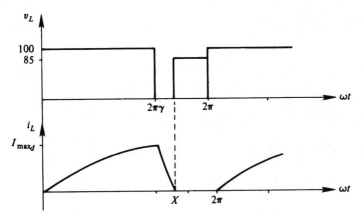

Fig. 5.16 Output waveforms of the d.c. chopper/d.c. motor drive of Example 5.5.

The average load voltage for discontinuous operation (5.33) gives

$$V_{L_{av}} = 100 \times \frac{2}{3} + (1 - 0.761) \times 85$$
$$= 66.67 + 20.32 = 87 \text{ V}$$

The corresponding average load current is given by (5.58)

$$I_{L_{av}} = \frac{1}{R_a}\left(\gamma V_s - \frac{X}{2\pi} E_b\right)$$
$$= \frac{1}{0.2}(66.67 - 64.69)$$
$$= 9.9 \text{ A}$$

The increase of back e.m.f. has resulted in a great reduction of average load current, compared with the situation of Example 5.4. To calculate the load power it is first necessary to determine the r.m.s. load current. The harmonic summation method of (5.60) is used. It is assumed that the motor back e.m.f. does not impede the flow of a.c. components of current. The impedances of the load circuit to the flow of the low order harmonic currents are given from Example 5.4

n	$\sin 2\pi n\gamma$	$\sin nX$	a_n	$\cos 2\pi n\gamma$	$\cos nX$	b_n	c_n
1	−0.866	−1	−0.5	−0.5	0.05	22.04	22.05
2	+0.866	−0.105	15.21	−0.5	−1	−3.18	15.54
3	0	0.99	−8.93	1.0	−0.156	−10.43	13.73

5.3 Worked examples

$Z_{L_1} = 2.1\,\Omega$, $Z_{L_2} = 4.193\,\Omega$, $Z_{L_3} = 6.282\,\Omega$

Equations (5.35), (5.36) are now used to calculate the harmonic load voltages.

The r.m.s. values of the harmonic load voltages are therefore

$$V_{L_1} = \frac{22.05}{\sqrt{2}} = 15.59\,\text{V}, \quad V_{L_2} = \frac{15.54}{\sqrt{2}} = 10.99\,\text{V}, \quad V_{L_3} = \frac{13.73}{\sqrt{2}} = 9.71\,\text{V}$$

It is therefore seen that the r.m.s. harmonic load currents are

$$I_{L_1} = \frac{15.59}{2.1} = 7.42\,\text{A}, \quad I_{L_2} = \frac{10.99}{4.193} = 2.62\,\text{A}, \quad I_{L_3} = \frac{9.71}{6.282} = 1.55\,\text{A}$$

Including the effect of the average value the r.m.s. load current, (5.60), is

$$I_{L_{\text{rms}}} = \sqrt{(9.9)^2 + (7.42)^2 + (2.62)^2 + (1.55)^2}$$
$$= 12.74\,\text{A}$$

The load power is therefore, from (5.64),

$$P_L = E_b I_{L_{\text{av}}} + I^2_{L_{\text{rms}}} R_a$$
$$= 85 \times 9.9 + 0.2(12.74)^2$$
$$= 841.5 + 32.46$$
$$= 874\,\text{W}$$

For the case of discontinuous load current, the average supply current is given by (5.69),

$$I_{S_{\text{av}}} = \frac{V_s - E_b}{R_a}\left[\gamma - \frac{\omega\tau}{2\pi}(1 - \varepsilon^{-2\pi\gamma/\omega\tau})\right]$$
$$= \frac{15}{0.2}\left[0.67 - \frac{10.472}{2}(1 - 0.67)\right]$$
$$= 75(0.67 - 0.55)$$
$$= 9\,\text{A}$$

The input power, is

$$P_{\text{in}} = V_s I_{s_{\text{av}}}$$
$$= 100 \times 9 = 900 \text{ W}$$

The drive efficiency, neglecting motor rotational losses, is

$$\eta = \frac{E_b I_{L_{\text{av}}}}{P_{\text{in}}}$$
$$= \frac{841.5}{900} = 93.5\%$$

Example 5.6
A class A transistor chopper transfers power from a 300 V battery to a load consisting of a 20 Ω resistor in series with a 10 mH inductor. A series choke of 10 µH is used in the supply line. The chopper operates at 5 kHz with a duty cycle of 67%. A snubber circuit of 15 Ω in series with 0.05 µF is connected across the power transistor, which has a conduction resistance of 0.01 Ω. The load impedance is shunted by a free-wheel diode that has a forward resistance of 0.2 Ω. If the load current is assumed to be smooth, calculate the output power.

Solution. The circuit diagram is shown in Fig. 5.11. Fig. 5.17 shows the load voltage waveform for which the average value, from (5.18), is

$$V_{L_{\text{av}}} = \gamma V_s = \frac{2}{3} \times 300 = 200 \text{ V}$$

If the load current is perfectly smooth the average voltage on the load inductor is zero. The average load current, (5.56), is

$$I_{L_{\text{av}}} = \frac{V_{L_{\text{av}}}}{R} = \frac{200}{20} = 10 \text{ A}$$

The output power is therefore

$$P_L = V_{L_{\text{av}}} I_{L_{\text{av}}} = 2 \text{ kW}$$

For power circuit calculations the snubber circuit current is negligibly small. When diode D conducts, capacitor C_s rings with inductor L_s at a natural frequency of 225 kHz. The action of a practical circuit would create switching spikes, ringing at 225 kHz, on the current waveforms of the diode and transistor at their switch-on points and on the transistor and capacitor voltages.

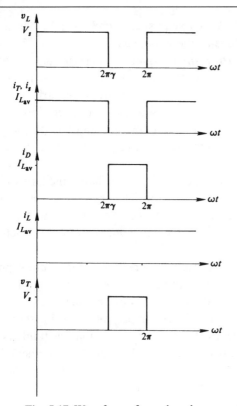

Fig. 5.17 Waveforms for a d.c. chopper with series R–L load in Example 5.6.

5.4 PROBLEMS

D.c. load control using d.c. choppers

5.1 Use the motor equations of Section 5.1 to show that, for a d.c. shunt motor, the torque–speed characteristics may be represented by the relationship (5.13).

Show that, using field resistance R_f as a parameter, the torque–speed characteristics are given by the form of Fig. 5.5.

5.2 Use the motor equations of Section 5.1 to show that, neglecting saturation, the torque–speed characteristics of a d.c. series motor may be represented by the relationship (5.16).

Deduce the effect on the torque–speed characteristics of changing the field resistance R_f.

5.3 Use the equations of Section 5.1 to show that, neglecting saturation, the torque–speed characteristic of a separately excited d.c. motor can be expressed as

$$T = \frac{K_T K_f E_f}{R_a R_f}\left(V - \frac{K_E K_f E_f}{R_f} \cdot N\right)$$

Sketch the form of the T–N characteristics represented by the above equation using armature voltage as the parameter.

The form of the above torque–speed equation suggests that torque is proportional to speed if $V = 0$. But, the physical reality is that no torque is developed if $V = 0$. How does this apparent contradiction arise?

5.4 Sketch the power circuit diagram for a class A SCR chopper. Show waveforms of the load voltages for the two duty-cycle conditions (i) $\gamma = \frac{1}{4}$, (ii) $\gamma = \frac{3}{4}$. For both conditions calculate the average value, r.m.s. value and ripple factor of the load voltage waveform.

5.5 In Problem 5.4, when $\gamma = \frac{3}{4}$, calculate the first three harmonic terms of the Fourier series for the load voltage waveform. Use these, together with the average value, to obtain an approximate value of the r.m.s. load voltage. Compare this approximate value with the value obtained from the defining integral.

5.6 For the class A chopper circuit derive an expression for the output power with resistive load. Calculate the per-unit value of this for (i) $\gamma = \frac{1}{4}$, (ii) $\gamma = \frac{3}{4}$.

5.7 A class A chopper circuit has a ratio of switch on-time/total period time or duty cycle defined by the symbol γ. If the supply voltage is V_s, sketch the load voltage waveform for a series R–L load and show that the fundamental component of this has the property $v_{L_1}(\omega t) = V_{L_1} \sin(\omega t + \psi_1)$ where

$$\hat{V}_{L_1} = \frac{V_s}{\pi}\sqrt{(1 - \cos 2\pi\gamma)^2 + \sin^2 2\pi\gamma} = \frac{2V_s}{\pi} \sin \pi\gamma$$

$$\psi_1 = \tan^{-1}\left(\frac{\sin 2\pi\gamma}{1 - \cos 2\pi\gamma}\right) = \pi\left(\frac{1}{2} - \gamma\right)$$

Calculate V_{L_1} and ψ_1, if $\gamma = \frac{1}{2}$, and superimpose a plot of $v_{L_1}(\omega t)$ on the corresponding load voltage waveform.

5.8 Show that the ripple factor of the load voltage for a d.c. chopper with passive impedance load is highest (i.e. worst) for the case of duty cycle $\gamma = \frac{1}{2}$.

5.9 A class A chopper is used to drive a separately excited d.c. motor. At a certain speed of operation and duty cycle the armature current becomes discontinuous. Sketch the form of load voltage that results and show that its average and r.m.s. values are given by equations (5.33) and (5.34) respectively.

5.10 In a class A chopper circuit with battery V_s, with resistive load and duty cycle γ, sketch waveforms of the voltages across the switch and the diode. Derive expressions for the average values of these.

5.11 Power is transferred from a battery with $V_s = 200$ V to a load consisting of resistor $R = 20\,\Omega$ in series with $L = 20$ mH via a class A chopper. Calculate the average value of the load voltage and current if the duty cycle is 75%. If the chopper switches at a frequency of 1000 Hz, calculate the values of the first and second harmonic components of the load current and hence calculate the load power dissipation.

5.12 For the chopper of Problem 5.11 sketch the waveform of the chopper current and calculate its maximum and minimum values. Determine the input power to the circuit and compare this with the output power previously obtained.

5.13 A d.c. supply with $V_s = 200$ V supplies power to a separately excited d.c. motor via a class A chopper. The motor has an armature circuit resistance of $0.33\,\Omega$ and inductance of $11\,\text{mH}$. The chopper is fully on at the rated motor speed of $1200\,\text{r.p.m.}$ when the armature current is $20\,\text{A}$. If the speed is to be reduced to $800\,\text{r.p.m.}$, with the load torque constant, calculate the necessary duty cycle. If the chopper frequency is $500\,\text{Hz}$, is the current continuous? If not, calculate the additional armature circuit inductance required to ensure continuity of the current.

5.14 For the d.c. motor of Problem 5.13 calculate the maximum and average values of the load current at $800\,\text{r.p.m.}$

5.15 A $100\,\text{V}$ battery supplies power to a d.c. separately excited motor, with $R_a = 0.2\,\Omega$ and $L_a = 1\,\text{mH}$, via a class A d.c. chopper operating at $300\,\text{Hz}$. With a duty cycle of $\frac{1}{3}$ the motor back e.m.f. is $25\,\text{V}$. Calculate the average values of the load voltage and current, the output power and the approximate efficiency of the motor.

5.16 For the d.c. motor of Problem 5.15 the duty cycle is increased to $\gamma = \frac{3}{4}$. This results in an increase of speed such that the back e.m.f. is now $70\,\text{V}$. Calculate the average load voltage and current and the drive efficiency.

5.17 A separately excited d.c. motor drives a constant torque load that requires an armature current of $25\,\text{A}$ from a $250\,\text{V}$ supply. The armature circuit resistance and inductance are $0.7\,\Omega$ and $2\,\text{mH}$ respectively. The armature voltage is chopper controlled at $1000\,\text{Hz}$ and it is necessary to use full conduction at the top speed of $1000\,\text{r.p.m.}$ At what value of speed will the armature current become discontinuous if $\gamma = \frac{1}{2}$?

5.18 For the chopper operation of Problem 5.17 calculate the necessary voltage and current ratings of the switch and the diode.

5.19 (a) Sketch the power circuit diagram for a class A SCR chopper which supplies the armature of a separately excited d.c. motor. Briefly explain the operation.

  (b) For a duty cycle $\gamma = \frac{3}{4}$, sketch consistent waveforms of the load voltage, SCR voltage, load current and supply current, for a typical steady-state cycle of operation, assuming continuous conduction of the load current.

  (c) Calculate the average value, r.m.s. value and ripple factor of the load voltage in terms of the battery voltage V_{dc}. Also calculate the first three harmonic terms of the Fourier series for the load voltage waveform and use these to provide a check calculation of the r.m.s. load voltage.

  (d) Derive an expression for the ratio of the average voltage across the SCR to the average voltage across the load in terms of duty cycle γ. For what value of γ is this ratio equal to unity?

6
Controlled bridge rectifiers with d.c. motor load

6.1 THE PRINCIPLES OF RECTIFICATION

The process of electrical rectification is where current from an a.c. supply is converted to a unidirectional form before being supplied to a load. Although unidirectional, the load current may pulsate in amplitude, depending on the load impedance. With resistive loads the load voltage polarity is fixed. The polarity of the voltage across series-connected load inductance elements may vary during the load current cycle.

In a rectifier circuit there are certain electrical properties that are of interest irrespective of circuit topology and impedance nature. These properties can be divided into two groups, (i) on the supply side, and (ii) on the load side of the rectifier, respectively. When the electrical supply system has a low (ideally zero) impedance, the sinusoidal supply voltages remain largely undistorted even when the rectifier action causes nonsinusoidal pulses of current to be drawn from the supply. For the purposes of circuit analysis one can assume that semiconductor rectifier elements, such as diodes and thyristor devices, are ideal in that they are dissipationless and have zero conducting voltage drop.

A study of rectifier circuits is basically a study of waveforms. No energy is stored within a rectifier so that there is a constant connection between the currents and voltages on the a.c. side and the current and voltage on the d.c. side. In rectifier calculations the essential requirement is to obtain an accurate physical picture of the operation and then establish circuit equations that are valid for the particular condition.

Equivalent circuits may be devised which correlate with each individual section of the corresponding nonsinusoidal supply current. Differential equations of the circuit currents may then be compared for consecutive parts of a current cycle and matched at the appropriate boundaries by current continuity criteria. This approach is followed here. An alternative approach is to establish circuit 'starting conditions' and perform a step-by-step digital analysis, incorporating the switching conditions of the circuit switches as logic steps.

Three-phase bridge rectifier circuits are very frequently used over a wide range of both electronic and electrical power applications. For example, they are widely used in brushless excitation systems for aircraft generators. There are many industrial applications where an individual three-phase generator is directly loaded with a full-wave bridge rectifier at its terminals. This raises some very interesting problems regarding the generator action that are outside the scope of the present book. The full-wave bridge rectifier is extensively used in static generator excitation schemes that are not of the rotating brushless kind, road vehicle generator systems, high voltage a.c.–d.c. power conversion and in a wide range of d.c. motor and a.c. motor speed control schemes. Because the three-phase bridge rectifier is so important and extensively used it is widely described in existing English language books on power electronics, electrical machines and electrical power supply systems. Some books are devoted largely or exclusively to rectifier circuits and deal extensively with three-phase bridge rectifiers. A relatively brief treatment is included in this book, assuming ideal three-phase supply, appropriate to the later chapters on motor control.

Three-phase controlled rectifiers invariably are naturally commutated by the cycling of the supply-side voltages. Normally there is no point in using gate turn-off devices and such rectifiers usually employ SCRs as switches. Only if the application results in a need for the converter to accept power regenerated from the load might the need arise to use gate turn-off switches such as power transistors or GTOs. In this chapter all semiconductor controlled switches are regarded as SCRs.

6.2 SEPARATELY EXCITED D.C. MOTOR WITH RECTIFIED SINGLE-PHASE SUPPLY

Each of the single-phase circuit configurations of Table 6.1 can be used to control the armature voltage and current of a separately excited d.c. motor. For the half-wave semi-converter and full converter connections the armature current i_a is unidirectional, whereas the double converter permits the flow of armature current in either direction. The polarity of the armature voltage v_a, defined by (5.10), is non-reversible for the half-wave and semi-converter circuits. The fully controlled converter enables positive or negative armature voltage to be applied while the double converter is completely comprehensive and permits operation in any of the four quadrants of the armature voltage–current plane.

The d.c. motor equations developed in Section 5.1 above are all valid for the present cases of rectified a.c. input. All of the converter circuits in Table 6.1 operate by natural commutation.

6.2.1 Single-phase semi-converter

The two versions of the semi-converter in Table 6.1 result in identical load-side performance. The freewheel diode (FWD) version of the converter is shown in Fig. 6.1 in which the d.c. motor is represented by its equivalent circuit. Consider the case in which the armature current is continuous, for which typical waveforms are shown in Fig. 6.2, when $\alpha = 60°$. While v_a is positive the freewheel diode is reverse biased and held in extinction.

While current flow is blocked in the SCRs, in the interval $0 \leq \omega t \leq \alpha$, for example, no supply current can flow and the load current freewheels through FWD. Each SCR carries one of the pulses of the supply current during each supply voltage cycle. The load voltages are therefore given by

$$v_a = i_a R_a + L_a \frac{di_a}{dt} + e_b, \alpha \leq \omega t \leq \pi \quad (6.1)$$

$$0 = i_a R_a + L_a \frac{di_a}{dt} + e_b, \pi \leq \omega t \leq \pi + \alpha \quad (6.2)$$

The time average value of the armature voltage is designated V or $V_{L_{av}}$ to be consistent with its use in Sections 5.1, 5.2 respectively

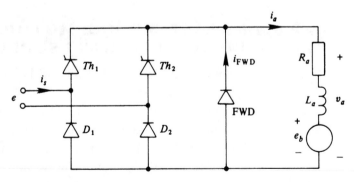

Fig. 6.1 Single-phase semi-converter (or half controlled converter) with freewheel diode, using SCR switches.

6.2 Separately excited d.c. motor with rectified single-phase supply 193

Table 6.1 *Single-phase naturally commutated controlled converter circuits.*

Type	Circuit	Operation
half-wave 1 pulse		
semi-converter 2 pulse		
full converter 2 pulse		
double converter 2 pulse		

$$V = V_{L_{av}} = \frac{1}{2\pi} \int_0^{2\pi} v_a(\omega t) \, d\omega t$$

$$= \frac{1}{\pi} \int_\alpha^\pi E_m \sin \omega t \, d\omega t \qquad (6.3)$$

$$\therefore V_{L_{av}} = \frac{E_m}{\pi} (1 - \cos \alpha)$$

The average load current may be obtained by putting (6.3) into (5.56),

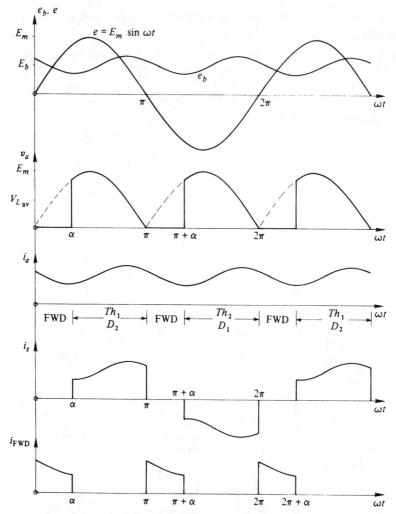

Fig. 6.2 Waveforms for single-phase semi-converter operation with continuous armature current. $\alpha = 60°$.

$$I_{L_{av}} = I_a = \frac{V_{L_{av}} - E_b}{R_a}$$
$$= \frac{E_m}{\pi R_a}(1 + \cos\alpha) - \frac{E_b}{R_a} \quad (6.4)$$

For a separately excited d.c. motor the instantaneous speed is proportional to the instantaneous back e.m.f. e_b. Curve e_b in Fig. 6.2 therefore also depicts the speed–time variation. The average value of the back e.m.f. E_b is used in (6.4) to calculate average armature current I_a.

The r.m.s. and harmonic properties of the current waveforms in Fig. 6.2 can only be accurately determined if analytical expressions are known for the instantaneous variables $i_a(\omega t)$, $i_s(\omega t)$ and $i_{FWD}(\omega t)$. Some approximation to the values of, for example, r.m.s. armature current and load power (using (5.64)) can be obtained by neglecting the ripple component of $i_a(\omega t)$ and assuming constant armature current. Alternatively one can calculate the harmonics of the load voltage and determine corresponding current harmonics by dividing by the appropriate harmonic impedance, as in Section 5.2.1.3.

If the motor is required to deliver an average torque T, the average speed N at which this torque is generated is obtained by substituting (6.3) into (5.12). Discontinuous armature current operation may result at some speeds with certain motors if the armature inductance is low and the SCR firing-angle is large.

6.2.2 Single-phase full converter

The two-quadrant, four-switch, full converter circuit is applied to a separately excited d.c. motor load, as shown in Fig. 6.3. Alternate switching of the pairs of SCRs Th_1, Th_4 or Th_2, Th_3 is used. If Th_1, Th_4 are conducting, for example, positive supply voltage is applied to the motor.

The applied voltage, from the ideal supply, is defined as

$$e = E_m \sin \omega t \tag{6.5}$$

This voltage is applied across the elements R_a and L_a in series with the back e.m.f. e_b so that

$$E_m \sin \omega t = i_a R_a + L_a \frac{di_a}{dt} + e_b(\omega t) \tag{6.6}$$

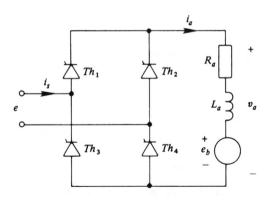

Fig. 6.3 Single-phase full converter with d.c. motor load, using SCR switches.

At an arbitrary instant of time defined by $\omega t = \alpha$, the net voltage impressed across the series R–L elements in Fig. 6.3 is

$$E_m \sin \alpha - e_b(\alpha) = i_a R_a + L_a \frac{di_a}{dt} \tag{6.7}$$

6.2.2.1 Continuous conduction

If $E_m \sin \alpha > e_b(\alpha)$ then current $i_a(\omega t)$ will flow continuously. The continuous current mode is illustrated in Fig. 6.4 for the case $\alpha = 60°$.

Instantaneous back e.m.f. variation $e_b(\omega t)$ follows the curve of instantaneous speed variation $n(\omega t)$. With a fully controlled converter, Fig. 6.3, conduction occurs in 180° pulses of supply current from $\alpha \le \omega t \le \pi + \alpha$. Each

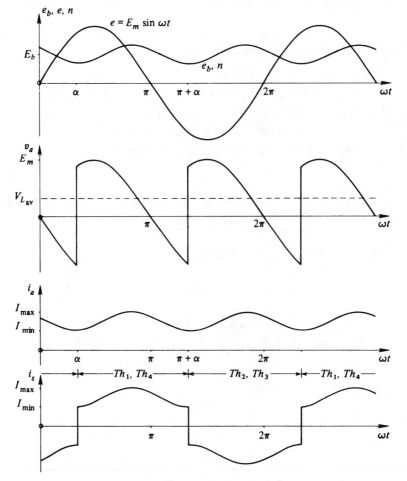

Fig. 6.4 Waveforms for single-phase full converter drive with continuous armature current. $\alpha = 60°$.

6.2 Separately excited d.c. motor with rectified single-phase supply

pair of thyristors conducts, in turn, for one-half of a supply cycle so that all four thyristors must be equally rated. When thyristors Th_2, Th_3 are triggered at $\omega t = \pi + \alpha$, negative voltage is applied across Th_1, Th_4 which causes them to commutate naturally.

From Fig. 6.4 it can be deduced that the time average value of the load voltage is given by

$$V = V_{L_{av}} = \frac{1}{2\pi} \int_0^{2\pi} v_a(\omega t) \, d\omega t, \text{ in general}$$

$$= \frac{2}{2\pi} \int_\alpha^{\pi+\alpha} E_m \sin \omega t = \frac{2E_m}{\pi} \cos \alpha \qquad (6.8)$$

For $\alpha = 60$, $V_{L_{av}} = 1/\pi$ per-unit and this is shown in Fig. 6.4.

The corresponding average load current is therefore found from

$$I = I_{L_{av}} = \frac{V_{L_{av}} - E_b}{R_a}$$

$$= \frac{2E_m}{\pi R_a} \cos \alpha - \frac{E_b}{R_a} \qquad (6.9)$$

The load current and voltage waveforms in Fig. 6.4 have a ripple frequency equal to twice the supply frequency. The a.c. harmonic components of v_a and i_a, therefore, have a lowest order of $n = 2$ (i.e. there is no supply frequency component).

The load voltage of Fig. 6.4 may be analysed in terms of its harmonic components, for $n \neq 1$, by

$$a_n = \frac{1}{\pi} \int_0^{2\pi} v_a(\omega t) \cos n\omega t \, d\omega t, \text{ in general}$$

$$= \frac{1}{\pi} \int_\alpha^{\pi+\alpha} E_m \sin \omega t \cos n\omega t \, d\omega t$$

$$+ \frac{1}{\pi} \int_{\pi+\alpha}^{2\pi+\alpha} E_m \sin(\omega t - \pi) \cos n\omega t \, d\omega t \qquad (6.10)$$

$$= \frac{E_m}{2\pi} \left[\frac{\cos(n-1)\alpha}{n-1} - \frac{\cos(n+1)\alpha}{n+1} \right]$$

$$b_n = \frac{1}{\pi} \int_0^{2\pi} v_a(\omega t) \sin n\omega t \, d\omega t, \text{ in general}$$

$$= \frac{1}{\pi} \int_\alpha^{\pi+\alpha} E_m \sin \omega t \sin n\omega t \quad (6.11)$$

$$+ \frac{1}{\pi} \int_{\pi+\alpha}^{2\pi+\alpha} E_m \sin(\omega t - \pi) \sin n\omega t \, d\omega t$$

$$= \frac{E_m}{2\pi} \left[\frac{\sin(n-1)\alpha}{n-1} - \frac{\sin(n+1)\alpha}{n+1} \right]$$

The peak value c_n of the nth harmonic load voltage is therefore given by (A.2), (see Appendix), for $n = 2, 4, 6, \ldots$, where a_n, b_n come from (6.10), (6.11).

An alternative form of expression for peak a.c. voltage component c_n is, for $n = 1, 2, 3, 4, \ldots$,

$$V_{L_n} = c_n = \sqrt{a_n^2 + b_n^2}$$

$$= \frac{E_m}{2\pi} \sqrt{\frac{1}{(2n-1)^2} + \frac{1}{(2n+1)^2} - \frac{2\cos 2\alpha}{(2n-1)(2n+1)}} \quad (6.12)$$

During the conduction interval $\alpha \leq \omega t \leq \pi + \alpha$ in Fig. 6.4 the armature circuit voltage equation is given by (6.6). If the speed ripple amplitude is small the back e.m.f. ripple will also be small and the average value of back e.m.f. E_b may be used in (6.6), to yield

$$i_a(\omega t) = \frac{E_m}{|Z_a|} \sin(\omega t - \phi_a) - \frac{E_b}{R_a} + K\varepsilon^{-\cot\phi_a(\omega t - \alpha)} \quad (6.13)$$

where K is a constant of integration,

$$|Z_a| = \sqrt{R_a^2 + \omega^2 L_a^2} \quad (6.14)$$

$$\cot \phi_a = \frac{R_a}{\omega L_a} = c \quad (6.15)$$

Now the oscillation of the motor current has its minimum value $I_{\min}$ at periodic time intervals such that

$$I_{\min} = i_a(\alpha) = i_a(\pi + \alpha) \quad (6.16)$$

If the substitutions $\omega t = \alpha$ and $\omega t = \pi + \alpha$ are made, in turn, in (6.13) and the identity (6.16) is used it is found that

$$K = \frac{2E_m \sin(\alpha - \phi_a)}{|Z_a|(\varepsilon^{-c\pi} - 1)} \quad (6.17)$$

6.2 Separately excited d.c. motor with rectified single-phase supply

Minimum current I_{min} is obtained by combining (6.13), (6.17) at $\omega t = \alpha$,

$$I_{min} = i_a(\alpha) = \frac{E_m}{|Z_a|} \sin(\alpha - \phi)\left(\frac{\varepsilon^{-c\pi} + 1}{\varepsilon^{-c\pi} - 1}\right) - \frac{E_b}{R_a} \tag{6.18}$$

The average value of the armature current can be found by the basic integration method

$$I = I_{L_{av}} = \frac{1}{\pi} \int_{\alpha}^{\pi + \alpha} i_a(\omega t) \, d\omega t \tag{6.19}$$

An evaluation of (6.19) by the substitution of (6.13) would be very tedious. Alternatively, the average current can be found by the harmonic summation method of Chapter 5, involving (5.60), (5.64). Fourier harmonics of the load current $i_a(\omega t)$ may be obtained by substituting (6.13) into (A.7) and (A.8), (see Appendix), to give

$$a_n = \frac{2K}{\pi(n^2 + c^2)} \left[(c \cos n\alpha - n \sin n\alpha)(\varepsilon^{-c\pi} - 1)\right] \tag{6.20}$$

$$b_n = \frac{2K}{\pi(n^2 + c^2)} \left[(n \cos n\alpha + c \sin n\alpha)(\varepsilon^{-c\pi} + 1)\right] \tag{6.21}$$

For the fundamental (supply frequency) components, $n = 1$, slightly more manageable forms are obtained

$$a_1 = -\frac{E_m}{|Z_a|} \sin \phi_a + \frac{2K}{\pi} \sin \phi_a \cos(\phi_a + \alpha)(\varepsilon^{-c\pi} + 1) \tag{6.22}$$

$$b_1 = \frac{E_m}{|Z_a|} \cos \phi_a + \frac{2K}{\pi} \sin \phi_a \sin(\phi_a + \alpha)(\varepsilon^{-c\pi} + 1) \tag{6.23}$$

The peak value of the fundamental load current is then given by

$$\hat{I}_{a_1} = c_1 = \sqrt{a_1^2 + b_1^2} \tag{6.24}$$

Calculation of the power dissipation requires use of the integral form (5.63) or accurate calculation of the r.m.s. current. This, in turn, becomes rather cumbersome since it is then necessary to square the three-term expression (6.13). The accurate calculation of circuit power and power factor is only straightforward analytically if the load inductance is large enough to make the load current constant. An approximation to the r.m.s. current can be obtained by the harmonic summation method.

It can be inferred from Fig. 6.4 that the r.m.s. values of the output current $i_a(\omega t)$ and input current $i_s(\omega t)$ are identical. Also, the load branch power dissipation in Fig. 6.3 is given by (5.65). This provides a method of calculating the power factor of the bridge operation, illustrated by Example 6.1.

6.2.2.2 Discontinuous conduction

Consider the conduction where the armature current $i_a(\omega t)$ falls to zero before the next pair of SCRs is switched in. In Fig. 6.5, for example, the conduction of armature current occurs between the limits $\alpha \leq \omega t \leq X$, where X is the extinction angle and $X < \pi + \alpha$. In the interval $X \leq \omega t \leq \pi + \alpha$, for example, all the SCRs are switched off and the load and supply currents are zero. During the current extinction intervals the back e.m.f. of the motor is

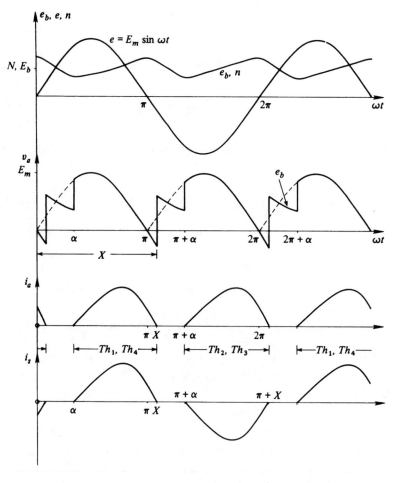

Fig. 6.5 Waveforms for single-phase full converter drive with discontinuous armature current. $\alpha = 60°$.

6.2 Separately excited d.c. motor with rectified single-phase supply

the only component of load voltage and this maps the time variation of the instantaneous speed. If the speed ripple is small,

$$v_a = E_m \sin \omega t \Big|_{\alpha,\pi+\alpha\ldots}^{X,\pi+X\ldots} + E_b \Big|_{X,\pi+X}^{\pi+\alpha,2\pi+\alpha} \qquad (6.25)$$

The average load voltage is then

$$V_{L_{av}} = \frac{E_m}{\pi}(\cos\alpha - \cos X) + \frac{E_b}{\pi}(\pi + \alpha - X) \qquad (6.26)$$

Using the now familiar expression it is found that

$$I_a = I_{L_{av}} = \frac{V_{L_{av}} - E_b}{R_a} = \frac{E_m}{\pi R_a}(\cos\alpha - \cos X) + \frac{E_b}{\pi R_a}(\alpha - X) \qquad (6.27)$$

The boundary between continuous and discontinuous conduction occurs when $X = \pi + \alpha$. At this condition (6.27) reduces to (6.9).

Equation (6.13) also represents the instantaneous discontinuous armature current, during its conduction intervals, e.g. $\alpha \leq \omega t \leq X$. But now current $i_a(\omega t) = 0$ at $\omega t = \alpha$. Substituting into (6.13) gives

$$0 = \frac{E_m}{|Z_a|}\sin(\alpha - \phi_a) - \frac{E_b}{R_a} + K \qquad (6.28)$$

Combining (6.28) and (6.13) gives, for $\alpha \leq \omega t \leq X$,

$$i_a(\omega t) = \frac{E_m}{|Z_a|}[\sin(\omega t - \phi_a) - \sin(\alpha - \phi_a)\varepsilon^{-c(\omega t - \alpha)}]$$
$$- \frac{E_b}{R_a}[1 - \varepsilon^{-c(\omega t - \alpha)}] \qquad (6.29)$$

With discontinuous operation it is necessary to evaluate the extinction angle X, which is found from the correlation $i_a(\omega t) = 0$ when $\omega t = X$. Therefore, in (6.29),

$$0 = \frac{E_m}{|Z_a|}[\sin(X - \phi_a) - \sin(\alpha - \phi_a)\varepsilon^{-c(X-\alpha)}]$$
$$- \frac{E_b}{R_a}[1 - \varepsilon^{-c(X-\alpha)}] \qquad (6.30)$$

Equation (6.30) is transcendental and must be solved by iteration.

As for the case of continuous conduction the r.m.s. values of the load and supply currents are identical. This value I_L can also be found by substituting from (6.29) into the defining integral

$$I_L^2 = \frac{1}{\pi} \int_\alpha^X i_a^2(\omega t)\, \mathrm{d}\omega t \tag{6.31}$$

Alternatively, I_L can be approximated by the harmonic summation method described for the continuous conduction mode.

6.2.2.3 Critical value of load inductance

For the circuit of Fig. 6.3 the forms of the motor armature current are shown in Figs. 6.4, 6.5 for continuous and discontinuous conduction respectively. The boundary between these two modes of operation occurs for the circuit condition when $i_a(\omega t)$ just reaches zero at $\omega t = \pi + \alpha = X$.

Putting $i_a(X) = 0$ and $\omega t = X = \pi - \alpha$ into (6.29) gives a magnitude criterion for continuity of the armature current

$$\frac{R_a}{|Z_a|} \sin(\alpha - \phi_a) \frac{(\varepsilon^{-c\pi} + 1)}{(\varepsilon^{-c\pi} - 1)} > \frac{E_b}{E_m} \tag{6.32}$$

Criterion (6.32) can also be deduced by putting $I_{\min} = i(\alpha) = 0$ in (6.18).

Now the maximum value of α for rectifier operation is 90°. Since continuous current is desirable the angle ϕ_a is usually greater than 80°. For most of the realistic control range $(\alpha - \phi_a)$ is negative, thereby cancelling the negative denominator term to give a positive LHS to inequality (6.32).

6.2.2.4 Power and power factor

The power dissipated in the load branch of Fig. 6.3 is given by (5.64), with appropriate changes of terminology,

$$P_a = I_L^2 R_a + E_b I \tag{6.33}$$

Component of power $E_b I_a$ represents the output power plus the motor friction and windage losses. In terms of motor developed torque T, from (5.9),

$$E_b I = TN = P_{\text{out}} \tag{6.34}$$

Neglecting the power loss in the rectifier switches and ignoring the motor core and rotational losses, the operating efficiency is

$$\eta = \frac{P_{\text{out}}}{P_{\text{in}}} = \frac{E_b I}{P_a} \tag{6.35}$$

6.2 Separately excited d.c. motor with rectified single-phase supply

Since the input current has an r.m.s. value equal to that of the motor current the operating power factor is

$$PF = \frac{P_a}{\frac{E_m}{\sqrt{2}} \times I_L} \tag{6.36}$$

6.2.3 Worked examples

Example 6.1
A separately excited d.c. motor is rated at 10 kW, 240 V, 1000 r.p.m. and is supplied with power from a fully controlled, single-phase bridge rectifier. The power supply is sinusoidal and rated at 240 V, 50 Hz. The motor armature resistance is $0.42\,\Omega$ and the motor constant is 2 volt seconds/radian. Some additional inductance is included in the armature circuit to ensure continuous conduction but its value is not known. Calculate the speed, power factor and efficiency of operation for SCR firing-angles $\alpha = 0°$ and $\alpha = 20°$ if the load torque is constant.

Solution. The average voltage applied to the load is, from (6.8),

$$V_{L_{av}} = \frac{2E_m}{\pi} \cos\alpha$$

$$= \frac{2\sqrt{2} \times 240}{\pi} \cos\alpha$$

$$= 216.1 \cos\alpha$$

At $\alpha = 0°$,

$V_{L_{av}} = 216.1$ V

At $\alpha = 20°$,

$V_{L_{av}} = 203$ V

Assume that the motor delivers its rated power at $\alpha = 0°$, then, from (6.34),

$$E_b I = 10\,000 \tag{a}$$

Also, from (6.9),

$$I = \frac{216.1 - E_b}{0.42} \tag{b}$$

Combining equations (a), (b) gives a quadratic equation in I

$$0.42I^2 - 216.1I + 10\,000 = 0$$

from which

$$I = \frac{216.1 \pm \sqrt{46\,699 - 16\,800}}{0.84}$$
$$= 51.43 \text{ A, taking the negative solution.}$$

From (a) above,

$$E_b = \frac{10\,000}{51.43} = 194.44 \text{ V}$$

From (5.6) the full-load motor torque is given by

$$T = K_T \Phi I$$
$$= 2 \times 51.43 = 102.86 \text{ N m}$$

From (6.34) the full-load speed is

$$N = \frac{E_b I}{T}$$
$$= \frac{194.44 \times 53.43}{102.86}$$
$$= 101 \text{ rad/s}$$
$$= 101 \times \frac{60}{2\pi} = 964.5 \text{ r.p.m.}$$

Since the load circuit inductance is not known it is not possible to accurately calculate the r.m.s. armature current either by the harmonic summation method or by the integral method (6.31). An approximation is made by assuming that the armature copper loss in the motor is given in terms of its average current by

armature copper loss $= I^2 R_a = (51.43)^2 0.42 = 1111$ W

The input power is therefore, approximately, given by (6.33),

$$P_a = 1111 + 10\,000 = 11\,111 \text{ W}$$

6.2 Separately excited d.c. motor with rectified single-phase supply

The efficiency is therefore, from (6.35),

$$\eta = \frac{10\,000}{11\,111} = 90\%$$

The above figure is optimistic because of the various losses neglected and the current approximation. The input power factor is therefore, (6.36),

$$PF = \frac{11\,111}{240 \times 51.43} = 0.9$$

Let the firing-angle be retarded to 20°, reducing the average armature voltage to 203 V. With a separately excited motor the speed regulation is small. If the no-load speed is assumed to be proportional to the average applied voltage, as implied in (5.17), then, at $\alpha = 20°$,

$$N = \frac{203}{216.1} \times 964.5 = 906 \text{ r.p.m.}$$

With constant torque, the output power varies proportionally with the speed and the current is constant

$$P_{out} = \frac{906}{60} \times 2\pi \times 102.86 = 9759 \text{ W}$$

$I = 51.43$ A, as before

From equation (a),

$$E_b = \frac{P_{out}}{I} = \frac{9759}{51.43} = 189.8 \text{ V}$$

The total power delivered to the motor is

$$P_a = I^2 R_a + E_b I$$
$$= (51.43)^2 0.42 + 9759$$
$$= 1111 + 9759 = 10\,870 \text{ W}$$

The efficiency is therefore

$$\eta = \frac{P_{out}}{P_a} = \frac{9759}{10\,870} = 89.8\%$$

The power factor is

$$PF = \frac{10\,870}{240 \times 51.43} = 0.88$$

Example 6.2
In the separately excited d.c. motor of Example 6.1 the armature circuit choke is removed, leaving the intrinsic armature inductance $L_a = 45\,\text{mH}$. Recalculate the speed and power factor when $\alpha = 0°$ if the motor is required to deliver its rated power.

Solution. It is first necessary to determine if the motor current is continuous:

$$c = \cot \phi_a = \frac{R_a}{\omega L_a} = \frac{0.42 \times 1000}{2\pi \times 50.45} = 0.0297$$

$$\phi_a = \cot^{-1}(0.0297) = 88.3°$$

$$\varepsilon^{-c\pi} = \varepsilon^{-0.0933} = 0.91$$

$$\frac{R_a}{|Z_a|} = \cos \phi_a = \cos 88.3° = 0.0297$$

$$\sin(\alpha - \phi_a) = \sin(-88.3°) = -0.999$$

In (6.32) the LHS is

$$(0.0297)(-0.999)\left(\frac{1.91}{-0.09}\right) = +0.63$$

The current will therefore be continuous if $E_b < 0.63 E_m$ or

$$E_b < 0.63 \times \sqrt{2} \times 240$$
$$< 213.8\,\text{V}$$

The value $E_b = 194.44\,\text{V}$ from Example 6.1 satisfies this criterion so that the current will be continuous even though no additional inductance is used. The values of average load voltage, average load current, torque and speed, at $\alpha = 0°$, remain true.

The second harmonic component of the load voltage (i.e. the lowest order a.c. harmonic component) is obtained from (6.12)

6.2 Separately excited d.c. motor with rectified single-phase supply

$$V_{L_2} = \frac{E_m}{2\pi} \sqrt{\frac{1}{9} + 1 - \frac{2}{3}}$$

$$= \frac{E_m}{2\pi} \sqrt{\frac{4}{9}}$$

$$= \frac{\sqrt{2} \times 240 \times 2}{2\pi \times 3} = 36 \text{ V (peak)}$$

The second harmonic component of load current is therefore

$$I_{L_2} = \frac{V_{L_2}}{\sqrt{R_a^2 + (2\omega L_a)^2}}$$

$$= \frac{36}{\sqrt{(0.42)^2 + \left(\frac{4\pi \times 50 \times 45}{1000}\right)^2}}$$

$$= \frac{36}{\sqrt{2} \times 9\pi} = 0.9 \text{ A (r.m.s.)}$$

The harmonic addition of I_{L_2} to the average armature current of 51.43 A will obviously make no significant difference. The values of power and power factor calculated in Example 6.1 are therefore still valid.

Example 6.3
A separately excited d.c. servomotor is rated at 500 W, 200 V, 1000 r.p.m. and has the armature circuit parameters $R_a = 0.15\,\Omega$, $L_a = 2.5\,\text{mH}$. The motor is supplied with power from an ideal sinusoidal supply of 240 V, 50 Hz via a fully controlled, single-phase, thyristor bridge rectifier. What external inductance (if any) must be included in the armature circuit to permit speed variation down to 500 r.p.m., with continuous current, if the load torque is constant at the rated value. Also, calculate the required thyristor firing-angle.

Solution. The rated motor torque, from (5.9), is

$$T = \frac{500}{1000 \times \frac{2\pi}{60}} = 4.775\,\text{N m},$$

Assume that at the rated speed of 1000 r.p.m. the motor operates with maximum applied voltage, so that $\alpha = 0°$. The motor average terminal voltage is, from (6.8), assuming continuous conduction,

$$V_{L_{av}}(\alpha = 0) = \frac{2E_m}{\pi}$$

$$= \frac{2\sqrt{2} \times 240}{\pi} = 216.074 \text{ V}$$

In calculating the motor average current one is often dealing with a small difference between two large voltages, $V_{L_{av}} - E_b$. Great accuracy of calculation is therefore necessary. From (6.9),

$$I = \frac{216.074 - E_b}{0.15} \tag{a}$$

Also, from (6.34),

$$E_b I = 500 \tag{b}$$

The simultaneous solution of (a) and (b) yields

$I = 2.32$ A,

$E_b = 215.72$ V

For this particular motor, at 1000 r.p.m. and $\alpha = 0°$,

$$c = \cot \phi_a = \frac{R_a}{\omega L_a} = \frac{0.15 \times 1000}{2\pi \times 50 \times 2.5} = 0.191$$

$\phi_a = \cot^{-1}(0.191) = 79.2°$

$\varepsilon^{-c\pi} = \varepsilon^{-0.6} = 0.55$

$R_a/|Z_a| = \cos \phi_a = 0.187$

$\sin(\alpha - \phi_a) = \sin(-79.2°) = -0.982$

In (6.32) it is seen that the LHS has the value

$$(0.187)(-0.982)\left(\frac{1.55}{-0.45}\right) = 0.6325$$

The RHS of (6.32) is

$$\frac{E_b}{E_m} = \frac{215.72}{\sqrt{2} \times 240} = 0.6356$$

6.2 Separately excited d.c. motor with rectified single-phase supply

It is seen that the motor is just on the margin between continuous and discontinuous operation. Obviously some additional inductance is likely to be needed at the proposed lower speed.

Let the motor now operate at 500 r.p.m., delivering rated torque. Then

$$P = TN = 250\,\text{W}$$

But, if the torque is unchanged then, from (5.6), the armature current is unchanged. Since the power output is halved, the back e.m.f. must be halved:

$$E_b(500\ \text{r.p.m.}) = \frac{215.72}{2} = 107.86\,\text{V}$$

The required motor applied voltage is deduced from (6.9), if continuous current has been maintained,

$$2.32 = \frac{V_{L_{\text{av}}} - 107.86}{0.15}$$

which gives

$$V_{L_{\text{av}}} = 108.21\,\text{V}$$

From (6.8) the required value of α is

$$\alpha = \cos^{-1}\left(\frac{\pi V_{L_{\text{av}}}}{2E_m}\right)$$
$$= \cos^{-1} 0.5 = 60°$$

The RHS of (6.32) is now given by

$$\frac{E_b}{E_m} = \frac{107.86}{\sqrt{2} \times 240} = 0.318$$

By iteration in (6.32) it is found that $\phi_a = 89.5°$. Therefore,

$$L_a = \frac{R_a}{\omega \cot \phi_a} = \frac{0.15 \times 1000}{2\pi \times 50 \times 0.00873} = 54.7\,\text{mH}$$

6.3 SEPARATELY EXCITED D.C. MOTOR WITH RECTIFIED THREE-PHASE SUPPLY

The theory and operation of the three-phase full wave controlled converter is fully discussed in Chapter 7 below, for the cases of passive impedance load. Three-phase converters are extensively used in adjustable speed d.c. drives from about 10 h.p. up to several thousand horsepower rating. The three-phase, half-wave circuit, Table 6.2, is not greatly used because of the d.c. components inherent in its line currents. The adoption of a full-wave bridge circuit not only eliminates the d.c. components in the supply lines but also permits optimum utilisation of the principal electrode ratings of the switches.

Table 6.2 *Three-phase naturally commutated controlled converter circuits.*

Type	Circuit	Operation
half-wave 3 pulse		
semi-converter 3 pulse		
full converter 6 pulse		
double converter 6 pulse		

6.3 Separately excited d.c. motor with rectified three-phase supply

With passive impedance loading, a three-phase full-wave bridge has twice the load voltage and four times the power capability, compared with a half-wave bridge. But the bridge switches in a full-wave circuit have to be rated for the peak line voltage compared with the peak phase voltage for a half-wave circuit. When reversible armature current is needed, to give four-quadrant operation, the double converter is used. This finds extensive use in the UK steel industry and other heavy engineering process applications.

When the application involves medium size motors, in the range 20–150 h.p., either the semi-converter or full converter is used. The condition of continuous armature current is invariably desired and, where necessary, additional armature inductance is included. The three-phase bridges are here assumed to be fed from an ideal three-phase sinusoidal supply defined below.

With a supply of zero impedance the three supply phase voltages for the circuit of Fig. 6.6 retain balanced sinusoidal form for any load condition. These voltages are defined by the equations

$$e_{aN} = E_m \sin \omega t \tag{6.37}$$

$$e_{bN} = E_m \sin(\omega t - 120°) \tag{6.38}$$

$$e_{cN} = E_m \sin(\omega t - 240°) \tag{6.39}$$

The corresponding line-to-line voltages at the supply point are

$$e_{ab} = e_{aN} + e_{Nb} = e_{aN} - e_{bN} = \sqrt{3} E_m \sin(\omega t + 30°) \tag{6.40}$$

$$e_{bc} = \sqrt{3} E_m \sin(\omega t - 90°) \tag{6.41}$$

$$e_{ca} = \sqrt{3} E_m \sin(\omega t - 210°) \tag{6.42}$$

Waveforms of the supply line voltages are given in Figs. 6.7 and 6.9. (*See* Sections 6.3.2.1 and 6.3.2.3, respectively.)

6.3.1 Three-phase semi-converter

The semi-converter circuit includes a freewheel diode FWD to assist in maintaining continuous load current. A cost advantage is obtained by the use of diodes in the lower half of the bridge, compared with the full converter. A further advantage is realised in that the semi-converter circuit absorbs less reactive voltamperes than the fully controlled converter. The average voltage $V_{L_{av}}$ at the load contains a contribution from the controlled upper half-bridge (or semi-converter) plus a contribution from the uncontrolled lower semi-converter. With continuous current (corresponding to high inductance operation with passive loads), for all firing-angles, one may represent the average load voltage as

$$V_{L_{av}} = \frac{3\sqrt{3}E_m}{2\pi} + \frac{3\sqrt{3}E_m}{2\pi}\cos\alpha$$
$$= \frac{3\sqrt{3}E_m}{2\pi}(1+\cos\alpha) \qquad (6.43)$$

When $\alpha = 0$ the average output voltage becomes identical to that of an uncontrolled three-phase bridge.

The average armature current is obtained by combining (6.43) with (5.56),

$$I = I_{L_{av}} = \frac{V_{L_{av}} - E_b}{R_a}$$
$$= \frac{3\sqrt{3}E_m}{2\pi R_a}(1+\cos\alpha) - \frac{E_b}{R_a} \qquad (6.44)$$

Equations (6.43), (6.44) are not valid for discontinuous current operation. The load voltage of a semi-converter is of three-pulse nature and has a higher ripple content than that of the fully controlled bridge. In addition, because only one-half of the bridge is controlled, the line currents are unsymmetrical and contain even harmonics. When $\alpha > 90°$ the average load voltage contribution of the upper half-bridge becomes negative. The result is that the overall load voltage v_a goes negative at some intervals during the cycle and the freewheel diode conducts.

6.3.2 Three-phase full converter

A circuit diagram is given in Fig. 6.6 in which the motor armature is represented by its equivalent circuit. With low armature inductance and large SCR

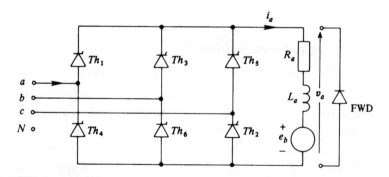

Fig. 6.6 Three-phase, full wave, controlled bridge converter with d.c. separately excited motor load.

6.3 Separately excited d.c. motor with rectified three-phase supply

firing-angle the armature current may become discontinuous especially if the d.c. motor speed (and therefore back e.m.f.) is high. If the motor armature circuit contains substantial series inductance and the firing-angle is small then the armature current is likely to be continuous, even with a large motor back e.m.f. The e.m.f. equations of the armature circuit are given by (5.10(a)) for instantaneous values and (5.11(a)) for average values.

6.3.2.1 Continuous conduction

Load voltage and current waveforms for continuous armature current operation are shown in Fig. 6.7 for a case when $E_b = E_m/4$. For all values of α the load voltage is defined by (7.7) and has the average value defined in (7.8), reproduced here,

$$V = V_{L_{av}} = \frac{3\sqrt{3}}{\pi} E_m \cos \alpha \tag{6.45}$$

The average armature current therefore has the value

$$I = \frac{V_{L_{av}} - E_b}{R_a}$$

$$= \frac{3\sqrt{3}}{\pi R_a} E_m \cos \alpha - \frac{E_b}{R_a} \tag{6.46}$$

The Fourier harmonic properties of the load voltage waveform, with phase voltage v_a as reference, and $n = 6, 12, 18 \ldots$, are found to be, after much manipulation,

$$\begin{aligned} a_n &= \frac{1}{\pi} \int_0^{2\pi} v_a(\omega t) \cos n\omega t \, d\omega t \\ &= \frac{6}{\pi} \int_{\alpha+30°}^{\alpha+90°} \sqrt{3} E_m \sin(\omega t + 30°) \cos n\omega t \, d\omega t \\ &= \frac{-12\sqrt{3} E_m}{\pi} (\cos \frac{n\pi}{6} \cdot \sin \frac{\pi}{6} [n \sin n\alpha \cdot \sin \alpha + \cos n \cdot \cos \alpha] \\ &= \frac{3\sqrt{3} E_m}{\pi} \left[\frac{2 \sin(n+1)\frac{\pi}{6} \cdot \cos(n+1)\alpha}{n+1} + \frac{2 \sin(n-1)\frac{\pi}{6} \cdot \cos(n-1)\alpha}{n-1} \right] \end{aligned}$$

$$\tag{6.47}$$

214 Controlled bridge rectifiers with d.c. motor load

Similarly

$$b_n = \frac{1}{\pi}\int_0^{2\pi} v_a(\omega t)\sin n\omega t \, d\omega t$$

$$= \frac{6}{\pi}\int_{\alpha+30°}^{\alpha+90°} \sqrt{3}E_m \sin(\omega t + 30°)\sin n\omega t \, d\omega t$$

$$= \frac{12\sqrt{3}E_m}{\pi(n^2-1)}\cos\frac{n\pi}{6}\cdot\sin\frac{\pi}{6}[n\cos n\alpha\cdot\sin\alpha - \sin n\alpha\cdot\cos\alpha]$$

$$= \frac{3\sqrt{3}E_m}{\pi}\left[\frac{2\sin(n+1)\frac{\pi}{6}\cdot\sin(n+1)\alpha}{n+1} + \frac{2\sin(n-1)\frac{\pi}{6}\cdot\sin(n-1)\alpha}{n-1}\right]$$

(6.48)

Since the full wave bridge is a six-pulse system, the load-side voltage contains a.c. ripple harmonics of order 6, 12, 18, ... etc. times the supply frequency. The peak value c_n of the load voltage may be expressed as

$$V_{L_n} = c_n = \sqrt{a_n^2 + b_n^2} \quad (n = 6, 12, 18, \ldots)$$

$$= \frac{3\sqrt{3}E_m}{\pi}\sqrt{\frac{1}{(n+1)^2} + \frac{1}{(n-1)^2} - \frac{2\cos 2\alpha}{(n-1)(n+1)}}$$

(6.49)

Alternatively, for $n = 1, 2, 3, \ldots$,

$$V_{L_n} = c_n = \frac{3\sqrt{3}E_m}{\pi}\sqrt{\frac{1}{(6n+1)} + \frac{1}{(6n-1)} - \frac{2\cos 2\alpha}{(6n+1)(6n-1)}}$$

(6.50)

Harmonic load currents can be obtained by dividing the harmonic voltage (6.50) by the corresponding harmonic impedance (5.61). The correspondence between (6.50) for a six-pulse system and (6.12) for a two-pulse system is because the respective load voltages, Fig. 6.4 and Fig. 6.7, have the same basic waveform.

For example, for a full converter, in the interval $\alpha + 30° \leq \omega t \leq \alpha + 90°$,

$$\sqrt{3}E_m \sin(\omega t + 30°) = i_a R_a + L_a \frac{di_a}{dt} + e_b$$

(6.51)

Assume that the motor speed ripple is small so that the back e.m.f. is constant at its average value E_b. From (6.51) it is found that

6.3 Separately excited d.c. motor with rectified three-phase supply

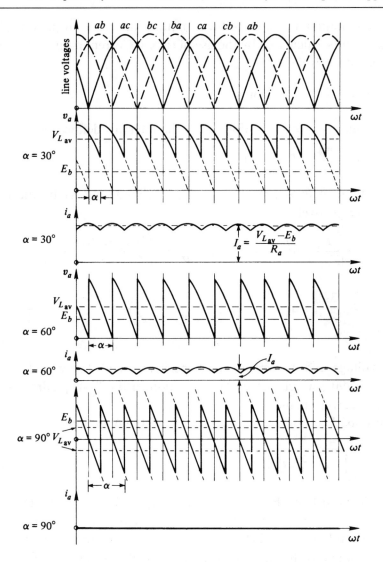

Fig. 6.7 Waveforms for the three-phase full converter drive with continuous armature current for a separately excited d.c. motor.

$$i_a(\omega t) = \frac{\sqrt{3}E_m}{|Z_a|} \sin\left(\omega t + \frac{\pi}{6} - \phi_a\right) - \frac{E_b}{R_a} + K_1 \varepsilon^{-c(\omega t - \pi/6 - \alpha)} \tag{6.52}$$

where $|Z_a|$, ϕ_a and c are defined by (6.14),(6.15) and K_1 is a constant of integration with the value

$$K_1 = \frac{\sqrt{3}E_m \sin(\alpha - \phi_a)}{|Z_a|(\varepsilon^{-c\pi/3} - 1)} \tag{6.53}$$

Maximum and minimum values of the armature current occur periodically in Fig. 6.7 and can be calculated by the use of (6.52). Current minima I_{min} occur at all of the intervals defined by

$$\omega t = \alpha - \frac{\pi}{6} + \frac{n\pi}{3}, \quad \text{for } n = 0, 1, 2 \ldots \tag{6.54}$$

Current maxima I_{max} occur in Fig. 6.7 at the intervals defined by

$$\omega t = \alpha + \frac{n\pi}{3}, \quad \text{for } n = 0, 1, 2 \ldots \tag{6.55}$$

The substitution of (6.54), (6.55) into (6.52), eliminating K_1 by the use of (6.53) gives, respectively,

$$I_{min} = \frac{\sqrt{3}E_m}{|Z_a|} \sin\left(\alpha - \phi_a + \frac{\pi}{3}\right) - \frac{E_b}{R_a} + \frac{\sqrt{3}E_m}{|Z_a|} \cdot \frac{\sin(\alpha - \phi_a)}{(\varepsilon^{-c\pi/3} - 1)} \tag{6.56}$$

$$I_{max} = \frac{\sqrt{3}E_m}{|Z_a|} \sin\left(\alpha - \phi_a + \frac{\pi}{2}\right) - \frac{E_b}{R_a} + \frac{\sqrt{3}E_m}{|Z_a|} \cdot \frac{\sin(\alpha - \phi_n) \cdot \varepsilon^{-c\pi/6}}{(\varepsilon^{-c\pi/3} - 1)} \tag{6.57}$$

The load voltage and current are both positive for all firing-angles in the range $0 \leq \alpha \leq 90°$, which represents operation in the first quadrant of the load voltage–current plane.

The r.m.s. load current I_L can be found from the basic integral expression

$$I_L = \frac{3}{\pi} \int_{\alpha+30°}^{\alpha+90°} i_a^2(\omega t) \, d\omega t \tag{6.58}$$

It may be deduced from Fig. 6.7 that each supply line current flows for a conduction interval of 120° or $2\pi/3$ rads every half-cycle of the supply voltage, as shown in Fig. 7.5, p. 241. The r.m.s. value of the supply line current I_s is therefore related to the r.m.s. load current by equation (7.24). Rewriting, using the terminology of this present chapter, gives

$$I_s = \sqrt{\frac{2}{3}} I_L \tag{6.59}$$

An expression for the steady-state torque–speed relationship can be obtained in terms of α by combining (5.14) and (6.45),

$$T = \frac{K_T K_f E_f}{R_a R_f} \left(\frac{3\sqrt{3}}{\pi} E_m \cos\alpha - \frac{K_E K_f E_f}{R_f} N \right) \tag{6.60}$$

or

6.3 Separately excited d.c. motor with rectified three-phase supply

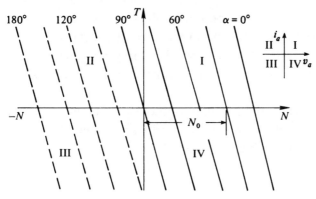

Fig. 6.8 Steady-state torque–speed characteristics of a separately excited d.c. motor drive. — full converter control, - - - double converter control.

$$T = N_0(\alpha) - KN \tag{6.61}$$

where

$$N_0 = \frac{3\sqrt{3} K_T K_f E_m E_f}{\pi R_a R_f} \cos\alpha = \text{no-load speed}$$

$$K = \frac{K_E K_T K_f^2 E_f^2}{R_a R_f^2} = \text{slope of the characteristic}$$

Typical characteristics are shown by the solid lines of Fig. 6.8. At $\alpha = 90°$ it is possible to sustain torque at low values of reverse speed. Performance in quadrants I and IV of the torque–speed plane corresponds to performance in quadrants I and II of the armature voltage–armature current plane.

6.3.2.2 Critical value of load inductance

At the boundary between discontinuous and continuous conduction the minimum armature current I_{min} has zero value. Putting $I_{min} = 0$ in (6.56) and rearranging gives a criterion for the maintenance of continuous armature current

$$\frac{R_a}{|Z_a|} \left[\sin\left(\alpha - \phi_a + \frac{\pi}{3}\right) + \frac{\sin(\alpha - \phi_a)}{(\varepsilon^{-c\pi/3} - 1)} \right] \geq \frac{E_b}{\sqrt{3} E_m} \tag{6.62}$$

6.3.2.3 Discontinuous conduction

Consider the condition shown in Fig. 6.9 where the same e.m.f. condition applies as in Fig. 6.7 (i.e. $E_b = E_m/4$) but now the motor operates with a much lower value of armature circuit inductance so that discontinuity of the armature current has occurred by the stage $\alpha = 60°$. Individual pulsations of current now occupy a conduction period θ_c which is less than the corresponding interval of 60° obtained with continuous operation. At $\alpha = 75°$ the average load voltage is $\cos \alpha = 0.26$ p.u., which is only marginally greater than the back e.m.f. The current pulsations are therefore very small and become zero when $V_{L_{av}} = E_b$.

Equation (6.52) is also valid for discontinuous conduction, in the intervals $\alpha \leq \omega t \leq \alpha + \theta_c$, where θ_c is the current conduction angle. But $i_a(\omega t)$ is then zero when $\omega t = \alpha + 30°$, in Fig. 6.9. Therefore

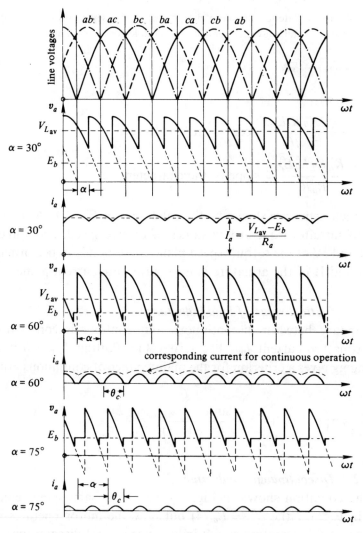

Fig. 6.9 Waveforms for the three-phase full converter drive with discontinuous armature current.

6.3 Separately excited d.c. motor with rectified three-phase supply

$$0 = \frac{\sqrt{3}E_m}{|Z_a|}\sin\left(\alpha + \frac{\pi}{3} - \phi_a\right) - \frac{E_b}{R_a} + K_1\varepsilon^{-c(-\pi/3)} \qquad (6.63)$$

Combining (6.52) and (6.63) gives

$$i_a(\omega t) = \frac{\sqrt{3}E_m}{|Z_a|}\left[\sin\left(\omega t + \frac{\pi}{6} - \phi_a\right) - \sin\left(\alpha + \frac{\pi}{3} - \phi_a\right)\varepsilon^{-c(\omega t + \pi/6 - \alpha)}\right]$$
$$- \frac{E_b}{R_a}[1 - \varepsilon^{-c(\omega t + \pi/6 - \alpha)}]$$

$$(6.64)$$

With discontinuous conduction it is necessary to evaluate the conduction angle θ_c. This can be done by noting that $i_a(\omega t) = 0$ when $\omega t = \alpha + \pi/6 + \theta_c$ in (6.64), giving a transcendental equation

$$0 = \frac{\sqrt{3}E_m}{|Z_a|}\left[\sin\left(\alpha + \frac{\pi}{3} + \theta_c - \phi_a\right) - \sin\left(\alpha + \frac{\pi}{3} - \phi_a\right)\varepsilon^{-c(\pi/3+\theta_c)}\right]$$
$$- \frac{E_b}{R_a}\left(1 - \varepsilon^{-c(\pi/3+\theta_c)}\right)$$

$$(6.65)$$

An iterative solution of (6.65) yields a value for θ_c. The r.m.s. value of the discontinuous armature current can be found from

$$I_L^2 = \frac{3}{\pi}\int_{\alpha+\pi/6}^{\alpha+\pi/6+\theta_c} i_a^2(\omega t)\, d\omega t \qquad (6.66)$$

Similarly, the average load current may be calculated from

$$I = \frac{3}{\pi}\int_{\alpha+\pi/6}^{\alpha+\pi/6+\theta_c} i_a(\omega t)\, d\omega t \qquad (6.67)$$

where $i_a(\omega t)$ is given by (6.64).

It may be deduced from Fig. 6.9 that the load voltage in the discontinuous condition is defined by the typical equation

$$v_a(\omega t) = \sqrt{3}E_m \sin(\omega t + 30°)\Big|_{\alpha+30°}^{\alpha+30°+\theta_c} + E_b\Big|_{\alpha+30°+\theta_c}^{\alpha+90°} \qquad (6.68)$$

220 Controlled bridge rectifiers with d.c. motor load

This has the average value

$$V_{L_{av}} = \frac{3}{\pi} \int_{\alpha+30°}^{\alpha+90°} v_a(\omega t)\, d\omega t$$

$$= \frac{3\sqrt{3}E_m}{\pi}[\cos(\alpha+60°) - \cos(\alpha+60°+\theta_c)] + \frac{3E_b}{\pi}\left(\frac{\pi}{3} - \theta_c\right)$$
(6.69)

When the boundary of full conduction is reached $\theta_c \to 60°$ and (6.69) reduces to (6.45). As an alternative to the use of (6.63), the average current can be found by combining (6.69) and (5.56).

6.3.2.4 Power and power factor

Equations (6.32)–(6.34) for the power and efficiency of a separately excited d.c. motor with single-phase supply are still valid for the case of three-phase supply. The power factor now has to be expressed in terms of phase quantities at the supply point. Equation (6.59) is valid for both continuous and discontinuous operation so that in terms of the load power P_a and r.m.s. load current I_L

$$PF = \frac{\dfrac{P_a}{3}}{\dfrac{\sqrt{3}E_m}{\sqrt{2}}\sqrt{\dfrac{2}{3}}I_L}$$

$$= \frac{P_a}{3E_m I_L}$$
(6.70)

6.3.2.5 Addition of freewheel diode

If the freewheel diode FWD in Fig. 6.6 is now connected across the armature the load voltage is prevented from going negative thereby restricting the performance to quadrant I operation.

For $0 \leq \alpha \leq 60°$, the load voltage is unchanged and has the average value given in (6.45). For $60° \leq \alpha \leq 120°$, the load voltage waveform is represented by the positive pulses only in Fig. 6.7. The average value is therefore given by

$$V_{L_{av}} = \frac{3}{\pi} \int_{\alpha+30°}^{150°} v_a(\omega t)\, d\omega t$$

$$= \frac{3}{\pi} \int_{\alpha+30°}^{150°} \sqrt{3} E_m \sin(\omega t + 30°)\, d\omega t \qquad (6.71)$$

$$= \frac{3\sqrt{3} E_m}{\pi} \left[-\cos(\omega t + 30°)\right]_{\alpha+30°}^{150°}$$

$$\therefore V_{L_{av}} = \frac{3\sqrt{3} E_m}{\pi} \left[1 + \cos(\alpha + 60°)\right]$$

When $\alpha > 60°$ the average voltage $V_{L_{av}}$ in (6.71) is slightly greater than the corresponding value (6.45) for standard bridge operation, and the control range now extends to 120°. The presence of the freewheel diode also causes a small improvement of operating power factor at the expense of increased input current harmonics.

6.3.3 Three-phase double converter

If control is required for both forward and reverse speeds and the production of positive and negative torques one can use a fully controlled bridge incorporating reversal of the applied armature voltage. A better solution is the use of the double or dual converter, Fig. 6.10, which avoids the need for change-over switches. The average motor voltage is required to be identical for both bridges, which sets up an ideal requirement that the firing-angles of the two sets of thyristors should sum to 180°. A series inductance is included in each of the four motor legs to accommodate the inevitable ripple voltage due to instantaneous inequalities. By controlled firing, negative armature voltage and current may be used to obtain reverse speed operation, as shown in

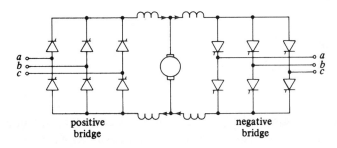

Fig. 6.10 Fully controlled double converter bridge rectifier with d.c. motor load.

Fig. 6.8. If the two bridges operate consecutively so that conduction occurs in only one bridge at a time, with the other completely blocked, the mode of control is called 'circulating current-free operation'. This mode of operation is readily realised provided that the motor current is continuous but serious difficulties arise if the current becomes discontinuous.

Alternatively, the two bridges can operate concurrently with a regulated degree of circulating current in a mode of control known as 'circulating current operation'. The circulating current acts to maintain armature current at all times with both converters in continuous conduction.

6.3.4 Worked examples

Example 6.4
A separately excited d.c. motor rated at 10 kW, 300 V, 1000 r.p.m. is supplied with power from a fully controlled, three-phase bridge rectifier. The ideal three-phase power supply is rated at 220 V, 50 Hz. This motor has an armature resistance $R_a = 0.2\,\Omega$ and sufficient added inductance to maintain continuous conduction. The motor constant is 1.38 volt seconds/radian, and it delivers rated power at $\alpha = 0°$. If the SCR firing-angle is retarded to $\alpha = 30°$, calculate the speed, power factor and efficiency of operation, if the load torque is constant.

Solution. An equivalent circuit is shown in Fig. 6.6. The average voltage applied to the armature at $\alpha = 0°$ is given, from (6.45), by

$$V_{L_{av}} = \frac{3\sqrt{3}}{\pi} E_m \cos\alpha$$
$$= \frac{3\sqrt{3}}{\pi} \times \frac{220\sqrt{2} \times 1}{\sqrt{3}}$$
$$= 297.1\,\text{V}$$

At $\alpha = 30°$,

$$V_{L_{av}} = 297.1 \cos 30°$$
$$= 257.3\,\text{V}$$

The steady-state speed–torque variation of a separately excited d.c. motor is described by (5.12). With constant flux the no-load speed $V/K_E\Phi$ is proportional to the (average) armature voltage.

6.3 Separately excited d.c. motor with rectified three-phase supply

Therefore, at $\alpha = 30°$, since $V = V_{L_{av}} = 257.3$ V, the speed is (assuming rated speed at $\alpha = 0°$)

$$N_{\alpha=30°} = \frac{257.3}{297.1} \times 1000 = 866 \text{ r.p.m.}$$

The motor-developed power is then, with constant T, assuming zero speed regulation

$$P_{out} = TN$$
$$= \frac{866}{60} \times 2\pi \times 95.5 = 8661 \text{ W}$$

The average armature current at $\alpha = 30°$ is obtained from two simultaneous equations. From (6.34),

$$E_b I = 8661 \qquad (a)$$

From (6.46),

$$I = \frac{257.3 - E_b}{0.2} \qquad (b)$$

The combination of (a) and (b) gives a quadratic equation in I,

$$0.2I^2 - 257.3I + 8661 = 0$$

which has the solution

$$I = \frac{257.3 \pm \sqrt{(257.3)^2 - 6929}}{0.4}$$
$$= \frac{257.3 \pm 243.5}{0.4} = 1252 \text{ A or } 34.5 \text{ A}$$

Taking the difference option, $I = 34.5$ A, gives a value of E_b from (a) above,

$$E_b = \frac{8661}{34.5} = 251 \text{ V}$$

An accurate calculation of the armature r.m.s. current is not possible, in this case, because the value of armature inductance L_a is not known. Neglecting the armature current ripple is, in effect, an assumption that the r.m.s. current is equal to the average current I. Then armature copper loss $= I^2 R_a = (34.5)^2 0.2 = 238$ W

The input power is therefore, from (6.33),

$$P_a = I^2 R_a + P_{out}$$
$$= 238 + 8661 = 8899 \text{ W}$$

The efficiency at $\alpha = 30°$ is therefore, approximately,

$$\eta = \frac{P_{out}}{P_a} = \frac{8661}{8899} = 97.3\%$$

This efficiency value is considerably optimistic due to the neglect of the field and rotational losses.

The r.m.s. line current, on the supply side, is found from (6.59). Once again, neglecting current ripple so that $I_L = I$

$$I_s = \sqrt{\frac{2}{3}} I = \sqrt{\frac{2}{3}} \times 34.5 = 28.2 \text{ A}$$

One may also proceed directly from (6.70)

$$PF = \frac{8899}{3 \times \frac{220}{\sqrt{3}} \times \sqrt{2} \times 34.5} = 0.48$$

This compares with a value $PF = (3/\pi) \cos \alpha = 0.827$ that would be obtained with passive inductive load.

Example 6.5
Calculate the voltage and current ratings required of the bridge thyristors in Example 6.4.

Solution. In a full wave bridge rectifier the peak inverse voltage (PIV) is the peak line-to-line voltage.

In this case

$$PIV = \sqrt{3} E_m = \sqrt{3} \times \frac{220}{\sqrt{3}} \times \sqrt{2} = 311 \text{ V}$$

6.3 Separately excited d.c. motor with rectified three-phase supply

The r.m.s. current of a thyristor is $1/\sqrt{2}$ times the r.m.s. supply line current.

In this case, neglecting current ripple,

$$I_T = \frac{1}{\sqrt{2}} I_s = \frac{1}{\sqrt{2}} 28.2 = 19.94 \text{ A}$$

Some degree of safety margin is desirable so that a reasonable choice of thyristor would be a rating of 400 V, 25 A.

Example 6.6

In the separately excited d.c. motor of Example 6.4 most of the armature circuit choke is removed leaving an armature inductance $L_a = 5$ mH. Calculate the speed and power factor for operation at $\alpha = 30°$ if the motor delivers its rated torque at all speeds.

Solution. Before the circuit equation can be applied it is necessary to know if the conduction is continuous or discontinuous.

At $\alpha = 30°$, $V_{L_{av}} = 257.3$ V, from Example 6.4. With constant torque load the armature current is also constant, equation (5.6), so that $I = 34.5$ A and $E_b = 251$ V.

The right-hand side (RHS) of the continuity criterion (6.62) is

$$\text{RHS} = \frac{E_b}{\sqrt{3}E_m} = \frac{251}{\sqrt{2} \times 220} = 0.807$$

Now

$$\omega L_a = \frac{2\pi \times 50 \times 5}{1000} = 1.57 \, \Omega$$

$$|Z_a| = \sqrt{0.2^2 + 1.57^2} = 1.58 \, \Omega$$

$$\phi_a = \tan^{-1} \frac{\omega L_a}{R_a} = \tan^{-1} 7.85 = 82.7°$$

$$\sin(\alpha - \phi_a) = \sin(-52.7°) = -0.795$$

$$\sin(\alpha - \phi_a + \pi/3) = \sin 7.3° = 0.127$$

$$c = \cot \phi_a = \cot 82.7° = 0.126$$

$c\pi/3 = 0.132$

$\varepsilon^{-c\pi/3} = 0.876$

The LHS of (6.62) is then

$$\frac{0.2}{1.57}\left[+0.127 + \frac{-0.795}{0.876-1}\right] = \frac{0.2}{1.57}[+0.127 + 6.41] = 0.837$$

It is seen that

LHS > RHS

so that the armature current is continuous. Therefore, as in Example 6.4, at $\alpha = 30°$,

$N = 866$ r.p.m.

$P_{\text{out}} = 8661$ W

The lowest armature current ripple frequency for a three-phase full wave bridge is the sixth harmonic. This has a harmonic impedance

$$|Z_a| = \sqrt{R_a^2 + (\omega_6 L_a)^2}$$
$$= \sqrt{(0.2)^2 + \left(\frac{6 \times 2\pi \times 50 \times 5}{1000}\right)^2}$$
$$= \sqrt{0.04 + (3\pi)^2} = 3\pi = 9.42\,\Omega$$

Similarly

$|Z_{a_{12}}| = 6\pi = 18.84\,\Omega$

For this bridge circuit, under continuous armature current, the harmonic armature voltage is given by (6.49) or (6.50)

$$V_{L_6} = \frac{3\sqrt{3}E_m}{\pi}\left(\frac{1}{7} + \frac{1}{5} - \frac{2\cos 60°}{7.5}\right)^{1/2}$$
$$= \frac{3}{\pi} \times \sqrt{3} \times \frac{220}{\sqrt{3}} \times \sqrt{2}\left(\frac{5+7-1}{35}\right)^{1/2} = 297.1\sqrt{\frac{11}{35}} = 166.6\,\text{V}$$

6.3 Separately excited d.c. motor with rectified three-phase supply

Therefore,

$$I_{L_6} = \frac{V_{L_6}}{Z_{a_6}} = \frac{166.6}{9.42} = 17.7\,\text{A}$$

For the 12th harmonic

$$V_{L_{12}} = \frac{3\sqrt{3}E_m}{\pi}\left(\frac{1}{13} + \frac{1}{11} - \frac{1}{13 \times 11}\right)^{1/2}$$

$$= 297.1\sqrt{\frac{23}{13 \times 11}} = 119.2\,\text{V}$$

Therefore,

$$I_{L_{12}} = \frac{119.2}{18.84} = 6.33\,\text{A}$$

An approximate value for the r.m.s. load current is obtained by taking the harmonic sum of the average value plus the two relevant harmonics,

$$I_L = \sqrt{I^2 + I_{L_6}^2 + I_{L_{12}}^2 + \ldots}$$

$$= \sqrt{(34.5)^2 + (17.7)^2 + (6.33)^2} = 39.29\,\text{A}$$

It is seen that I_L is only slightly greater than the average armature current. Use of the more correct value I_L in the power equation (6.33) gives

$$P_a = (39.29)^2 0.2 + 8661$$
$$= 308.7 + 8661 = 8970\,\text{W}$$

The corrected (approximate) efficiency is therefore

$$\eta = \frac{P_{\text{out}}}{P_a} = \frac{8661}{8970} = 96.6\%$$

(compared with 97.3% previously).

The corrected power factor becomes, from (6.70),

$$PF = \frac{8970}{3 \times \frac{220}{\sqrt{3}} \times \sqrt{2} \times 35.06} = 0.474$$

(compared with 0.48 previously).

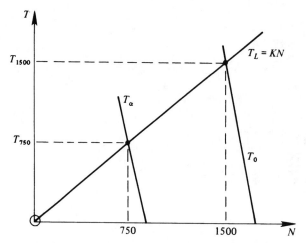

Fig. 6.11 Torque–speed characteristics of a fully-controlled converter d.c. drive (Example 6.7).

Example 6.7

The load torque characteristic of a certain drive application is proportional to drive speed. This torque is provided by a separately excited d.c. motor rated at 25 kW, 600 V, 1500 r.p.m. with armature resistance $R_a = 0.4\,\Omega$ and inductance $L_a = 6.3$ mH. Speed control is required from rated speed to one-half speed. The armature voltage is to be controlled by a three-phase, fully controlled thyristor bridge rectifier supplied from an ideal three-phase supply of 440 V, 60 Hz. Is supplementary inductance required to maintain continuous conduction?

Solution. An equivalent circuit of the bridge and motor is shown in Fig. 6.6. The torque–speed diagrams of the motor and load are shown in Fig. 6.11. At the upper speed of 1500 r.p.m., with thyristor firing-angle $\alpha = 0°$, the motor delivers torque T_0 and its rated power. From (5.9),

$$T_{1500} = \frac{25\,000}{\frac{1500}{60}2\pi} = 159.2\,\text{N m}$$

At the lower speed of 750 r.p.m. the torque T_{750} is seen, Fig. 6.11, to be one-half the value at 1500 r.p.m.

$$T_{750} = \frac{T_{1500}}{2} = 79.6\,\text{N m}$$

6.3 Separately excited d.c. motor with rectified three-phase supply

This is given by the intersection of the load line T_L with the motor line T_α at firing-angle α. It is seen that

$$P_{750} = T_{750} \times N_{750} = \frac{P_{1500}}{4} = \frac{25\,000}{4} = 6.25\,\text{kW}$$

At full speed, with $\alpha = 0°$, from (6.45), assuming continuous conduction,

$$V = V_{L_{av}} = \frac{3\sqrt{3}E_m}{\pi} \cos \alpha$$

$$= \frac{3}{\pi} \times \sqrt{3} \times \frac{440\sqrt{2}}{\sqrt{3}} \times 1 = 594.2\,\text{V}$$

The average armature current and back e.m.f. are given by the solution of two simultaneous equations.

From (6.34),

$$E_b I = 25\,000 \qquad (a)$$

From (6.46),

$$I = \frac{594.2 - E_b}{0.4} \qquad (b)$$

It is found that

$$I = \frac{-594.2 \pm 559.4}{0.8} = 43.5\,\text{A or } 1441\,\text{A}$$

The lower value is obviously the correct one.

From (a),

$$E_b = \frac{25\,000}{43.5} = 574.7\,\text{V}$$

In separately excited d.c. motor the field flux is constant. Neglecting armature reaction and saturation effects, the motor torque (5.6) is proportional to the average armature current. Therefore,

$$\frac{I_{750}}{I_{1500}} = \frac{T_{750}}{T_{1500}}$$

so that

$$I_{750} = \frac{1}{2} I_{1500} = 21.75 \text{ A}$$

Therefore

$$E_{b_{750}} = \frac{P_{750}}{I_{750}} = \frac{6250}{21.75} = 287.4 \text{ V}$$

If the current is continuous at 750 r.p.m. then, from (6.46),

$$21.75 = \frac{V_{L_{av}} - 287.4}{0.4}$$

which gives, at 750 r.p.m.,

$$V_{L_{av}} = 296.1 \text{ V}$$

This is one-half the value for operation at 1500 r.p.m.

In (6.45) it is seen that

$$296.1 = \frac{3}{\pi} \times \sqrt{3} \times \frac{400\sqrt{2}}{\sqrt{3}} \cos \alpha$$

which gives

$$\cos \alpha = 0.498$$
$$\alpha = 60°$$

The RHS of the armature current continuity criterion (6.62), for 750 r.p.m. operation, is

$$\text{RHS} = \frac{E_b}{\sqrt{3} E_m} = \frac{287.4}{\sqrt{2} \times 440} = 0.462$$

Now,

$$\phi_a = \tan^{-1} \frac{\omega L_a}{R_a} = \tan^{-1} \left(\frac{2\pi \times 60 \times 6.3}{0.4 \times 1000} \right) = 80.44°$$

$$\sin(\alpha - \phi_a) = \sin(-20.44°) = -0.349$$

$$\sin(\alpha - \phi_a + \pi/3) = \sin(39.56°) = 0.637$$

6.3 Separately excited d.c. motor with rectified three-phase supply

$$c = \cot \phi_a = \cot 80.44° = 0.168$$

$$c\frac{\pi}{3} = 0.176$$

$$\varepsilon^{-c\pi/3} - 0.84$$

The LHS of (6.62) is given by

$$\frac{0.4}{2.41}\left[0.637 + \frac{-0.349}{0.84 - 1}\right]$$

$$= \frac{0.4}{2.41}\left[0.637 + \frac{0.349}{0.16}\right]$$

$$= \frac{0.4}{2.41}[2.818] = 0.468$$

With the intrinsic armature inductance, at $\alpha = 60°$ and 750 r.p.m., it is seen that LHS > RHS in (6.62). The continuity criterion is satisfied and no additional inductance is needed.

Example 6.8
A double converter bridge circuit supplies power to a 560 V, 50 A separately excited d.c. motor with an armature resistance of 1.2 ohms. The voltage drop on the bridge thyristors is 20 V at rated motor current. Power is supplied by an ideal three-phase source with an r.m.s. line voltage of 415 V. Find the necessary firing-angle and motor back e.m.f. for:

(i) motoring operation at rated load current with motor terminal voltage of 500 V,

(ii) regeneration operation at rated load current with terminal voltage of 500 V,

(iii) motor plugged at rated load current with terminal voltage of 500 V and a current limiting resistor of 10 ohms.

Solution. The appropriate circuit diagram is given in Fig. 6.10 with corresponding steady-state torque–speed characteristics in Fig. 6.8. With a 415 V supply the maximum average voltage at the motor is

$$V = V_{L_{av}} = \frac{3\sqrt{3}E_m}{\pi} = \frac{3\sqrt{3}\,415\sqrt{2}}{\pi\sqrt{3}} = 560 \text{ V}$$

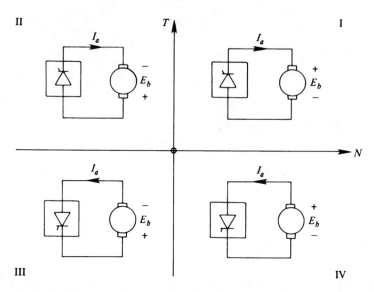

Fig. 6.12 Polarity and current direction for four-quadrant operation of a double converter drive

Due to the voltage drop on the conducting thyristors the armature circuit voltage equation (5.11(a)) is modified to

$$E_b + I_a R_a = V_{L_{av}} - V_T = V_{motor}$$

Operation in the four quadrants of the torque–speed plane is illustrated in Fig. 6.12.

(i) With $V_{motor} = 500$ V, $I_a = 55$ A, in quadrant I,

$$E_b = 500 - 50 \times 1.2 = 440 \text{ V}$$
$$V_{L_{av}} = 500 + 20 = 520 \text{ V}$$

From (6.45),

$$\cos \alpha = \frac{V_{L_{av}}}{3\sqrt{3} E_m} = \frac{520\pi}{3\sqrt{2} \times 415} = 0.928$$
$$\alpha = \cos^{-1} 0.928 = 21.9°$$

(ii) For regeneration operation the motor is delivering power to the supply

$$V_{motor} = -500\text{V}, \ I_a = 50 \text{ A, in quadrant II}$$

$$E_b = -500 - 60 = -560\,\text{V}$$

$$V_{L_{av}} = -500 + 20 = -480\,\text{V}$$

$$\therefore \cos\alpha = \frac{480\pi}{3\sqrt{2}\times 415} = -0.865$$

$$\alpha = \cos^{-1}(0.865) = 180 - 31 = 149°$$

(iii) If the motor is plugged through a 10 ohm resistance, operation still proceeds in quadrant II with an additional voltage drop of $10I_a$ in series with the converter.

$$V_{\text{motor}} = -500\,\text{V},\ I_s = 50\,\text{A}$$

$$E_b = -500 - 60 = -560\,\text{V}$$

$$V_{L_{av}} = -500 + 20 + 10\times 50 = 20\,\text{V}$$

$$\therefore \cos\alpha \frac{20\pi}{3\sqrt{2}\times 415} = 0.036$$

$$\alpha = \cos^{-1}(0.036) = 88°$$

6.4 PROBLEMS

Separately excited d.c. motor with rectified single-phase supply

6.1 Sketch the variation of the average output voltage versus firing-angle for (i) semi-converter, (ii) full converter with continuous conduction.

6.2 For a single-phase, fully controlled bridge rectifier with continuous load current, show that the peak amplitude of the nth harmonic load voltage is given by (6.1). Calculate and sketch the amplitudes of harmonics $n = 2, 3, 4, 5, 6$ when $\alpha = 0°$, $30°$, $60°$ and $90°$.

6.3 For the full-wave bridge rectifier of Problem 6.2 calculate the per-unit average load voltage and hence the per-unit r.m.s. load voltage at each of $\alpha = 0°$, $30°$, $60°$ and $90°$.

6.4 A single-phase supply rated at 240 V, 50 Hz supplies power to a separately excited d.c. motor load with $R_a = 1\,\Omega$, $L_a = 20\,\text{mH}$. At a certain speed of operation the back e.m.f. E_b is 100 V. When the thyristor firing-angle $\alpha = 30°$, calculate the average and r.m.s. values of the armature current, the power delivered to the motor and the power factor of operation.

6.5 Repeat Problem 6.4 if the armature circuit inductance is now reduced to 2 mH.

6.6 A separately excited d.c. motor is rated at 10 kW, 200 V, 1000 r.p.m. It is supplied with power from a fully controlled, single-phase, SCR converter with ideal supply 230 V, 50 Hz. The motor armature resistance is 0.25 Ω and a large inductor choke is included to ensure continuous armature current. If the motor constant $K\Phi$ has a value of 0.2 V/r.p.m. calculate the speed, torque, power factor and efficiency of operation if the motor delivers its rated power at maximum motor voltage. If the motor speed is unchanged when $\alpha = 30°$, calculate the new values of power factor and efficiency.

6.7 An ideal single-phase supply 230 V, 50 Hz, provides power for a fully controlled bridge converter supplying a separately excited d.c. motor rated at 1000 W, 200 V, 800 r.p.m. The motor has the armature parameters $R_a = 0.4\,\Omega$, $L_a = 5\,\text{mH}$. If the motor delivers rated power at rated speed, with $\alpha = 0°$, calculate the power factor and efficiency of operation. What is the speed of operation at $\alpha = 30°$, if rated torque is delivered?

6.8 For the motor of Problem 6.7 it is required to operate with full torque at two-thirds rated speed. Calculate the thyristor firing-angle required and the additional armature inductance (if any) required to maintain continuous current.

6.9 A fully controlled, single-phase, bridge rectifier with four identical SCRs transfers power to a separately excited d.c. motor with armature resistance R_a and inductance L_a. If the supply voltage is given by $e = E_m \sin \omega t$ show that the instantaneous armature current, at firing-angle α, is given by (6.13).

6.10 For the controlled d.c. motor of Problem 6.9 calculate the average armature current for continuous current operation by use of the integral method (6.19). Show that the value thus obtained is consistent with (6.9).

Separately excited d.c. motor with rectified three-phase supply

6.11 A separately excited d.c. motor is fed from a three-phase, fully controlled bridge rectifier. The peak value of the supply phase voltage is E_m and the motor armature has resistance R_a. Use the load voltage waveforms to show that the average continuous armature current at firing-angle α is given by (6.9).

6.12 For the three-phase d.c. motor drive of Problem 6.11 plot the per-unit values of the load voltage harmonics versus firing-angle for $n = 6, 12, 18$ and 24. How would the corresponding load current harmonics vary with firing-angle?

6.13 Plot comparative curves of the variation of the average load voltage, versus thyristor firing-angle, for the three-phase semi-converter, the full converter and the full converter with freewheel diode.

6.14 Show that the expression for the average armature current given in Problem 6.11 is obtained if the instantaneous current equation is substituted into the defining integral (of form (6.19)).

6.15 A separately excited d.c. motor rated at 50 kW, 300 V, 1000 r.p.m. is supplied with power from a fully controlled, three-phase bridge rectifier. The rectifier is energised from an ideal three-phase supply rated at 225 V, 50 Hz. The motor has an armature resistance of 0.15 Ω. Series inductance is included in the armature circuit to ensure continuous conduction. Speed adjustment is required in the range 700–1000 r.p.m. while delivering rated torque. Calculate the necessary retardation of the firing-angle and the range of current variation.

6.16 In Problem 6.15 the external armature inductance is removed from the circuit leaving $L_a = 27$ mH. Is it possible to operate at 700 r.p.m. with continuous armature current? If not, calculate the additional inductance needed to just maintain continuity of the current.

6.17 The 50 kW motor of Problem 6.15 is operated with supplementary armature circuit inductance such that the total $L_a = 10$ mH. It is required to deliver its rated torque at one-half rated speed. Calculate the power factor and efficiency at one-half rated speed.

6.18 Calculate the voltage and current ratings required of the bridge thyristors in Problem 6.17.

6.19 The load torque–speed characteristic of a 20 kW, pump motor is given by $T_L = KN^2$. The motor delivers its rated power at 1250 r.p.m. The 600 V motor is supplied from a three-phase supply of 440 V, 50 Hz via a fully controlled, six-pulse bridge rectifier. Speed control is required from full speed to one-half speed. Calculate the power factor and the efficiency of the extremes of operation.

7

Three-phase naturally commutated bridge circuit as a rectifier or inverter

7.1 THREE-PHASE CONTROLLED BRIDGE RECTIFIER WITH PASSIVE LOAD IMPEDANCE

With an ideal supply of zero impedance the three supply phase voltages for the circuit of Fig. 7.1 retain balanced sinusoidal form for any load condition. These voltages are defined by the equations

$$e_{aN} = E_m \sin \omega t \tag{7.1}$$

$$e_{bN} = E_m \sin(\omega t - 120°) \tag{7.2}$$

$$e_{cN} = E_m \sin(\omega t - 240°) \tag{7.3}$$

The corresponding line-to-line voltages at the supply point are

$$e_{ab} = e_{aN} + e_{Nb} = e_{aN} - e_{bN} = \sqrt{3} E_m \sin(\omega t + 30°) \tag{7.4}$$

$$e_{bc} = \sqrt{3} E_m \sin(\omega t - 90°) \tag{7.5}$$

$$e_{ca} = \sqrt{3} E_m \sin(\omega t - 210°) \tag{7.6}$$

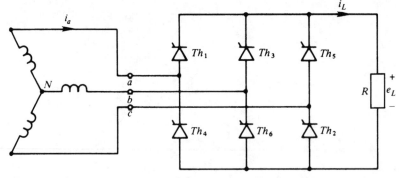

Fig. 7.1 Three-phase, full-wave, controlled bridge rectifier circuit, using SCR switches.

Waveforms of the supply voltages are given in Fig. 7.2 and Fig. 7.4.

The device numbering notation shown in the SCR bridge rectifier circuit of Fig. 7.1 is standard for the three-phase, full-wave bridge in both its rectifier and inverter modes of operation. To provide a current path from the supply side to the load side requires the simultaneous conduction of two appropriate switches. When one element of the upper group of SCRs and one of the lower group conducts, the corresponding line-to-line voltage is applied directly to the load. For example, if Th_1 and Th_6 conduct simultaneously then line voltage e_{ab} is applied across the load. Clearly there are some conduction patterns which are not permissible. If, for example, any two SCRs conduct simultaneously from either the top half or the bottom half of the bridge this would represent a short circuit on the a.c. supply. In order to provide load current of the maximum possible continuity and smoothness appropriate bridge switches must conduct in pairs sequentially, for conduction intervals up to 120° or $\pi/3$ rads of the supply voltage. The average load voltage and current are controlled by the firing-angle of the bridge SCRs, each measured from the crossover point of its respective phase voltages.

7.1.1 Resistive load and ideal supply

When the SCR firing-angle α is 0° the bridge operates like a diode rectifier circuit, with the waveforms given in Fig. 7.2. The corresponding conduction sequence of the circuit devices is given in Fig. 7.2(a), in which the upper tier represents the upper half of the bridge. Supply line current $i_a(\omega t)$, for the first cycle, Fig. 7.2(f), is made up of four separate components contributed by the four separate circuits shown in Fig. 7.3. An alternative representation of the device conduction pattern is shown in Fig. 7.2(b) in which the upper and lower tiers do not represent the upper and lower halves of the bridge but the pattern shows that the thyristors conduct in numerical order.

The circuit operation possesses two different modes of operation depending on the value of the firing-angle α. In the range $0 \leq \alpha \leq \pi/3$, the load voltage and current are continuous, Fig. 7.4, and an on-coming switch will instantly commutate an off-going device. In the range $\pi/3 \leq \alpha \leq 2\pi/3$, the load current becomes discontinuous because an off-going SCR extinguishes before the corresponding on-going SCR is fired. For resistive loads, with negligible supply reactance, both the load current and the supply current are always made up of parts of sinusoids, patterned from the line voltages. For all firing-angles the sequence order of switch conduction in the circuit of Fig. 7.1 is always that shown in Fig. 7.2(a). However, the onset of conduction

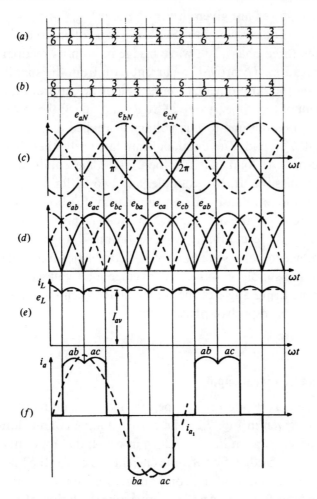

Fig. 7.2 Waveforms for three-phase, full-wave, controlled bridge rectifier circuit with resistive load. $\alpha = 0°$: (a), (b) switching sequence, (c) supply phase voltages, (d) supply line voltages, (e) load current and voltage, (f) supply line current.

is delayed, after the phase voltage crossover at $\omega t = 30°$, until the appropriate forward biased SCRs are gated and fired.

Consider operation at $\alpha = 30°$, for example. In Fig. 7.4 forward bias voltage occurs on switches Th_1 and Th_6 at $\omega t = 30°$. But if the firing-angle is set at $\alpha = 30°$ conduction via Th_1 and Th_6, Fig. 7.3(a), does not begin until $\omega t = \alpha + 30° = 60°$ and then continues for 60°. At $\omega t = \alpha + 90° = 120°$ the dominant line voltage is e_{ac}, SCR Th_6 is reverse biassed and conduction continues via the newly fired SCR Th_2, Fig. 7.3(b), for a further 60°. At $\omega t = 180°$, SCR Th_1 is commutated off by the switching in of Th_3 and line

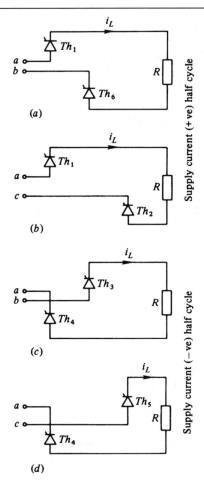

Fig. 7.3 Equivalent circuits of conduction for resistive load and ideal supply. $\alpha = 0°$. (a) $30° \leq \omega t \leq 90°$ (Th_1, Th_6 on), (b) $90° \leq \omega t \leq 150°$ (Th_1, Th_2 on), (c) $210° \leq \omega t \leq 270°$ (Th_3, Th_4 on), (d) $270° \leq \omega t \leq 330°$ (Th_5, Th_4 on).

current i_a, Fig. 7.5(c), becomes zero so that the load current path is provided by Th_2 and Th_3 for a further 60° interval. When $\omega t = 240°$, the dominant line voltage is e_{ba}. Firing switch Th_4 transfers the load current from Th_2, Fig. 7.3(c), and supply current resumes in phase 'a' in the opposite direction. After a further 60°, at $\omega t = 300°$, line voltage e_{ca} is dominant, Fig. 7.5(a), and the switching in of Th_5 causes the commutation of Th_3. Switches Th_4, Th_5 then provide the load current path which is fed from phase 'c' to phase 'a', as shown in Fig. 7.3(d).

240 Three-phase naturally commutated bridge circuit

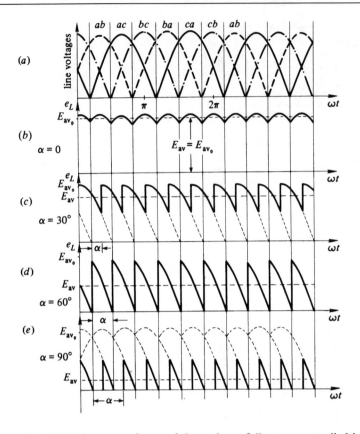

Fig. 7.4 Voltage waveforms of three-phase, full-wave, controlled bridge rectifier. R load: (*a*) supply line voltages, (*b*) load voltage, $\alpha = 0°$, (*c*) load voltage, $\alpha = 30°$, (*d*) load voltage, $\alpha = 60°$, (*e*) load voltage, $\alpha = 90°$.

7.1.1.1 Load-side quantities

The sequence of SCR firing creates the load voltage (and current) waveforms shown in Fig. 7.4. In Mode 1 operation, where $0 \le \alpha \le 60°$, the average voltage can be obtained by taking any 60° interval of $e_L(\omega t)$.

For $\alpha + 30° \le \omega t \le \alpha + 90°$,

$$e_L(\omega t) = \sqrt{3} E_m \sin(\omega t + 30°)\Big|_{\alpha+30°}^{\alpha+90°} \tag{7.7}$$

The average value of (7.7) in terms of peak phase voltage E_m is

7.1 Rectifier with passive load impedance

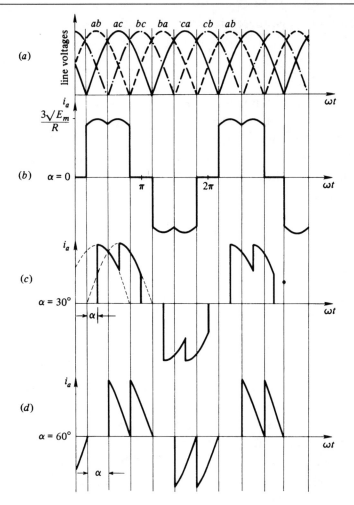

Fig. 7.5 Waveforms of three-phase, full-wave, controlled bridge rectifier. R load: (a) supply line voltages, (b) supply line current, $\alpha = 0°$, (c) supply line current, $\alpha = 30°$, (d) supply line current, $\alpha = 60°$.

$$
\begin{aligned}
E_{av} &= \frac{3}{\pi} \int_{\alpha+30°}^{\alpha+90°} \sqrt{3} E_m \sin(\omega t + 30°) \, d\omega t \\
&= \frac{3\sqrt{3}}{\pi} E_m \cos \alpha \\
&= E_{av_0} \cos \alpha
\end{aligned}
\tag{7.8}
$$

where

$$
E_{av_0} = \frac{3\sqrt{3}}{\pi} E_m = 1.654 E_m \tag{7.9}
$$

The average current I_{av} is, therefore, a function of α

$$I_{av} = \frac{E_{av}}{R} = \frac{E_{av_0}}{R} \cos\alpha \tag{7.10}$$

with resistive load the instantaneous load voltage is always positive. When the anode voltage of an SCR goes negative then extinction occurs. At a firing angle $\alpha > 60°$ the load voltage and current therefore become discontinuous, as shown in Fig. 7.4. This represents a different mode of operation from $\alpha < 60°$ and the load voltage is then described by the following equation: for $60° \leq \alpha \leq 120°$,

$$e_L(\omega t) = \sqrt{3}E_m \sin(\omega t + 30°)|_{\alpha+30°}^{150°} \tag{7.11}$$

The average value of (7.11) is

$$E_{av} = \frac{3\sqrt{3}E_m}{\pi}[1 + \cos(\alpha + 60°)] \tag{7.12}$$

When $\alpha = 60°$, (7.8) and (7.12) give identical results. At $\alpha = 120°$, the average load voltage becomes zero.

The waveforms of Fig. 7.4 show that the load voltage waveform has a repetition rate six times that of the phase voltage. This means that the lowest ripple frequency is six times the fundamental frequency. If a Fourier analysis is performed on the load voltage waveform the two lowest order harmonics are the d.c. level (i.e. the average value) followed by the sixth harmonic.

Load power dissipation can be found from the r.m.s. (not the average) load current. The r.m.s. or effective load current, I_L, is defined as

$$I_L = \sqrt{\frac{1}{2\pi}\int_0^{2\pi} i_L^2(\omega t)\, d\omega t} \tag{7.13}$$

where $i_L = e_L/R$ from (7.7) or (7.11).

Comparing the waveforms of the supply and load currents at a given firing-angle, Figs. 7.4, 7.5, one would anticipate that $I_L > I_a$ because $i_L(\omega t)$ has a greater area under the curve than does $i_a(\omega t)$ and therefore $i_L^2(\omega t)$ is likely to be greater than $i_a^2(\omega t)$. The substitution of (7.7) or (7.11) respectively into (7.13) gives

$$I_L|_{0\leq\alpha\leq 60°} = \frac{\sqrt{3}E_m}{2R}\sqrt{\frac{2\pi + 3\sqrt{3}\cos 2\alpha}{\pi}} \tag{7.14}$$

7.1 Rectifier with passive load impedance 243

$$I_L|_{60 \leq \alpha \leq 120°} = \frac{\sqrt{3}E_M}{2R} \sqrt{\frac{4\pi - 6\alpha - 3\sin(2\alpha - 60°)}{\pi}} \quad (7.15)$$

The ideal output current from a rectifier circuit is a unidirectional current of constant value (as shown in Fig. 7.8(*b*)). A measure of the nonconstancy or ripple of any actual load current is defined by the ripple factor,

$$\text{ripple factor} = \sqrt{\left(\frac{I_L}{I_{av}}\right)^2 - 1} \quad (7.16)$$

With an ideal $i_L(\omega t)$ waveform, $I_L = I_{av}$ and the ripple factor is zero.

Power dissipation in the bridge circuit of Fig. 7.1 is assumed to occur entirely in the load resistor. This may be obtained from the r.m.s. (not the average) load current

$$P_L = I_L^2 R \quad (7.17)$$

Combining equations (7.14), (7.15) with (7.17) gives

$$P_L|_{0 \leq \alpha \leq 0°} = \frac{3E_m^2}{4\pi R}(2\pi + 3\sqrt{3}\cos 2\alpha) \quad (7.18)$$

$$P_L|_{60 \leq \alpha \leq 120°} = \frac{3E_m^2}{4\pi R}[4\pi - 6\alpha - 3\sin(2\alpha - 60°)] \quad (7.19)$$

The load-side properties of the bridge are summarised in Table 7.1.

7.1.1.2 Supply-side quantities

Waveforms of the currents on the supply side of the bridge are shown in Fig. 7.5 for Mode 1 operation. The instantaneous supply currents for the two modes of operation are defined by:

For $0 \leq \alpha \leq 60°$,

$$i_a(\omega t) = \frac{\sqrt{3}E_m}{R}\sin(\omega t + 30°)|_{\alpha+30°}^{\alpha+90°} \cdots + \frac{\sqrt{3}E_m}{R}\sin(\omega t - 30°)|_{\alpha+90°}^{\alpha+150°} \cdots \quad (7.20)$$

For $60° \leq \alpha \leq 120°$,

$$i_a(\omega t) = \frac{\sqrt{3}E_m}{R}\sin(\omega t + 30°)|_{\alpha+30°}^{150°} + \frac{\sqrt{3}E_m}{R}\sin(\omega t - 30°)|_{\alpha+90°}^{120°} \quad (7.21)$$

The r.m.s. values of the supply line currents may be obtained via the defining integral (7.13). Substituting (7.20), (7.21) into the form of (7.13) gives

Table 7.1 *Three-phase, full-wave, controlled bridge rectifier with ideal supply: load side properties.*

	Resistive load	Highly inductive load
Instantaneous load voltage	$0 \leqslant \alpha \leqslant 60°$ $\sqrt{3}\, E_m \sin(\omega t + 30°)\vert_{\alpha+30}^{\alpha+90}$ $60° \leqslant \alpha \leqslant 120°$ $\sqrt{3}\, E_m \sin(\omega t + 30)\vert_{\alpha+30}^{150°}$	$E_{avo} \cos \alpha$
Average load voltage	$0 \leqslant \alpha \leqslant 60°$ $E_{avo} \cos \alpha$ $60° \leqslant \alpha \leqslant 120°$ $E_{avo}[1 + \cos(\alpha + 60°)]$	$E_{avo} \cos \alpha$
R.m.s. load current	$0 \leqslant \alpha \leqslant 60°$ $\dfrac{\sqrt{3}\, E_m}{2R}\sqrt{\dfrac{2\pi + 3\sqrt{3}\cos 2\alpha}{\pi}}$ $60° \leqslant \alpha \leqslant 120°$ $\dfrac{\sqrt{3}\, E_m}{2R}\sqrt{\dfrac{4\pi - 6\alpha - 3\sin(2\alpha - 60°)}{\pi}}$	$\dfrac{E_{avo}}{R}\cos\alpha$
Load power	$0 \leqslant \alpha \leqslant 60°$ $\dfrac{3 E_m^2}{4\pi R}(2\pi + 3\sqrt{3}\cos 2\alpha)$ $60° \leqslant \alpha \leqslant 120°$ $\dfrac{3 E_m^2}{4\pi R}[4\pi - 6\alpha - 3\sin(2\alpha - 60°)]$	$\dfrac{E_{avo}^2}{R}\cos^2\alpha$

$$I_a\vert_{0 \leq \alpha \leq 60°} = \frac{E_m}{\sqrt{2}R}\sqrt{\frac{2\pi + 3\sqrt{3}\cos 2\alpha}{\pi}} \qquad (7.22)$$

$$I_a\vert_{60 \leq \alpha \leq 120°} = \frac{E_m}{\sqrt{2}R}\sqrt{\frac{4\pi - 6\alpha - 3\sin(2\alpha - 60°)}{\pi}} \qquad (7.23)$$

At $\alpha = 60°$, (7.22) and (7.23) are found to be identical.

Comparison of the r.m.s. supply and load currents gives, for both modes of operation,

$$I_L = \sqrt{\frac{3}{2}} I_a \qquad (7.24)$$

7.1.1.3 Operating power factor

The power dissipated at the load must be equal to the power at the supply point. This provides a method of calculating the operating power factor

$$PF = \frac{P_L}{3E_a I_a} \qquad (7.25)$$

Substituting for P_L ((7.18) or (7.19)) and I_a ((7.22) or (7.23)) into (7.25), noting that $E_a = E_m/\sqrt{2}$, gives

$$PF|_{0 \leq \alpha \leq 60°} = \sqrt{\frac{2\pi + 3\sqrt{3}\cos 2\alpha}{4\pi}} \qquad (7.26)$$

$$PF|_{60° \leq \alpha \leq 120°} = \sqrt{\frac{4\pi - 6\alpha - 3\sin(2\alpha - 60°)}{4\pi}} \qquad (7.27)$$

When $\alpha = 0$, (7.26) has the value

$$PF|_{\alpha=0} = \sqrt{\frac{2\pi + 3\sqrt{3}}{4\pi}} = 0.956 \qquad (7.28)$$

which is a well-known result for the uncontrolled (diode) bridge.

The power factor of the three-phase rectifier bridge circuit, as for any circuit, linear or nonlinear, with sinusoidal supply voltages, can be represented as the product of a 'distortion factor' and a 'displacement factor'. The distortion factor is largely related to load impedance nonlinearity; in this case, the switching action of the SCRs. The displacement factor is the cosine of the phase-angle between the fundamental components of the supply voltage and current. This angle is partly due to the load impedance phase-angle but mainly due here to the delay angle of the current, introduced by the switching.

Both the displacement factor and the distortion factor are functions of the fundamental component of the supply current. This is calculated in terms of the Fourier coefficients a_1, b_1, quoted from the Appendix for the order $n = 1$.

$$a_1 = \frac{1}{\pi} \int_0^{2\pi} i(\omega t) \cos \omega t \, d\omega t \qquad (A.9)$$

$$b_1 = \frac{1}{\pi}\int_0^{2\pi} i(\omega t)\sin\omega t\, d\omega t \tag{A.10}$$

Expressions for the coefficients a_1, b_1 are given in Table 7.2.
The displacement factor and distortion factor are given by

$$\text{displacement factor of supply current} = \cos\psi_1 = \cos\left(\tan^{-1}\frac{a_1}{b_1}\right) \tag{7.29}$$

$$\text{distortion factor of supply current} = \frac{I_{a_1}}{I_a} = \frac{I_1}{I} \tag{7.30}$$

where

$$I_{a_1} = c_1/\sqrt{2} = \frac{\sqrt{a_1^2 + b_1^2}}{\sqrt{2}} \tag{7.31}$$

Expressions for the displacement factor and distortion factor, for both modes of operation, are also given in Table 7.2. The product of these is seen to satisfy the defining relation

$$PF = (\text{displacement factor})(\text{distortion factor}) \tag{7.32}$$

7.1.1.4 Shunt capacitor compensation

Some degree of power factor correction can be obtained by connecting equal lossless capacitors C across the supply terminals, Fig. 7.6. The bridge voltages and currents are unchanged and so is the circuit power dissipation. The capacitor current $i_c(\omega t)$ is a continuous function unaffected by the SCR switching. In phase 'a', for example, capacitor current $i_{ca}(\omega t)$ is given by

$$i_{ca}(\omega t) = \frac{E_m}{X_c}\sin(\omega t + 90°) = \frac{E_m}{X_c}\cos\omega t \tag{7.33}$$

where $X_c = 1/2\pi fC$.
The instantaneous supply current $i_{s_a}(\omega t)$ in phase 'a' is now

$$i_{s_a}(\omega t) = i_a(\omega t) + i_{c_a}(\omega t) \tag{7.34}$$

Therefore,

$$i_{s_a}|_{0\leq\alpha\leq 60°} = \frac{E_m}{X_c}\cos\omega t + \frac{\sqrt{3}E_m}{R}\sin(\omega t + 30°)\Big|_{\alpha+30°}^{\alpha+90°}$$
$$+ \frac{\sqrt{3}E_m}{R}\sin(\omega t - 30°)\Big|_{\alpha+90°}^{210°} \tag{7.35}$$

7.1 Rectifier with passive load impedance

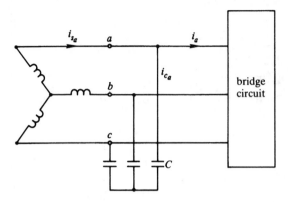

Fig. 7.6 Three-phase bridge circuit with supply side capacitors.

$$i_{s_a}|_{60°\leq\alpha\leq120°} = \frac{E_m}{X_c}\cos\omega t + \frac{\sqrt{3}E_m}{R}\sin(\omega t + 30)|_{\alpha+30°}^{150°}$$
$$+ \frac{\sqrt{3}E_m}{R}\sin(\omega t - 30°)|_{\alpha+90°}^{210°} \quad (7.36)$$

The substitution of (7.35), (7.36) respectively into (7.13) gives modified expressions for the r.m.s. supply current,

$$I_{s_a}|_{0\leq\omega t\leq60°} = \frac{E_m}{\sqrt{2}R}\sqrt{\frac{R^2}{X_c^2} + \frac{2\pi + 3\sqrt{3}\cos 2\alpha}{\pi} - \frac{3\sqrt{3}R}{\pi X_c}\sin 2\alpha} \quad (7.37)$$

$$I_{s_a}|_{60°\leq\alpha\leq120°}$$
$$= \frac{E_m}{\sqrt{2}R}\sqrt{\frac{R^2}{X_c^2} + \frac{4\pi - 6\alpha - 3\sin(2\alpha - 60°)}{\pi} - \frac{3R}{\pi X_c}[\sin(2\alpha + 30°) + 1]}$$
$$(7.38)$$

Since the system voltages and power are unchanged by the presence of the capacitor the power factor will be improved if the r.m.s. supply current with the capacitor is reduced below the level of the bridge r.m.s. current (which is the supply r.m.s. current in the absence of the capacitor).

If the compensated power factor is denoted by PF_c the ratio of compensated to uncompensated power factor is found to be

Table 7.2 Three-phase, full-wave, controlled bridge rectifier with ideal supply: supply-side properties.

	Resistive load		Highly inductive load
	Mode 1 $0 \leq \alpha \leq 60°$	Mode 2 $0 \leq \alpha \leq 120°$	
Fourier coefficients of (a_1) supply current	$-\dfrac{3\sqrt{3}E_m}{2\pi R}\sin 2\alpha$	$-\dfrac{3E_m}{2\pi R}[1+\cos(2\alpha-60)]$	$-\dfrac{9}{\pi^2}\dfrac{E_m}{R}\sin 2\alpha$
(b_1)	$\dfrac{E_m}{2\pi R}(2\pi+3\sqrt{3}\cos 2\alpha)$	$\dfrac{E_m}{2\pi R}[4\pi-6-3\sin(2\alpha-60)]$	$\dfrac{9}{\pi^2}\dfrac{E_m}{R}(1+\cos 2\alpha)$
Fundamental supply current (I_1)	$\dfrac{1}{\sqrt{2}}\sqrt{a_1^2+b_1^2}$	$\dfrac{1}{\sqrt{2}}\sqrt{a_1^2+b_1^2}$	$\dfrac{1}{\sqrt{2}}\sqrt{a_1^2+b_1^2}$
r.m.s. current (I)	$\dfrac{E_m}{\sqrt{2}R}\sqrt{\dfrac{2\pi+3\sqrt{3}\cos 2\alpha}{\pi}}$	$\dfrac{E_m}{\sqrt{2}R}\sqrt{\dfrac{4\pi-6\alpha-3\sin(2\alpha-60)}{\pi}}$	$\dfrac{3\sqrt{2}E_m\cos\alpha}{\pi R}$
Displacement factor ($\cos\psi_1$)	$\dfrac{2\pi+3\sqrt{3}\cos 2\alpha}{\sqrt{27+4\pi^2+12\sqrt{3}\pi\cos 2\alpha}}$	$\dfrac{4\pi-6\alpha-3\sin(2\alpha-60)}{\sqrt{9[1+\cos(2\alpha-60)]^2+[4\pi-6\alpha-3\sin(2\alpha-60)]^2}}$	$\cos\alpha$
Distortion factor (I_1/I)	$\sqrt{\dfrac{27+4\pi^2+12\sqrt{3}\pi\cos 2\alpha}{4\pi(2\pi+3\sqrt{3}\cos 2\alpha)}}$	$\sqrt{\dfrac{9[1+\cos(2\alpha-60)]^2+[4\pi-6\alpha-3\sin(2\alpha-60)]^2}{4\pi[4\pi-6\alpha-3\sin(2\alpha-60)]}}$	$\dfrac{3}{\pi}$
Power factor	$\sqrt{\dfrac{2\pi+3\sqrt{3}\cos 2\alpha}{4\pi}}$	$\sqrt{\dfrac{4\pi-6\alpha-3\sin(2\alpha-60)}{4\pi}}$	$\dfrac{3}{\pi}\cos\alpha$

$$\left.\frac{PF_c}{PF}\right|_{0\le\alpha\le 60°} = \sqrt{\frac{\frac{1}{\pi}(2\pi + 3\sqrt{3}\cos 2\alpha)}{\frac{R^2}{X_c^2} + \frac{1}{\pi}(2\pi + 3\sqrt{3}\cos 2\alpha) - \frac{3\sqrt{3}R}{\pi X_c}\sin 2\alpha}} \quad (7.39)$$

$$= \frac{I_a}{I_{S_a}}$$

$$\left.\frac{PF_c}{PF}\right|_{60°\le\alpha\le 120°}$$

$$= \sqrt{\frac{\frac{1}{\pi}[4\pi - 6\alpha - 3\sin(2\alpha - 60°)]}{\frac{R^2}{X_c^2} + \frac{1}{\pi}[4\pi - 6\alpha - 3\sin(2\alpha - 60°)] - \frac{3R}{\pi X_c}[\sin(2\alpha + 30) + 1]}}$$

$$(7.40)$$

The ratio PF_c/PF will be greater than unity, indicating that power factor improvement has occurred, when the following inequalities are true:
For $0 \le \alpha \le 60°$,

$$\frac{R}{X_c}\left(\frac{R}{X_c} - \frac{3\sqrt{3}}{\pi}\sin 2\alpha\right) < 0 \quad (7.41)$$

For $60° \le \alpha \le 120°$,

$$\frac{R}{X_c}\left[\frac{R}{X_c} - \frac{3}{\pi}\{\sin(2\alpha + 30) + 1\}\right] < 0 \quad (7.42)$$

For the limiting values of firing-angle α, being zero in (7.41) and 120° in (7.42) it is found that R/X_c would need to be negative to cause power factor improvement. In other words, when $\alpha = 0$ the use of capacitance does not give improvement but actually makes the power factor worse.

The use of supply point capacitance aims to reduce the displacement angle ψ_{s_1} to zero so that displacement factor $\cos\psi_{s_1} = 1.0$, which is the highest realisable value. From (7.29) it is seen that $\psi_{s_1} = 0$ when $a_{s_1} = 0$. If (7.35), (7.36) are substituted into (A.9) it is found that

$$\left.a_{s_1}\right|_{0\le\alpha\le 60°} = \frac{E_m}{X_c} - \frac{3\sqrt{3}E_m}{2\pi R}\sin 2\alpha \quad (7.43)$$

$$\left.a_{s_1}\right|_{60°\le\alpha\le 120°} = \frac{E_m}{X_c} - \frac{3E_m}{2\pi R}[1 + \cos(2\alpha - 60)] \quad (7.44)$$

Unity displacement factor and maximum power factor compensation are therefore obtained by equating (7.43), (7.44) to zero:
For $0 \le \alpha \le 60°$,

$$\frac{R}{X_c} - \frac{3\sqrt{3}}{2\pi} \sin 2\alpha = 0 \qquad (7.45)$$

For $60° \le \alpha \le 120°$,

$$\frac{R}{X_c} - \frac{3}{2\pi}[1 + \cos(2\alpha - 60°)] = 0 \qquad (7.46)$$

When the conditions of (7.45), (7.46) are satisfied the power factor has attained its maximum value realisable by capacitor compensation:
For $0 \le \alpha \le 60°$,

$$PF_{c|\cos\psi=1} = \frac{2\pi + 3\sqrt{3}\cos 2\alpha}{\sqrt{4\pi(2\pi + 3\sqrt{3}\cos 2\alpha) - 27\sin^2 2\alpha}} \qquad (7.47)$$

For $60° \le \alpha \le 120°$,

$$PF_{c|\cos\psi=1}$$
$$= \frac{4\pi - 6\alpha - 3\sin(2\alpha - 60°)}{\sqrt{4\pi[4\pi - 6\alpha - 3\sin(2\alpha - 60°)] - 9[1 + \cos(2\alpha - 60°)]^2}} \qquad (7.48)$$

The degree of power factor improvement realisable by capacitor compensation is zero at $\alpha = 0$ and is small for small firing-angles. For firing-angles in the mid-range $30° \le \alpha \le 60°$ significant improvement is possible. Note that the criteria of (7.45), (7.46) are not the same as the criteria of (7.41), (7.42) because they do not refer to the same constraint.

7.1.1.5 Worked examples

Example 7.1
A three-phase full-wave controlled bridge has a resistive load, $R = 100\,\Omega$. The three-phase supply 415 V, 50 Hz may be considered ideal. Calculate the average load voltage and the power dissipation at (i) $\alpha = 45°$, (ii) $\alpha = 90°$.

Solution. At $\alpha = 45°$, from (7.8),

$$E_{av} = \frac{3\sqrt{3}E_m}{\pi} \cos \alpha$$

where E_m is the peak value of the phase voltage. Assuming that 415 V represents the r.m.s. value of the line voltage,

$$E_m = 415 \frac{\sqrt{2}}{\sqrt{3}}$$

Therefore,

$$E_{av} = \frac{3\sqrt{3}}{\pi} \times 415 \frac{\sqrt{2}}{\sqrt{3}} \times \frac{1}{\sqrt{2}} = 396.3 \text{ V}$$

The power is given by (7.18),

$$P_L = \frac{3E_m^2}{4\pi R}(2\pi + 3\sqrt{3}\cos 2\alpha)$$
$$= \frac{3}{4\pi} \times \frac{415^2}{100} \times \frac{2}{3} \times 2\pi$$
$$= 1722 \text{ W}$$

At $\alpha = 90° = \pi/2$, from (7.12),

$$E_{av} = \frac{3\sqrt{3}E_m}{\pi}[1 + \cos(\alpha + 60°)] = \frac{3\sqrt{3}}{\pi} \times 415 \times \frac{\sqrt{2}}{\sqrt{3}}\left(1 - \frac{\sqrt{3}}{2}\right) = 75.1 \text{ V}$$

The power is now given by (7.19),

$$P_L = \frac{3E_m^2}{4\pi R}[4\pi - 6\alpha - 3\sin(2\alpha - 60°)]$$
$$= \frac{3}{4\pi} \times \frac{415^2}{100} \times \frac{2}{3}\left(4\pi - 3\pi - \frac{3\sqrt{3}}{2}\right)$$
$$= \frac{415^2}{200\pi}(\pi - 2.589)$$
$$= 149 \text{ W}$$

Example 7.2
For a three-phase full-wave controlled bridge with resistive load and ideal supply, obtain a value for the load current ripple when $\alpha = 60°$, compared with uncontrolled operation.

Solution. The r.m.s. values of the load current in the two modes of operation are given by (7.14), (7.15). The average values are given in (7.8), (7.10) and (7.12). Taking the ratio I_L/I_{av} it is found that:
For $0 \leq \alpha \leq 60°$,

$$\frac{I_L}{I_{av}} = \frac{\frac{\sqrt{3}E_m}{2R}\sqrt{\frac{2\pi + 3\sqrt{3}\cos 2\alpha}{\pi}}}{\frac{3\sqrt{3}E_m}{\pi R}\cos\alpha}$$

$$= \frac{\pi}{6}\sqrt{\frac{2\pi + 3\sqrt{3}\cos 2\alpha}{\pi \cos^2 \alpha}} \qquad (i)$$

For $60 \leq \alpha \leq 120°$,

$$\frac{I_L}{I_{av}} = \frac{\frac{\sqrt{3}E_m}{2R}\sqrt{\frac{4\pi - 6\alpha - 3\sin(2\alpha - 60)}{\pi}}}{\frac{3\sqrt{3}E_m}{\pi R}[1 + \cos(\alpha + 60°)]}$$

$$= \frac{\pi}{6}\sqrt{\frac{4\pi - 6\alpha - 3\sin(2\alpha - 60°)}{\pi[1 + \cos(\alpha + 60°)]^2}} \qquad (ii)$$

From each of the relations (i), (ii) above, at $\alpha = 60°$, it is found that

$$\frac{I_L}{I_{av}} = 1.134$$

The ripple factor is, from (7.10),

$$RF = \sqrt{\left(\frac{I_L}{I_{av}}\right)^2 - 1}$$

$$= 0.535$$

From relation (i) above, at $\alpha = 0$,

$$\frac{I_L}{I_{av}} = \frac{\pi}{6}\sqrt{\frac{2\pi + 3\sqrt{3}}{\pi}}$$

$$= \frac{\pi}{6} \times 1.912$$

$$= 1.0$$

The ripple factor at $\alpha = 0$, is therefore, very small but it becomes significant as the firing-angle is retarded.

Example 7.3
Calculate the operating power factor for the three-phase full-wave bridge of Example 7.1 at (i) $\alpha = 45°$ and (ii) $\alpha = 90°$. If the maximum possible compensation by capacitance correction is realised, calculate the new values of power factor and the values of capacitance required.

7.1 Rectifier with passive load impedance

Solution. At $\alpha = 45°$, from (7.26),

$$PF = \sqrt{\frac{2\pi + 3\sqrt{3}\cos 90°}{4\pi}}$$

$$= \sqrt{\tfrac{1}{2}} = 0.707$$

At $\alpha = 90° = \pi/2$, from (7.27),

$$PF = \sqrt{\frac{4\pi - 3\pi - 3\sin 120°}{4\pi}}$$

$$= \sqrt{\frac{\pi - \frac{3\sqrt{3}}{2}}{4\pi}} = 0.2$$

If the maximum realisable compensation is achieved the power factor is then given by (7.47), (7.48).

At $\alpha = 45°$, from (7.47),

$$PF_c = \frac{2\pi + 3\sqrt{3}\cos 90°}{\sqrt{4\pi(2\pi + 3\sqrt{3}\cos 90°) - 27\sin^2 90°}}$$

$$= \frac{2\pi}{\sqrt{8\pi^2 - 27}} = \frac{2\pi}{7.21} = 0.87$$

At $\alpha = 90°$, from (7.48),

$$PF_c = \frac{4\pi - 3\pi - 3\sin 120°}{\sqrt{4\pi(4\pi - 3\pi - 3\sin 120°) - 9(1 + \cos 120°)^2}}$$

$$= \frac{\pi - \frac{3\sqrt{3}}{2}}{\sqrt{4\pi\left(\pi - \frac{3\sqrt{3}}{2}\right) - 9\left(1 - \tfrac{1}{2}\right)^2}}$$

$$= \frac{0.544}{\sqrt{6.83 - 2.25}}$$

$$= 0.254$$

The criteria for zero displacement factor are given in (7.45), (7.46).

At $\alpha = 45°$, from (7.45),

$$\frac{1}{X_c} = 2\pi f C = \frac{3\sqrt{3}}{2\pi R}\sin 2\alpha$$

$$C = \frac{1}{2\pi \times 50} \times \frac{3\sqrt{3}}{2\pi \times 100} = 52.5\,\mu\text{F}$$

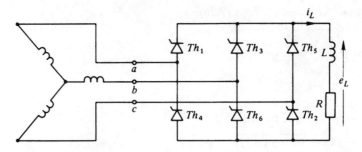

Fig. 7.7 Three-phase, full-wave, controlled bridge rectifier circuit with series R–L load.

At $\alpha = 90°$, from (7.46),

$$\frac{1}{X_c} = \frac{3}{2\pi R}[1 + \cos(2\alpha - 60°)]$$

$$C = \frac{3}{2 \times 50} \times \frac{3}{2\pi \times 100}(1 - \tfrac{1}{2})$$

$$= 15.2\,\mu F$$

7.1.2 Highly inductive load and ideal supply

7.1.2.1 Load-side quantities

The three-phase controlled bridge rectifier is most commonly used in applications where the load impedance is highly inductive. The load inductance is often introduced in the form of a large choke in series with the load resistor, Fig. 7.7. If the load-side inductance smooths the load current to make it, very nearly, a pure direct current as shown in Fig. 7.8(b), then

$$i_L(\omega t) = I_{av} = I_L = I_m \qquad (7.49)$$

With a smooth load current there is zero average voltage on the smoothing inductor and the average load voltage falls entirely on the load resistor so that (7.10) remains true. The pattern of the load current and supply currents is shown in Fig. 7.8 for firing-angles up to $\alpha = 60°$. Unlike the case with resistive load, the load current is continuous for all values of α in the control range and only one mode of operation occurs. The average voltage, for all firing-angles, is identical to that derived in (7.8) with the corresponding average current in (7.10). With constant load resistance the heights of the current pulses reduce as the firing-angle is retarded because the a.c. driving voltage is proportional to $\cos \alpha$, Table 7.1.

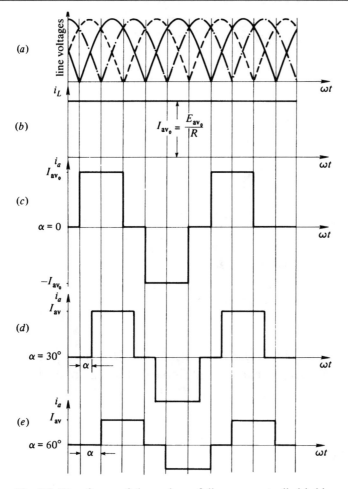

Fig. 7.8 Waveforms of three-phase, full-wave, controlled bridge rectifier with highly inductive load and ideal supply: (*a*) supply line voltages, (b) load current, $\alpha = 0°$, (c) supply line current i_a, $\alpha = 0°$, (d) supply line current, $\alpha = 30°$, (e) supply line current, $\alpha = 60°$.

It is seen from (7.10) that the average load current becomes zero at $\alpha = 90°$. The controlled range with highly inductive load is therefore smaller than with resistive load, as shown in Fig. 7.9. With a smooth load current there is no ripple component at all and the current ripple factor has the ideal value of zero.

For $\alpha \leq 60°$, the instantaneous load voltage, with highly inductive load, is the same as for resistive load. At $\alpha = 75°$, $e_L(\omega t)$ contains a small negative component for part of the cycle. When $\alpha = 90°$, the instantaneous load voltage has positive segments identical to those in Fig. 7.4(e) but these are

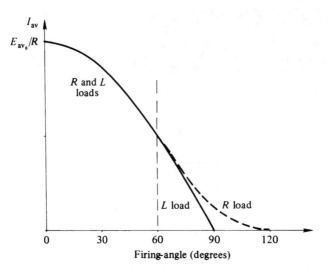

Fig. 7.9 Average load current versus SCR firing-angle for three-phase, full-wave, controlled bridge rectifier circuit.

balanced by corresponding negative segments, Fig. 9.2(d), to give an average value of zero. Although the load current ripple factor is zero, the load voltage ripple factor is determined by the ratio E_L/E_{av}. From (7.8) and (7.14), with $\alpha \leq 60°$,

$$\frac{E_L}{E_{av}} = \frac{\pi}{6}\sqrt{\frac{2 + 3\sqrt{3}\cos 2\alpha}{\pi \cos^2 \alpha}} \qquad (7.50)$$

The load power dissipation is proportional to the square of the load r.m.s. current and therefore, substituting (7.10) into (7.17)

$$P_L = I_L^2 R = I_{av}^2 R = \left(\frac{E_{av_0}}{R}\cos\alpha\right)^2 R$$
$$= \frac{E_{av_0}^2}{R}\cos^2\alpha = \frac{27 E_m^2}{\pi^2 R}\cos^2\alpha \qquad (7.51)$$

The load-side properties are summarised in Table 7.1, p. 244.

7.1.2.2 Supply-side quantities

The supply current $i_a(\omega t)$ shown in Fig. 7.8 is defined by the equation

$$i_a(\omega t) = \frac{E_{av_0}}{R}\cos\alpha \Big|_{\alpha+30°}^{\alpha+150°} - \frac{E_{av_0}}{R}\cos\alpha \Big|_{\alpha+210°}^{\alpha+330°} \qquad (7.52)$$

Since the r.m.s. values of the negative and positive parts of the wave are identical the r.m.s. supply current is given by

$$I_a = \sqrt{\frac{1}{\pi}\int_{\alpha+30°}^{\alpha+150°}\left(\frac{E_{av_0}}{R}\cos\alpha\right)^2 d\omega t}$$

$$= \frac{E_{av_0}}{R}\cos\alpha\sqrt{\frac{1}{\pi}[\omega t]_{\alpha+30°}^{\alpha+150°}}$$

$$= \sqrt{\frac{2}{3}}\frac{E_{av_0}}{R}\cos\alpha \qquad (7.53)$$

$$= \frac{3\sqrt{2}E_m\cos\alpha}{\pi R}$$

$$= \sqrt{\frac{2}{3}}I_{av}$$

The value in (7.53) is found to be $\sqrt{2}$ times the corresponding value for a half-wave rectifier, and is identical to (7.24) for the case of resistive load.

The operating power factor of the bridge can be obtained by substituting (7.51) and (7.52) into (7.25), noting that $E_a = E_m/\sqrt{2}$.

$$PF = \frac{P_L}{3E_a I_a}$$

$$= \frac{\frac{27 E_m^2}{\pi^2 R}\cos^2\alpha}{3\frac{E_m}{\sqrt{2}}\frac{3\sqrt{2}E_m\cos\alpha}{\pi R}} \qquad (7.54)$$

$$= \frac{3}{\pi}\cos\alpha$$

The power factor is also found to be $\sqrt{2}$ times the value for a three-phase, half-wave, controlled bridge and has the well-known value of $3/\pi$ or 0.955 for $\alpha = 0°$ (or diode bridge) operation.

A Fourier analysis of the supply current $i_a(\omega t)$ shows that the coefficients a_1, b_1 below are valid for the fundamental component:

$$a_1 = \frac{1}{\pi}\int_0^{2\pi} i_a(\omega t)\cos\omega t\, d\omega t$$

$$= \frac{2}{\pi} \frac{E_{av_0}}{R} \cos\alpha [\sin\omega t]_{\alpha+30°}^{\alpha+150°}$$

$$= -\frac{2\sqrt{3}}{\pi} \frac{E_{av_0}}{R} \sin\alpha \cos\alpha \tag{7.55}$$

$$= -\frac{9}{\pi^2} \frac{E_m}{R} \sin 2\alpha$$

$$b_1 = \frac{1}{\pi} \int_0^{2\pi} i_a(\omega t) \sin\omega t \, d\omega t$$

$$= \frac{2}{\pi} \frac{E_{av_0}}{R} \cos\alpha [-\cos\omega t]_{\alpha+30°}^{\alpha+150°}$$

$$= \frac{2\sqrt{3}}{\pi} \frac{E_{av_0}}{R} \cos^2\alpha \tag{7.56}$$

$$= \frac{9}{\pi^2} \frac{E_m}{R} (1 + \cos 2\alpha)$$

Equations (7.55) and (7.56) can be used to obtain a very important relationship

$$\frac{a_1}{b_1} = \frac{-\sin 2\alpha}{(1+\cos 2\alpha)} = -\tan\alpha = \tan\psi_1 \tag{7.57}$$

From (7.57) it can be seen that the displacement angle ψ_1 of the input current is equal to the firing-angle (the negative sign representing delayed firing),

$$\alpha = \psi_1 \tag{7.58}$$

The displacement factor $\cos\psi_1$ is therefore equal to the cosine of the delayed firing-angle,

$$\cos\psi_1 = \cos\alpha \tag{7.59}$$

The relationship of (7.59) is true for both half-wave and full-wave bridges with highly inductive load. It is not true for bridges with purely resistive loading. The distortion factor of the input current is obtained by combining equations (7.30), (7.31), (7.53), (7.55) and (7.56):

distortion factor of supply current $= \dfrac{I_{a_1}}{I_a}$

$$= \dfrac{\dfrac{1}{\sqrt{2}} \dfrac{9}{\pi^2} \dfrac{E_m}{R} \sqrt{\sin^2 2\alpha + (1 + \cos 2\alpha)^2}}{\dfrac{3\sqrt{2} E_m}{\pi R} \cos \alpha}$$

$$= \dfrac{3}{\pi} \qquad (7.60)$$

The product of the displacement factor (7.59) and the distortion factor (7.60) is seen to give the power factor (7.54). Some of the supply-side properties of the inductively loaded bridge are included in Table 7.2, located in the previous section.

For any balanced three-phase load with sinusoidal supply voltage the real or active input power P is given by

$$P = 3EI_1 \cos \psi_1 \qquad (7.61)$$

where I_1 is the r.m.s. value of the fundamental component of the supply current and $\cos \psi_1$ is the displacement factor (not the power factor). Substituting (7.55), (7.56) and (7.59) into (7.61) gives

$$P = \dfrac{27}{\pi^2} \dfrac{E_m^2}{R} \cos^2 \alpha \qquad (7.62)$$

which is seen to be equal to the power P_L dissipated in the load resistor, equation (7.51).

7.1.2.3 Shunt capacitor compensation

If equal capacitors C are connected in star at the supply point, Fig. 7.6, the instantaneous supply current is given by

$$i_{s_a}(\omega t) = \dfrac{E_m}{X_c} \cos \omega t + \dfrac{3\sqrt{3} E_m}{\pi R} \cos \alpha \Big|_{\alpha+30°}^{\alpha+150°} - \dfrac{3\sqrt{3} E_m}{\pi R} \cos \alpha \Big|_{\alpha+210°}^{\alpha+330°}$$

$$(7.63)$$

The substitution of (7.63) into (7.13) gives an expression for the r.m.s. supply current

$$I_{S_a} = \sqrt{\frac{1}{\pi} \int \left[\left(\frac{E_m}{X_c} \cos \omega t\right)_0^\pi + \left(\frac{3\sqrt{3}E_m}{\pi R} \cos \alpha\right)_{\alpha+30°}^{\alpha+150°} \right]^2 d\omega t} \qquad (7.64)$$

$$= \frac{E_m}{\sqrt{2}R} \sqrt{\frac{R^2}{X_c^2} - \frac{18}{\pi^2} \frac{R}{X_c} \sin 2\alpha + \frac{36}{\pi^2} \cos^2 \alpha}$$

When the capacitance is absent, X_c becomes infinitely large and (7.64) reduces to (7.53). The power flow and the terminal voltage are unaffected by the connection of the capacitors. The compensated power factor is given by combining (7.62) and (7.64),

$$PF_c = \frac{P}{3E_a I_{S_a}}$$

$$= \frac{\frac{9}{\pi^2} \cos^2 \alpha}{\sqrt{\frac{R^2}{4X_c^2} - \frac{9}{2\pi^2} \frac{R}{X_c} \sin 2\alpha + \frac{9}{\pi^2} \cos^2 \alpha}} \qquad (7.65)$$

The ratio of the compensated power factor to the uncompensated power factor is given by the ratio of the load current to the supply current,

$$\frac{PF_c}{PF} = \frac{I_a}{I_{S_a}} = \sqrt{\frac{\frac{18}{\pi^2} \cos^2 \alpha}{\frac{R^2}{2X_c^2} - \frac{9}{\pi^2} \frac{R}{X_c} \sin 2\alpha + \frac{18}{\pi^2} \cos^2 \alpha}} \qquad (7.66)$$

The power factor is therefore improved when $PF_c/PF > 1$, which occurs when

$$\frac{R}{2X_c} \left(\frac{R}{X_c} - \frac{18}{\pi^2} \sin 2\alpha \right) < 0 \qquad (7.67)$$

Examination of the inequality (7.67) shows that power factor improvement occurs when $0 < C < 18 \sin 2\alpha / \omega \pi^2 R$.

Fourier coefficient a_{s_1} for the fundamental component of the compensated supply current is given by

$$a_{s_1} = \frac{1}{\pi} \int_0^{2\pi} i_{s_a}(\omega t) \cos \omega t \, d\omega t$$

$$= \frac{2}{\pi} \int \left[\left(\frac{E_m}{X_c} \cos^2 \omega t \right)_0^\pi + \left(\frac{3\sqrt{3} E_m}{\pi R} \cos \alpha \cos \omega t \right)_{\alpha+30°}^{\alpha+150°} \right] d\omega t$$

$$= \frac{E_m}{X_c} - \frac{9}{\pi^2} \frac{E_m}{R} \sin 2\alpha \qquad (7.68)$$

When $C = 0$, (7.68) reduces to (7.55).

To obtain the maximum displacement factor, coefficient a_{s_1} must be zero. The condition for maximum realisable capacitor compensation is therefore, from (7.68),

$$C = \frac{9}{\pi^2} \frac{1}{\omega R} \sin 2\alpha \qquad (7.69)$$

Equating (7.68) to zero and substituting into (7.65) gives the maximum power factor achievable by terminal capacitor compensation

$$PF_{c_{\max}} = \frac{\frac{3}{\pi} \cos \alpha}{\sqrt{1 - \frac{9}{\pi^2} \sin^2 \alpha}} \qquad (7.70)$$

For any non-zero value of α it is seen that the uncompensated power factor $(3/\pi) \cos \alpha$ is improved due to optimal capacitor compensation, as illustrated in Fig. 7.10. Over most of the firing-angle range the degree of power factor improvement is substantial. A disadvantage of power factor compensation by the use of capacitors is that, for fixed load resistance, the value of the optimal capacitor varies with firing-angle.

7.1.2.4 Worked examples

Example 7.4

A three-phase, full-wave controlled bridge rectifier contains six ideal SCRs and is fed from an ideal three-phase voltage source of 240 V, 50 Hz. The load resistor $R = 10\,\Omega$ is connected in series with a large smoothing inductor. Calculate the average load voltage and the power dissipation at (i) $\alpha = 30°$, (ii) $\alpha = 60°$.

Solution. If 240 V represents the r.m.s. value of the line voltage, then the peak phase voltage E_m is given by

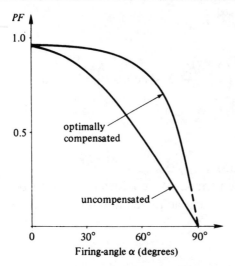

Fig. 7.10 Power factor versus firing-angle for three-phase, full-wave, controlled bridge rectifier with highly inductive load.

$$E_m = \frac{\sqrt{2}}{\sqrt{3}} \times 240$$

From (7.8)

$$E_{av} = \frac{3 \times \sqrt{3}}{\pi} \times \frac{\sqrt{2}}{\sqrt{3}} \times 240 \cos \alpha$$

$$= 324 \cos \alpha$$

At $\alpha = 30°$, $E_{av} = 280.6$ V
At $\alpha = 60°$, $E_{av} = 162$ V
The power dissipation is given by (7.51),

$$P = I_{av}^2 R = \frac{E_{av}^2}{R}$$

At $\alpha = 30°$, $P = 7.863$ kW
At $\alpha = 60°$, $P = 2.625$ kW

Example 7.5
For the three-phase bridge of Example 7.4 calculate the displacement factor, the distortion factor and the power factor at (i) $\alpha = 30°$, (ii) $\alpha = 60°$.

Solution. From (7.59) it is seen that the displacement factor is given by

displacement factor $= \cos \psi_1 = \cos \alpha$

At $\alpha = 30°$, $\cos \alpha = 0.866 = \sqrt{3}/2$

At $\alpha = 60°$, $\cos \alpha = 0.5$

Because the waveshape of the supply current is not affected by the firing-angle of the bridge SCRs the supply distortion factor is constant. From (7.60),

$$\text{distortion factor} = \frac{3}{\pi} = 0.955$$

For loads with sinusoidal supply voltage the power factor, seen from the supply point, is the product of the displacement factor and the distortion factor,

$$PF = \frac{3}{\pi} \cos \alpha$$

At $\alpha = 30°$, $PF = 0.827$
At $\alpha = 60°$, $PF = 0.478$

Example 7.6
For the three-phase bridge of Example 7.4 calculate the required voltage and current ratings of the bridge SCRs.

Solution. In a three-phase full-wave bridge the maximum voltage on a switch is the peak value of the line voltage,

$$E_{max} = \sqrt{2} E_{line}$$

where E_{line} is the r.m.s. value of the line voltage. Therefore

$$E_{max} = \sqrt{2} \times 240 = 339.4 \text{ V}$$

(Note that E_{max} is $\sqrt{3}$ times the peak value E_m of the phase voltage.)
From (7.53) the r.m.s value of the supply current is

$$I_a = \sqrt{\frac{2}{3}} I_{av}$$

$$= \frac{3\sqrt{2} E_m}{\pi R} \cos \alpha$$

But each SCR conducts only one (positive) pulse of current every supply voltage cycle. In Fig. 7.7 for example, SCR Th_1 conducts only the positive pulses of current $i_a(\omega t)$ in Fig. 7.8. Therefore,

$$I_{Th_1} = \sqrt{\frac{1}{2\pi} \int_{\alpha+30°}^{\alpha+150°} i_a^2(\omega t) \, d\omega t}$$

The defining expression for I_{Th}, above, is seen to have the value $1/\sqrt{2}$ that of I_a in (7.53),

$$I_{Th_1} = \frac{1}{\sqrt{2}} I_a$$

$$= \frac{3E_m}{\pi R} \text{ at } \alpha = 0°$$

Therefore

$$I_{Th_1} = \frac{3}{\pi} \times \frac{\sqrt{2}}{\sqrt{3}} \times 240 \times \frac{1}{10}$$

$$= 18.7 \text{ A}$$

Example 7.7

The three-phase full-wave rectifier of Example 7.4 is to have its power factor compensated by the connection of equal, star-connected capacitors at the supply point. Calculate the maximum value of capacitance that will result in power factor improvement and the optimum capacitance that will give the maximum realisable power factor improvement at (i) $\alpha = 30°$, (ii) $\alpha = 60°$. In each case compare the compensated power factor with the corresponding uncompensated value.

Solution. The criterion for power factor improvement is defined by (7.67), which shows that

$$C_{max} = \frac{18 \sin 2\alpha}{\omega \pi^2 R}$$

At $\alpha = 30°$,

$C_{max} = 503 \, \mu\text{F}$

At $\alpha = 60°$,

$C_{max} = 503 \, \mu\text{F}$

The optimum value of capacitance that will cause unity displacement factor and maximum power factor is given in (7.69),

$$C_{opt} = \frac{9}{\omega \pi^2 R} \sin 2\alpha$$

(Note that $C_{opt} = C_{max}/2$.)
For both firing-angles

$C_{opt} = 251.5 \, \mu\text{F}$

In the presence of optimum capacitance the power factor is obtained from (7.70),

$$PF_{C_{max}} = \frac{3 \cos \alpha}{\sqrt{\pi^2 - 9 \sin^2 \alpha}}$$

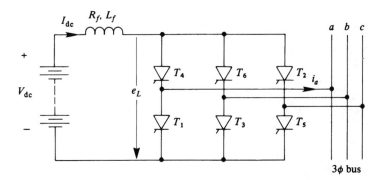

Fig. 7.11 Three-phase, naturally commutated basic bridge inverter.

At $\alpha = 30°$, $PF_c = 0.941$, which compares with the uncompensated value $PF = 0.827$ (Example 7.4).

At $\alpha = 60°$, $PF_c = 0.85$, which compares with the uncompensated value $PF = 0.478$ (Example 7.4).

7.2 THREE-PHASE CONTROLLED BRIDGE RECTIFIER–INVERTER

7.2.1 Theory of operation

The three-phase bridge rectifier of Fig. 7.1 can be used as an inverter if the passive load is replaced by a d.c. supply with reversed polarity voltage as shown in Fig. 7.11. The current direction on the d.c. side is unchanged. For inverter operation the voltage level, frequency and waveform on the a.c. side are set by the bus and cannot be changed. As with rectifier operation, described in Section 7.1, the anode voltages of the switches undergo cyclic variation and are therefore switched off by natural commutation. There is no advantage to be gained here by the use of gate turn-off devices.

In the operation of the inverter circuit of Fig. 7.11 certain restrictions must be imposed on the switching sequence of the switches. For example, both switches of any inverter arm, such as T_1 and T_4, cannot conduct simultaneously. Similarly, only one switch of the upper bridge (T_2, T_4, T_6) and one switch of the lower bridge (T_1, T_3, T_5) can conduct simultaneously. If sequential firing is applied to the six SCRs the three phase currents are identical in form but mutually displaced in phase by 120°. The detailed operation of the circuit for rectifier operation, described in Section 7.1, is again relevant here, except that the range of firing-angles is now $90° \leq \alpha \leq 180°$.

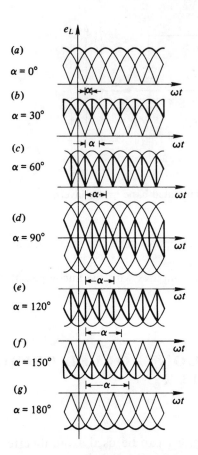

Fig. 7.12 Instantaneous voltage e_L on the d.c. side of a bridge rectifier–inverter circuit, assuming ideal a.c. supply (no overlap).

The instantaneous voltage e_L on the d.c. side of the bridge is shown in Fig. 7.12, assuming an ideal a.c. supply. For $\alpha \leq 60°$ the waveforms are identical to those of rectifier operation with passive load, Fig. 7.4, and $e_L(\omega t)$ is always positive with a positive average value. At $\alpha = 90°$ the waveform, Fig. 7.12(d), has an average value of zero and is identical to that of rectifier operation with a highly inductive load.

When $\alpha > 90°$ the average value of $e_L(\omega t)$ is negative. With passive load, Fig. 7.7, this negative voltage cannot circulate negative current and therefore rectifier operation ceases at $\alpha = 90°$. For $\alpha \geq 120°$ voltage $e_L(\omega t)$ is negative at all times. It is seen that the polarity of $e_L(\omega t)$ (and, by implication, its average value E_{av}) in Fig. 7.11 will then be opposite to that of Fig. 7.7. Since $e_L(\omega t)$ is time varying it does not, in general, coincide with the constant

7.2 Rectifier–inverter

voltage V_{dc} from the d.c. source. For this reason a filter inductor L_f must be included to absorb the ripple voltage e_{L_f}.

$$e_{L_f} = V_{dc} - e_L(\omega t) = L_f \frac{dI_{dc}}{dt} + R_f I_{dc} \tag{7.71}$$

If the d.c. side current is very smooth, which occurs when L_f is large, then $dI_{dc}/dt \to 0$ and the ripple voltage falls largely on the resistance R_f of the filter inductor.

The waveform of the current on the a.c. side of the bridge depends largely on the magnitude of the filter inductor. If this inductor is large the line current assumes a rectangular waveform similar to that obtained for rectifier operation with a highly inductive load, Fig. 7.8. The current pulses have a conduction period of 120° followed by a dwell period of 60°. In the circuit of Fig. 7.11 the average value of the bridge voltage is, from (7.8),

$$\begin{aligned}
\text{average value of } e_L(\omega t) = E_{av} &= E_{av_0} \cos \alpha \\
&= \frac{3\sqrt{3}E_m}{\pi} \cos \alpha \\
&= \frac{3}{\pi} \sqrt{2} \left(\frac{\sqrt{3}E_m}{\sqrt{2}} \right) \cos \alpha \\
&= 1.35 E \cos \alpha
\end{aligned} \tag{7.72}$$

where E_m is the peak phase voltage and E is the r.m.s. line voltage.

The average value of the current on the d.c. side is seen, from Fig. 7.11, to be satisfied by the relation, for rectifier operation,

$$\begin{aligned}
I_{dc} &= \frac{V_{dc} + E_{av}}{R_f} \\
&= \frac{1}{R_f} (V_{dc} + 1.35 E \cos \alpha) \\
&= \frac{1}{R_f} (V_{dc} + 2.34 V_1 \cos \alpha)
\end{aligned} \tag{7.73}$$

where V_1 is the r.m.s. phase voltage.

For inverter operation, $\alpha > 90°$ and $\cos \alpha$ is negative, so that the filter inductor voltage is $V_{dc} - E_{av}$. At $\alpha = 90°$, $E_{av} = 0$ and the inverter presents a short circuit to the direct current.

In order to maintain constant I_{dc} in the presence of adjustable SCR firing-angle α it is necessary to simultaneously vary R_f or V_{dc} or both. If direct voltage V_{dc} is constant, then increase of the retardation angle α would result in a decrease of I_{dc} due to growth of the bucking voltage E_{av}. Current I_{dc} in Fig. 7.11 will become zero, from (7.73), when

$$\cos\alpha = \frac{\pi V_{dc}}{3\sqrt{3}E_m} = \frac{V_{dc}}{1.35E} \tag{7.74}$$

Since the switches of the rectifier–inverter bridge are presumed to be lossless there must be a power balance on each side of the circuit. If the r.m.s. value of the fundamental component of the inverter line current is I_{a_1} and the displacement angle is ψ_1, then

$$P = 3\frac{E_m}{\sqrt{2}}I_{a_1}\cos\psi_1 \text{ (a.c. side)}$$
$$= E_{av}I_{dc} \text{ (d.c. side)} \tag{7.75}$$

The power dissipation $I_{dc}^2 R_f$ in the filter comes from the battery but this is external to the bridge.

The real power P becomes zero if $E_{av} = 0$ or $I_{dc} = 0$ or both. To make $P = 0$ the SCR firing conditions are therefore that $\alpha = 90°$, from (7.72), or $\alpha = \cos^{-1}(V_{dc}/1.35E)$, from (7.74).

Now the important relationship $\alpha = \psi_1$ of (7.58) remains true and combining this with (7.72) and (7.74) gives

$$I_{a_1} = \frac{\sqrt{2}\times\sqrt{3}}{\pi}I_{dc} = \frac{\sqrt{6}}{\pi}I_{dc} = 0.78\,I_{dc} \tag{7.76}$$

The fundamental line current therefore has a peak value $\sqrt{2}\times 0.78\,I_{dc} = 1.1 I_{dc}$ and is, by inspection, in time-phase with $i_a(\omega t)$. An example of this is shown in Fig. 7.13(a).

Although the real power in watts transferred through the inverter comes from the d.c. source, the reactive voltamperes has to be provided by the a.c. supply. The a.c. side current can be thought of as a fundamental frequency component lagging its corresponding phase-voltage by $\psi_1(=\alpha)$ radians plus a series of odd higher harmonics. In terms of the fundamental frequency component, the inverter action can be interpreted either as drawing lagging current from the a.c. system or, alternatively, as delivering leading current to the a.c. system. Rectifier action and inverter action of the bridge circuit are depicted in the equivalent circuits of Fig. 7.14. A notation I_1 is used (rather than I_{a_1}) because the circuits are true for any phase.

The fundamental a.c. side components may be represented in phasor form as, for example, in Fig. 7.15. With rectifier operation, both the in-phase component of current $I_1\cos\alpha$ and the quadrature component $I_1\sin\alpha$ are drawn from the a.c. bus. With inverter operation, the in-phase component

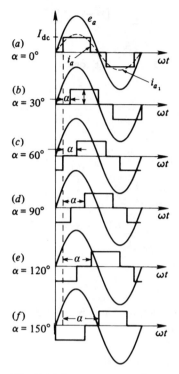

Fig. 7.13 Instantaneous phase voltage and current on the a.c. side of a bridge rectifier–inverter, assuming ideal a.c. supply. (Note – it is assumed that V_{dc} is adjusted proportionately to $\cos\alpha$ to maintain I_{dc} constant.)

of current is opposite in sign to the case of rectifier operation and represents active or real power delivered to the a.c. bus, but the quadrature component still represents reactive voltamperes drawn from the a.c. bus. As α increases from 90°, with constant current, an increased amount of power is transferred to the a.c. system and the reactive voltampere requirement is reduced.

Even for the condition of zero power transfer there may still be currents flowing in the inverter. At $\alpha = 90°$, Fig. 7.13(d), the inverter current is in quadrature lagging its respective phase-voltage. The in-phase component of current $I_1 \cos\alpha$ in Figs. 7.14, 7.15 is then zero and so is the power. But the quadrature component of current $I_1 \sin\alpha$ is finite and therefore, from (7.76), a component of the inverter phase-current acts as a 'magnetising' current even though there is no magnetic field and no capability of storing energy. In Fig. 7.14, the reactive voltamperes Q is given by an expression complementary to the expression for real power P,

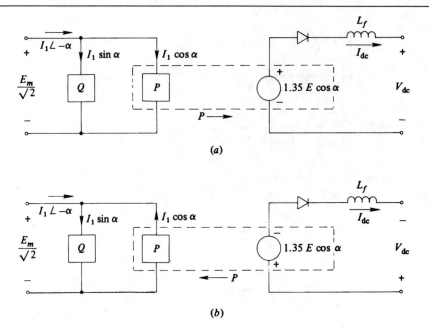

Fig. 7.14 Equivalent per-phase circuits for the three-phase bridge rectifier–inverter: (*a*) rectifier operation, (*b*) inverter operation.

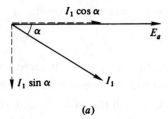

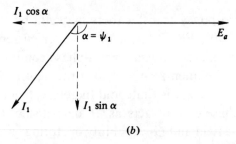

Fig. 7.15 Phasor diagrams for the fundamental frequency components of the a.c. side voltage and current: (*a*) rectifier operation, (*b*) inverter operation.

$$Q = \frac{3E_m}{\sqrt{2}} I_{a_1} \sin \alpha = 3 \frac{E_m}{\sqrt{2}} I_{a_1} \sin \psi_1 \tag{7.77}$$

Combining (7.72) and (7.76) with (7.77) gives

$$Q = E_{av_0} I_{dc} \sin \alpha \tag{7.78}$$

The combination of voltampere components P and Q thus gives

$$\sqrt{P^2 + Q^2} = E_{av_0} I_{dc} \tag{7.79}$$

Since only fundamental components are being considered the ratio

$$\frac{P}{\sqrt{P^2 + Q^2}} = \frac{E_{av}}{E_{av_0}} = \cos \alpha = \text{displacement factor} \tag{7.80}$$

It can be seen that (7.80) confirms the earlier result of (7.59). The r.m.s. value of the total a.c. side line current is given by (7.53) if the filter inductance is large. Combining (7.76) with (7.53) shows that the a.c. side line current distortion factor I_{a_1}/I has the value $3/\pi$, as in (7.60).

7.2.2 Worked examples

Example 7.8
Power is transferred from a 300 V battery to a three-phase, 230 V, 50 Hz a.c. bus via a controlled SCR inverter. The inverter switches may be considered lossless and a large filter inductor with resistance 10 Ω is included on the d.c. side. Calculate the power transferred and the power factor if (a) $\alpha = 90°$, (b) $\alpha = 120°$, (c) $\alpha = 150°$.

Solution. The circuit is represented in Fig. 7.11 with the a.c. side current waveform shown in Fig. 7.13. The average voltage of the inverter is given by (7.22),

$$E_{av} = \frac{3\sqrt{3}E_m}{\pi} \cos \alpha$$

In this case the peak phase-voltage E_m is

$$E_m = \sqrt{2} \times \frac{230}{\sqrt{3}} = 187.8 \text{ V}$$

Therefore

$$E_{av} = \frac{3\sqrt{3}}{\pi} \times 187.8 \cos\alpha = 310.6 \cos\alpha$$
$$= 0, \text{ at } \alpha = 90°$$
$$= -155.3 \text{ V, at } \alpha = 120°$$
$$= -269 \text{ V, at } \alpha = 150°$$

The negative sign indicates that E_{av} opposes the current flow that is created by the connection of V_{dc}. The current on the d.c. side is given by (7.73),

$$I_{dc} = \frac{V_{dc} + E_{av}}{R_f} = \frac{|V_{dc}| - |E_{av}|}{R_f}$$

At $\alpha = 90°$,

$$I_{dc} = \frac{300}{10} = 30 \text{ A}$$

At $\alpha = 120°$,

$$I_{dc} = \frac{300 - 155.3}{10} = 14.47 \text{ A}$$

At $\alpha = 150°$,

$$I_{dc} = \frac{300 - 269}{10} = 3.1 \text{ A}$$

The power transferred through the inverter into the a.c. system is the battery power minus the loss in R_f.

$$P = V_{dc}I_{dc} - I_{dc}^2 R_f$$

At $\alpha = 90°$,

$$P = 300 \times 30 - (30)^2 \times 10 = 0$$

At $\alpha = 120°$,

$$P = 300 \times 14.47 - (14.47)^2 \times 10$$
$$= 4341 - 2093.8 = 2247.2 \text{ W}$$

At $\alpha = 150°$,

$$P = 300 \times 3.1 - (3.1)^2 \times 10$$
$$= 930 - 96.1 = 833.9 \text{ W}$$

The peak height of the a.c. side current is also the battery current I_{dc}. The fundamental a.c. side current has an r.m.s. value given by (7.76),

$$I_1 = 0.78 I_{dc}$$

At $\alpha = 90°$,

$$I_1 = 23.4 \text{ A}$$

At $\alpha = 120°$,

$I_1 = 11.3$ A

At $\alpha = 150°$,

$I_1 = 2.42$ A

At $\alpha = 120°$, for example, the power given by the a.c. side equation (7.75) is

$$P = 3 \times \frac{230}{\sqrt{3}} \times 11.3 \cos 120° = -2251 \text{ W}$$

This agrees very nearly with the calculated power on the d.c. side.
The displacement factor is defined directly from (7.59) and (7.80),

displacement factor $= |\cos \psi_1| = |\cos \alpha|$

At $\alpha = 90°$,

$DF = 0$

At $\alpha = 120°$,

$DF = |\cos 120°| = 0.5$

At $\alpha = 150°$,

$DF = |\cos 150°| = 0.866$

The distortion factor for a waveform such as the current in Fig. 7.13 was shown in Section 7.1.2.2 above to have the value $3/\pi$. Now

$PF = $ displacement factor $\times$ distortion factor

At $\alpha = 90°$,

$PF = 0$

At $\alpha = 120°$,

$$PF = \frac{3}{\pi} \times 0.5 = 0.477$$

At $\alpha = 150°$,

$$PF = \frac{3}{\pi} \times 0.866 = 0.827$$

Example 7.9
Calculate the switch ratings for the operation of the inverter in Example 7.8.

Solution. Operation of the inverter of Fig. 7.11 requires that the peak line voltage fall sequentially on each SCR

$V_{T_{max}} = \sqrt{2} \times 230 = 325.3$ V

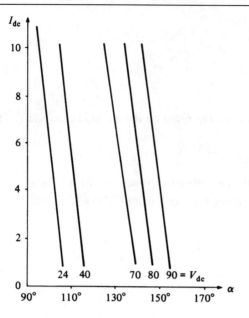

Fig. 7.16 Measured d.c. side current (amps) versus firing-angle for a three-phase naturally commutated inverter.

Each SCR current has one rectangular pulse (positive going only), as in Fig. 7.13, per cycle. This has an r.m.s. value

$$I_T = \sqrt{\frac{1}{2\pi}\int_{\pi/6}^{5\pi/6} I_{dc}^2 \, d\omega t} = \sqrt{\frac{I_{dc}^2}{3}}$$

Therefore

$$I_T = \frac{I_{dc}}{\sqrt{3}}$$

The maximum value of I_{dc} occurs at $\alpha = 90°$. In this case

$$I_T = \frac{30}{\sqrt{3}} = 17.32 \, \text{A}$$

The expression above for I_T is confirmed by (7.53), which defines the r.m.s. value of the line current as $\sqrt{2}I_T$.

The practically selected devices might be rated at 400 V, 20 A.

Example 7.10
The current–firing-angle characteristics of Fig. 7.16 were measured from a battery powered SCR inverter feeding into a three-phase transformer. Deduce the transformer terminal voltage.

Solution. A point on one of the characteristics is chosen arbitrarily. Consider the point $V_{dc} = 70\,\text{V}$, $I_{dc} = 6\,\text{A}$, when $\alpha = 130°$. The power leaving the battery is $P = 70 \times 6 = 420\,\text{W}$. Neglecting any power loss in the filter inductor all of this power reappears on the a.c. side of the inverter. The r.m.s. value of the fundamental inverter current is, from (7.76),

$$I_1 = 0.78\,I_{dc} = 4.68\,\text{A}$$

In (7.75),

$$420 = \frac{3E_m}{\sqrt{2}} \times 4.68 \times \cos 130°$$

which gives

$$E_m = 65.8\,\text{V}$$

The r.m.s. line voltage on the a.c. side was therefore

$$E = \sqrt{3} \times \frac{65.8}{\sqrt{2}} = 80\,\text{V}$$

7.3 PROBLEMS

Three-phase, full-wave controlled bridge with resistive load and ideal supply

7.1 A three-phase rectifier bridge of six ideal switches is connected to provide full-wave controlled rectifier current to a resistive load. The ideal three-phase supply provides balanced sinusoidal voltages at the input terminals. Show that the average load voltage E_{av} is given by equations (7.8), (7.12) in the two respective modes of operation. Sketch E_{av} versus firing-angle α over the full operating range.

7.2 A three-phase, full-wave bridge rectifier containing six ideal SCRs supplies a resistive load $R = 100\,\Omega$. The ideal supply 240 V, 50 Hz provides balanced sinusoidal voltages. Calculate the average load current and power dissipation at (i) $\alpha = 30°$, (ii) $\alpha = 60°$, (iii) $\alpha = 90°$.

7.3 For the three-phase bridge circuit of Problem 7.2 deduce and sketch the voltage waveform across a switch at $\alpha = 30°$.

7.4 For the three-phase bridge circuit of Problem 7.1 show that the r.m.s. values of the supply current are given by equations (7.22) and (7.23).

7.5 For a three-phase, full-wave bridge circuit with resistive load, show that, for both modes of operation, the r.m.s. supply current I_a is related to the r.m.s. load current I_L by the relation (7.24).

7.6 Derive expressions for the fundamental component of supply current into a three-phase, full-wave, controlled bridge rectifier supplying a resistive load. Calculate the r.m.s. value of this fundamental component with a supply of 230 V, 50 Hz and a load resistor $R = 100\,\Omega$.

7.7 The power input to a three-phase, full-wave, controlled bridge rectifier is given by the relation

$$P = 3EI_{a_1} \cos \psi_1$$

where E is the r.m.s. phase voltage, I_{a_1} is the r.m.s. value of the fundamental component of the supply current and $\cos \psi_1$ is the displacement factor (not the power factor!). Calculate P for the bridge circuit of Problem 7.6 and check that the value obtained agrees with the power dissipation calculated on the load side.

7.8 Use Fourier series to obtain general expressions for the higher harmonic components of the supply current to a three-phase, full-wave, controlled bridge rectifier with resistive load. Calculate the values of the harmonics up to the seventh harmonic and check that the harmonics sum (very nearly) to the r.m.s. value of the input current.

7.9 Derive expressions for the displacement factor $\cos \psi_1$ and distortion factor I_1/I for a three-phase, full-wave controlled bridge rectifier with resistive load. Show that the respective products of these are consistent with the expressions (7.26), (7.27) for the power factor.

7.10 Calculate and sketch the power factor of a three-phase, full-wave bridge rectifier with resistive load over the operating range of SCR firing-angles.

7.11 Use the information of Problem 7.7 to derive expressions for the reactive voltamperes Q into a three-phase, full-wave bridge rectifier with resistive load, where

$$Q = 3EI_{a_1} \sin \psi_1$$

Does a knowledge of real power P and reactive voltamperes Q account for all the apparent voltamperes $S(=3EI_a)$ at the bridge terminals?

7.12 Three equal capacitors C are connected in star across the terminals of a full-wave, three-phase bridge rectifier with resistive load. If $X_c = R$, sketch waveforms of a capacitor current, a bridge input current and the corresponding current at $\alpha = 30°$. Does the waveform of the supply current seem to represent an improvement compared with the uncompensated bridge?

7.13 For the three-phase bridge circuit of Problem 7.2 what will be the minimum value of supply point capacitance per-phase that will cause power factor improvement at (i) $\alpha = 30°$, (ii) $\alpha = 60°$ and (iii) $\alpha = 90°$?

7.14 For the three-phase bridge circuit of Problem 7.2 what must be the respective values of the compensating capacitors to give the highest realisable power factor (by capacitor correction) at the three values of firing-angle?

7.15 For the three-phase bridge circuit of Problem 7.2 calculate the operating power factor at each value of firing-angle. If optimum compensation is now achieved by the use of the appropriate values of supply point capacitance, calculate the new values of power factor.

7.3 Problems

Three-phase, full-wave controlled bridge rectifier with highly inductive load and ideal supply

7.16 A three-phase, full-wave controlled bridge rectifier contains six ideal SCRs and is fed from an ideal three-phase supply of balanced sinusoidal voltages. The load consists of a resistor R in series with a large filter inductor. Show that, for all values of SCR firing-angle α, the average load voltage is given by (7.8).

Sketch E_{av} versus α and compare the result with that obtained for purely resistive load.

7.17 For the three-phase inductively loaded bridge of Problem 7.16 calculate the Fourier coefficients a_1, b_1 of the fundamental component of the supply current. Use these to show that the displacement angle $\psi_1(\tan^{-1} a_1/b_1)$ is equal to the SCR firing-angle α.

7.18 A three-phase, full-wave controlled bridge rectifier is supplied from an ideal three-phase voltage source of 415 V, 50 Hz. The load consists of resistor $R = 100$ ohms in series with a very large filter inductor. Calculate the load power dissipation at (i) $\alpha = 30°$, (ii) $\alpha = 60°$ and compare the values with those that would be obtained in the absence of the load filter inductor.

7.19 Show that for the inductively loaded bridge of Problem 7.16 the distortion factor of the supply current is independent of the SCR firing-angle.

7.20 Show that the waveform of the supply current to a three-phase, full-wave controlled bridge rectifier with highly inductive load is given by

$$i(\omega t) = \frac{2\sqrt{3}I_{av}}{\pi} [\sin(\omega t - \alpha) - \tfrac{1}{5}\sin 5(\omega t - \alpha) - \tfrac{1}{7}\sin 7(\omega t - \alpha) \ldots]$$

where I_{av} is the average load current.

7.21 For the three-phase bridge rectifier of Problem 7.16 show that the power input is equal to the load power dissipation.

7.22 Derive an expression for the load voltage ripple factor, RF, for a three-phase inductively loaded bridge rectifier and show that this depends only on the SCR firing-angle. Obtain a value for the case $\alpha = 0$ and thereby show that the RF is zero within reasonable bounds of calculation.

7.23 For the inductively loaded bridge rectifier of Problem 7.16 show that the r.m.s. supply current is given by

$$I = \frac{3\sqrt{2}E_m}{\pi R} \cos \alpha$$

Calculate this value for the cases specified in Problem 7.18.

7.24 For the inductively loaded bridge of Problem 7.18 calculate the r.m.s. current and peak reverse voltage ratings required of the bridge switches.

7.25 Show that the average load voltage of a three-phase, full-wave controlled bridge circuit with highly inductive load can be obtained by evaluating the integral

$$E_{\text{av}} = \frac{6}{2\pi} \int_{\alpha-30°}^{\alpha+30°} E_m \cos\omega t \, d\omega t$$

Sketch the waveform of the instantaneous load voltage $e_L(\omega t)$ for $\alpha = 75°$ and show that it satisfies the above relationship.

7.26 A three-phase, full-wave SCR bridge is fed from an ideal three-phase supply and transfers power to a load resistor R. A series choke on the load side gives current smoothing that may be considered ideal. Derive an expression for the r.m.s. value of the fundamental component of the supply current. Use this expression to show that the reactive voltamperes entering the bridge is given by

$$Q = \frac{27 E_m^2}{4\pi^2 R} \sin 2\alpha$$

7.27 For the three-phase bridge rectifier of Problem 7.18 calculate the power factor. If equal capacitors C are now connected in star at the supply, calculate the new power factor when $X_c = R$. What is the minimum value of firing-angle α at which compensation to the degree $X_c = R$ renders a power factor improvement?

7.28 For the bridge rectifier circuit of Problem 7.16 derive an expression for the maximum terminal capacitance that will give power factor improvement.

7.29 The bridge rectifier circuit of Problem 7.18 is compensated by the use of equal capacitors C connected in star at the supply terminals. Calculate the values of capacitance that will give unity displacement factor at (i) $\alpha = 30°$, (ii) $\alpha = 60°$. In each case calculate the degree of power factor improvement compared with uncompensated operation.

7.30 For the bridge circuit of Problem 7.28 sketch, on squared paper, consistent waveforms of the bridge line current, the capacitor current and the supply line current. Does the waveform of the supply current appear less distorted than the rectangular pulse waveform of the bridge current?

7.31 A three-phase, full-wave bridge rectifier circuit, Fig. 7.7, supplies power to load resistor R in the presence of a large load filter inductor. Equal capacitors are connected at the supply terminals to give power factor improvement by reducing the displacement angle ψ_1 to zero at the fixed SCR firing-angle α. Derive a general expression for the supply current distortion factor in the presence of supply capacitance. For the case when C has its optimal value so that the displacement factor is increased to unity is the distortion factor also increased?

Three-phase controlled bridge rectifier–inverter

7.32 A naturally commutated, three-phase inverter contains six ideal SCRs and transfers energy into a 440 V, 50 Hz, three-phase supply from an 800 V d.c. battery. The battery and the inverter are linked by a smoothing inductor with a resistance of 12.4 ohms. Calculate the power transferred at $\alpha = 90°$, 120°, 150° and 170°.

7.33 For the inverter application of Problem 7.32 calculate the voltage and r.m.s. current ratings required of the switches.

7.34 A large solar energy installation utilises a naturally commutated, three-phase SCR inverter to transfer energy into a power system via a 660 V transformer. The collected solar energy is initially stored in an 800 V battery which is linked to the inverter through a large filter choke of resistance 14.2 Ω. What is the maximum usable value of the SCR firing-angle? Calculate the power transferred at the firing-angle of 165°. What is the necessary transformer rating?

7.35 Calculate the necessary SCR voltage and r.m.s. current ratings for the inverter application of Problem 7.34.

7.36 Use the inverter characteristics of Fig. 7.16 to deduce the form of the corresponding V_{dc}–I_{dc} characteristics with $\cos \alpha$ as the parameter. If the maximum d.c. side voltage is 100 V, what is the firing-angle required to give a direct current of 10 A if $R_f = 1\,\Omega$?

7.37 Sketch the main circuit of a naturally commutated, three-phase, controlled bridge inverter. If the a.c. side r.m.s. line voltage V is fixed, sketch the variation of inverter power transfer with SCR firing-angle α and d.c. side voltage V_{dc}. If $\alpha = 120°$, what is the minimum value of ratio V_{dc}/V that will permit inversion?

Sketch the waveform of the current passing between the inverter and the supply and give a phasor diagram interpretation to explain the inverter operation. Why is it often necessary to connect capacitance across the terminals of a naturally commutated inverter of high kVA rating?

8
Single-phase voltage controllers

Smooth control of the voltage level across a single-phase load can be obtained by the use of a bidirectional electronic switch. As with the rectifier circuit applications of Chapter 6 and Chapter 7, the alternating current source provides the means of natural commutation of the conducting switches. Gate turn-off switches are not necessary.

A pair of silicon controlled rectifiers connected in inverse-parallel or a triac can perform the function of an electronic switch suitable for use with a.c. supply. The basic single-phase arrangement is shown in Fig. 8.1. If suitable gating pulses are applied to the SCRs while their respective anode voltages are positive, current conduction is initiated. The conduction angle depends on the firing-angle (often also known as the triggering angle), measured from anode voltage zero, and the phase-angle of the load to sinusoidal currents of

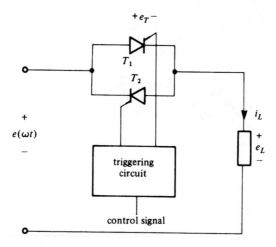

Fig. 8.1 Basic circuit arrangement of single-phase, inverse-parallel connected SCR pair.

supply frequency. With the arrangement of Fig. 8.1 the effective load voltage can be varied from zero, corresponding to extinction of both devices, to almost full supply voltage, corresponding to full conduction of both devices.

Two different modes of switching operation are mainly used:

(a) If triggering is used to permit complete cycles of load current followed by complete cycles of extinction, the load voltage and current waveform is variously described as 'burst firing', 'zero voltage switching', 'cycle selection', 'on–off control' or 'integral-cycle switching'. Some interesting properties and applications of this mode of control are described in Section 8.3 below.

(b) If each switch is triggered at some non-zero point on its respective anode voltage cycle, the load voltage waveform is described as 'phase-angle controlled'. With reactive loads the load voltage and current have different nonsinusoidal waveforms. When each SCR is triggered at an identical point on its respective anode voltage waveform (i.e. symmetrical triggering) the load voltage and current have identical positive and negative alternations with frequency spectra containing only odd harmonics.

By Kirchhoff's loop law the instantaneous thyristor voltage and load voltage in Fig. 8.1 always sum to the instantaneous value of the supply voltage.

$$e = e_T + e_L \tag{8.1}$$

When a switch is conducting, its forward voltage drop is of the order of 1 volt and this constitutes a reverse voltage on the reverse-connected device which is held in extinction. A current flowing in SCR T_1 of Fig. 8.1, for example, serves to reverse-bias SCR T_2. This cannot switch on, regardless of triggering condition, until the current in T_1 has fallen below its holding value (a few milliamperes).

8.1 RESISTIVE LOAD WITH SYMMETRICAL PHASE-ANGLE TRIGGERING

8.1.1 Harmonic properties

With resistive load a single gating pulse of magnitude 1–3 V is usually sufficient to switch on an SCR. From the point of view of power transfer the rise of current and consequent load voltage may be considered instantaneous. A set of theoretical waveforms for sinusoidal supply voltage with an arbitrary triggering angle $\alpha = 60°$ is given in Fig. 8.2, which is consistent with (8.1).

282 Single-phase voltage controllers

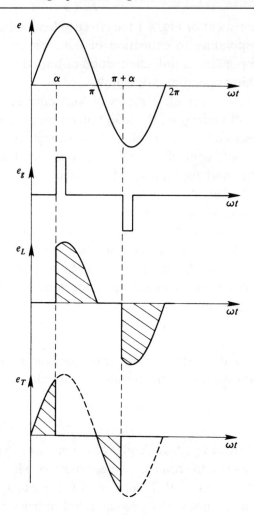

Fig. 8.2 Theoretical voltage waveforms for the circuit of Fig. 8.1. R load; $\alpha = 60°$.

Oscillograms obtained with actual circuits are very similar to the waveforms of Fig. 8.2, differing only in slight rounding of some of the discontinuities. The voltage waveforms are defined in terms of the peak supply phase voltage E_m by

$$e = E_m \sin \omega t \tag{8.2}$$

$$e_L = E_m \sin \omega t \big|_{\alpha, \pi+\alpha, \ldots}^{\pi, 2\pi, \ldots} \tag{8.3}$$

$$e_T = E_m \sin \omega t \big|_{0, \pi, \ldots}^{\alpha, \pi+\alpha, \ldots} \tag{8.4}$$

8.1 Resistive load with symmetrical phase-angle triggering

In (8.2)–(8.4) it is assumed that the supply voltage remains sinusoidal in the presence of the nonsinusoidal current drawn through any supply impedance.

The Fourier coefficients of the load voltage waveform are obtained by combining (8.3) with equations (A.6)–(A.10) from the Appendix

$$\frac{a_0}{2} = \frac{1}{2\pi} \int_0^{2\pi} e_L(\omega t)\, d\omega t \tag{A.6}$$

$$= \frac{1}{2\pi} \int_{\alpha,\pi+\alpha}^{\pi,2\pi} E_m \sin\omega t\, d\omega t$$

$$= 0 \tag{8.5}$$

It is seen from (8.5) that the time average or d.c. value of the function represented in (8.3) is zero. In any cycle of 2π radians the area under the positive wave is equal to the area under the negative wave.

For the fundamental (supply frequency) components it is seen that, in the present case,

$$a_1 = \frac{1}{\pi} \int_{\alpha,\pi+\alpha}^{\pi,2\pi} E_m \sin\omega t \cos\omega t\, d\omega t$$

$$= \frac{E_m}{2\pi} (\cos 2\alpha - 1) \tag{8.6}$$

$$b_1 = \frac{1}{\pi} \int_{\alpha,\pi+\alpha}^{\pi,2\pi} E_m \sin^2\omega t\, d\omega t$$

$$= \frac{E_m}{2\pi} [\sin 2\alpha + 2(\pi - \alpha)] \tag{8.7}$$

Coefficients a_1 and b_1 may be combined to give the peak amplitude $\hat{E}_{L_1}$ and phase-angle ψ_1 of the fundamental component of the load voltage as follows:

$$\hat{E}_{L_1} = c_1 = \sqrt{a_1^2 + b_1^2} = \frac{E_m}{2\pi} \sqrt{(\cos 2\alpha - 1)^2 + [\sin 2\alpha + 2(\pi - \alpha)]^2} \tag{8.8}$$

$$\psi_1 = \tan^{-1}\frac{a_1}{b_1} = \tan^{-1}\left[\frac{\cos 2\alpha - 1}{\sin 2\alpha + 2(\pi - \alpha)}\right] \tag{8.9}$$

When $\alpha = 90°$, for example, it is found that $a_1 = -E_m/\pi$, $b_1 = E_m/2$, $c_1 = 0.59 E_m$ and $\psi_1 = -32.5°$. The fundamental component $e_1(\omega t)$ of the discontinuous load voltage is therefore defined by the relationship

$$e_1(\omega t) = 0.59 E_m \sin(\omega t - 32.5°) \tag{8.10}$$

284 Single-phase voltage controllers

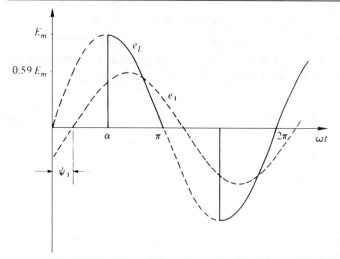

Fig. 8.3 Fundamental component of load voltage for the single-phase controller. R load; $\alpha = 90°$.

This waveform is shown in Fig. 8.3 together with the time variation of the actual load voltage $e_L(\omega t)$. It is important to note that the Fourier component $e_1(\omega t)$ of Fig. 8.3 and (8.12) does not exist physically. Component $e_1(\omega t)$ exists only as a mathematical entity but it is very useful in understanding the action and properties of the circuit. Variation of the per-unit amplitude c_1 with α is shown in Fig. 8.4.

Substitution of $e_L(\omega t)$ from (8.3) into (A.7)–(A.8), Appendix gives, for the nth Fourier harmonic,

$$a_n = \frac{E_m}{2\pi} \left[\frac{1}{n+1} \{1 + (-1)^{n+1}\}\{\cos(n+1)\alpha - 1\} \right. \\ \left. - \frac{1}{n-1}\{1 + (-1)^{n-1}\}\{\cos(n-1)\alpha - 1\} \right] \quad (8.11)$$

$$b_n = \frac{E_m}{2\pi} \left[\frac{\sin(n+1)\alpha}{n+1}\{1+(-1)^{n+1}\} - \frac{\sin(n-1)\alpha}{n-1}\{1+(-1)^{n-1}\} \right] \quad (8.12)$$

For odd harmonic terms the values of $(-1)^{n+1}$ and $(-1)^{n-1}$ are unity. Equations (8.11), (8.12) then reduce to

$$a_{n=3,5,7\ldots} = \frac{E_m}{2\pi}\left[\frac{2}{n+1}\{\cos(n+1)\alpha - 1\} - \frac{2}{n-1}\{\cos(n-1)\alpha - 1\}\right] \quad (8.13)$$

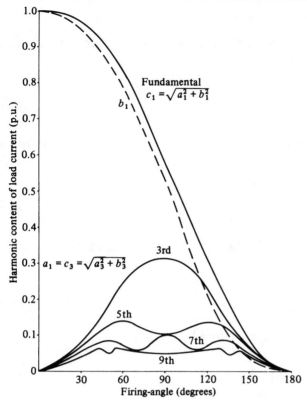

Fig. 8.4 Harmonic components of the load voltage for the single-phase controller. R load.

$$b_{n=3,5,7\ldots} = \frac{E_m}{2\pi}\left[\frac{2}{n+1}\sin(n+1)\alpha - \frac{2}{n-1}\sin(n-1)\alpha\right] \tag{8.14}$$

$$\hat{E}_{L_n} = c_n = \sqrt{a_n^2 + b_n^2} \tag{8.15}$$

$$\psi_n = \tan^{-1}\frac{a_n}{b_n} \tag{8.16}$$

For n = even and $n = 0$ (i.e. the d.c. component), coefficients a_n, b_n are zero. The Fourier spectrum of the load voltage therefore contains only odd harmonics with per-unit magnitudes as shown in Fig. 8.4. At $\alpha = 0°$, the load voltage is sinusoidal and therefore contains no higher harmonic components. Coefficient b_1 is predominant for small triggering angles and it is found that $a_1 = c_3$. At $\alpha = 90°$ the third harmonic component is about one-half the value of the fundamental component.

8.1.2 R.m.s. voltage and current

Any function $e_L(\omega t)$ that is periodic in 2π radians has a root mean square (r.m.s.) or effective value defined by

$$E_L = \sqrt{\frac{1}{2\pi} \int_0^{2\pi} e_L^2(\omega t)\, d\omega t} \qquad (8.17)$$

The function $e_L(\omega t)$ is defined by (8.3) for the circuit of Fig. 8.1. Substituting (8.3) into (8.17) gives

$$E_L = E_m \sqrt{\frac{1}{2\pi}\left[(\pi - \alpha) + \frac{\sin 2\alpha}{2}\right]} \qquad (8.18)$$

The r.m.s. value of the sinusoidal supply voltage is given by the standard relationship

$$E = \frac{E_m}{\sqrt{2}} \qquad (8.19)$$

The r.m.s. load voltage E_L can therefore be written

$$E_L = E\sqrt{\frac{1}{2\pi}[2(\pi - \alpha) + \sin 2\alpha]} \qquad (8.20)$$

The evaluation of (8.18) for successive values of α gives the relationship shown in Fig. 8.5. Over much of the anticipated working range the r.m.s. load voltage is 10–20% greater than the fundamental value.

With resistive load the instantaneous current is given by

$$i = \frac{e_L}{R} = \frac{E_m}{R}\sin \omega t \Big|_{\alpha,\pi+\alpha\ldots}^{\alpha\pi,2\pi\ldots} \qquad (8.21)$$

The r.m.s. load current I can therefore be written directly from (8.18)

$$I = \frac{E_m}{R}\sqrt{\frac{1}{2\pi}\left[(\pi - \alpha) + \frac{\sin 2\alpha}{2}\right]} \qquad (8.22)$$

In terms of the r.m.s. supply voltage E the r.m.s. current is given by

$$I = \frac{E}{R}\sqrt{\frac{1}{2\pi}[2(\pi - \alpha) + \sin 2\alpha]} \qquad (8.23)$$

In terms of its r.m.s. harmonic components, the circuit r.m.s. current may also be written

$$I^2 = I_1^2 + I_3^2 + I_5^2 + \ldots \qquad (8.24)$$

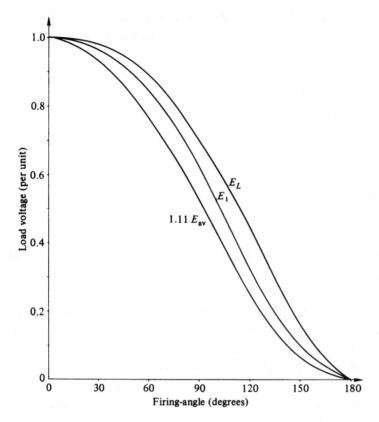

Fig. 8.5 Variation of r.m.s., fundamental and average load voltages with firing-angle. R load.

At $\alpha = 90°$, for example, the per-unit harmonic magnitudes are $I_1 = 0.59$, $I_3 = 0.33$, $I_5 = 0.10$, $I_7 = 0.10$, $I_9 = 0.05$, as seen from Fig. 8.4. Taking these five components only, gives $I = 0.692$ p.u. compared with the true value $I = 0.707$ p.u. obtained, in effect, by summing infinite harmonic components via equation (8.24).

The measurement of r.m.s. voltage and current in power electronic a.c. circuits raises problems. Neither a moving-iron instrument nor a rectifier moving-coil instrument (such as an Avometer) will give a reliable indication of the true r.m.s. value.

In a rectifier moving-coil instrument, for a.c. measurement, the torque on the instrument movement is proportional to the half-wave average current. The instrument scale is calibrated to read 1.11 times the average current, which then gives the r.m.s. value of a sinusoidal current. When delayed triggering is employed the form factor (i.e. the ratio r.m.s./average values) of the waveform is no longer 1.11 and the instrument indication can be wildly inaccurate, being very different from either the half-wave average or the r.m.s. values.

The mean or time average value of the load voltage is zero over any complete number of cycles. The half-cycle average value of the waveform $e_L(\omega t)$ of Fig. 8.2 is given by

$$\begin{aligned} E_{\text{av}} &= \frac{1}{\pi} \int_\alpha^\pi E_L(\omega t)\, \mathrm{d}\omega t \\ &= \frac{1}{\pi} \int_\alpha^\pi E_m \sin \omega t\, \mathrm{d}\omega t \\ &= \frac{E_m}{\pi}(1 + \cos \alpha) \end{aligned} \quad (8.25)$$

The indication on a rectifier moving-coil voltmeter is 1.11 times the p.u. value of (8.25) and is shown in Fig. 8.5. There is a large discrepancy from the true r.m.s. value E_L.

Standard types of moving-iron instrument have nominal frequency ranges of the order 25–100 Hz. In the circuit of Fig. 8.1 they are found to record a value close to the r.m.s. value of the fundamental component for 50 Hz operation. Such instruments thus act as a low-pass filter in which the lowest order higher harmonic (in this case the third harmonic, of frequency 150 Hz) is above the filter cut-off frequency. If it is required to accurately measure the r.m.s. value a 'true r.m.s.' instrument must be used, based on either the integration principle of (8.17) or the summation of harmonics principle of (8.24). Instruments such as the ironless type of dynamometer, electrostatic, thermal or some types of digital instruments, are appropriate.

8.1.3 Power and power factor

8.1.3.1 Average power

Power is defined as the time rate of energy transfer and it is appropriate to nominate as 'power' only the two variables instantaneous power and average power which have the dimension voltamperes and the physical nature of power.

The quantity often referred to as 'apparent power' in books on electrical power systems has the dimension voltamperes but does not have any physical nature at all. It is a figure of merit representing the energy-transfer capability of a system. To distinguish this from the real physical quantities it is hereafter referred to as 'apparent voltamperes'.

8.1 Resistive load with symmetrical phase-angle triggering

Similarly, the quantity usually known as 'reactive power' has the dimension voltamperes but no absolute physical form. It is not associated with energy dissipation and may or may not be associated with the storage of energy in fields of force.

One may thus distinguish by terminology between the real physical quantities instantaneous power and average power and the figures of merit known as apparent voltamperes and reactive voltamperes. This distinction has a number of basic properties especially relevant in circuit analysis. For example, it follows from the *Principle of Conservation of Energy* that the instantaneous power and average power at the terminals of an electric circuit are equal to the respective sums of the instantaneous and average powers in the individual circuit components. The principle of conservation of energy does not apply to apparent voltamperes and it may be shown that for any circuit, other than purely resistive or purely reactive, irrespective of supply voltage waveform, no simple analytical relationship exists between the apparent voltamperes at the terminals and the apparent voltamperes of individual circuit components. There is no such thing as the conservation of apparent voltamperes and no such thing as the conservation of reactive voltamperes.

The instantaneous current in the circuit of Fig. 8.1 is given by (8.21). The combination of equations (8.2) and (8.21) gives the instantaneous power

$$ei = \frac{E_m^2}{R} \sin^2 \omega t \Big|_{\alpha, \pi+\alpha, \ldots}^{\pi, 2\pi, \ldots} \tag{8.26}$$

Time variation of the instantaneous power is shown in Fig. 8.6 for an arbitrary triggering angle $\alpha = 60°$. The double supply frequency pulsation is positive for all time regardless of triggering angle. Average power P into the circuit of Fig. 8.1 is given by the basic defining integral

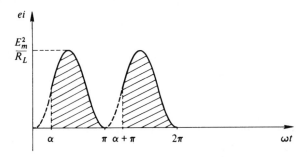

Fig. 8.6 Instantaneous terminal power for the single-phase controller. R load; $\alpha = 60°$.

$$P = \frac{1}{T}\int_0^T ei\,dt \tag{8.27}$$

where T is the period of the ei product. In the present case one can use the repetitive period π or 2π. Combining (8.26) and (8.27) gives

$$P = \frac{1}{\pi}\int_\alpha^\pi \frac{E_m^2}{R}\sin^2\omega t\,d\omega t$$

$$= \frac{E_m^2}{4\pi R}[2(\pi - \alpha) + \sin 2\alpha] \tag{8.28}$$

In terms of the r.m.s. supply voltage $E(= E_m/\sqrt{2})$ the power dissipation is

$$P = \frac{E^2}{2\pi R}[2(\pi - \alpha) + \sin 2\alpha] \tag{8.29}$$

The average power dissipation is equal to b_1 (p.u.) times the power dissipation E^2/R for a sinusoidal resistive circuit. Variation of the per-unit power with SCR firing-angle is therefore given by curve b_1 of Fig. 8.4.

In terms of the r.m.s. current I and r.m.s. load voltage E_L the power is given by

$$P = I^2 R = \frac{E_L^2}{R} \tag{8.30}$$

In terms of harmonic r.m.s. components

$$P = R(I_1^2 + I_3^2 + I_5^2 + \ldots)$$

$$= \frac{1}{R}(E_{L_1}^2 + E_{L_3}^2 + E_{L_5}^2 + \ldots) \tag{8.31}$$

Now, in any circuit, average power is dissipated only by those combinations of voltage and current that have the same frequencies. Also, in order to dissipate average power, these like-frequency voltage and current terms must have components in time-phase with each other. Since the terminal voltage is sinusoidal, in the present case, and of supply frequency, average power is dissipated by the supply voltage in combination with the in-phase component of the fundamental current,

$$P = EI_1 \cos\psi_1 \tag{8.32}$$

Average power P can be measured by the connection of an ironless dynamometer wattmeter. Such an instrument has a response that is practically independent of waveform and frequency, so that it indicates the average power accurately with nonsinusoidal voltage and/or current. The connection of a wattmeter at the terminals of the circuit of Fig. 8.1 would place the sinusoidal supply voltage across the potential coil. If the power loss in the switches and their heat sinks (which is of such order as to produce a potential difference of about one or two volts across an essentially dissipative device) is negligible, then a wattmeter with its potential coil across the supply will indicate the dissipation of load power fairly accurately.

8.1.3.2 Power factor

Irrespective of waveform, the power factor of a circuit is the factor by which the apparent voltamperes must be multiplied in order to give the average power

$$\text{power factor} = \frac{\text{average power}}{\text{apparent voltamperes}} = \frac{P}{EI} \qquad (8.33)$$

The most reliable way to determine the power factor in a nonsinusoidal circuit is to measure P, E and I separately and to use (8.33).

In a linear sinusoidal circuit of phase-angle Φ, with sinusoidal applied voltage, the power factor has the value $\cos\Phi$. The value $\cos\Phi$ does not represent the power factor when the current or voltage or both are nonsinusoidal. In the metering of nonsinusoidal systems it is essential to avoid labelling as 'power factor' the reading of (say) an induction-type power factor meter. If the system is nonsinusoidal such an instrument will only read $\cos\Phi$ when the fundamental component is very dominant; with a dominant third harmonic, for example, an induction-type power factor meter could give an indication equivalent to $\cos\Phi_3$.

The phase-angle between the voltage and current of a load at any harmonic frequency is a function only of the load itself, not of the excitation-voltage waveform. Because one is mostly concerned, in practice, with the phase-angle ψ_1, which represents the phase displacement between the fundamental components of voltage and current at the circuit terminals, the factor $\cos\psi_1$ is given the special name 'displacement factor' or fundamental power factor

$$\text{displacement factor} = \cos\psi_1 \qquad (8.34)$$

For circuits with sinusoidal supply voltage but nonsinusoidal current the distortion factor of the current may be significant

$$\text{distortion factor} = \frac{\text{power factor}}{\text{displacement factor}}$$

$$= \frac{EI_1 \cos \psi_1}{EI \cos \psi_1} \quad (8.35)$$

$$= \frac{I_1}{I}$$

Substituting from (8.29) into (8.33) gives

$$PF = \sqrt{\frac{1}{2\pi}[2(\pi - \alpha) + \sin 2\alpha]} \quad (8.36)$$

Comparison of (8.36) with (8.23) and (8.29) shows that

$$PF = I(\text{p.u.}) = \sqrt{P(\text{p.u.})} \quad (8.37)$$

Variation of power factor, displacement factor and distortion factor for the entire range of triggering angles is shown in Fig. 8.7. The power factor curve is identical to the r.m.s. voltage curve of Fig. 8.5 which is, in turn, proportional to the square root of the average power curve.

8.1.3.3 Reactive voltamperes and power factor correction

The load voltage waveform for an SCR firing-angle $\alpha = 90°$ is shown in Fig. 8.3. For this value of α the fundamental phase-angle ψ_1 is $-32.5°$. Also, from (8.8) the peak magnitude c_1 of the fundamental current is 0.59 per unit, as shown in Fig. 8.4. Fundamental current is shown in Fig. 8.3 lagging the supply voltage by $32.5°$.

Now a fundamental lagging component of current is usually associated with energy storage in a magnetic field. But in the circuit of Fig. 8.1, with resistive load, no storage of energy is possible. For this reason the pulsations of instantaneous power (Fig. 8.6) are always positive. Since the load is not capable of storing energy, no return of energy can occur from the load to the supply and the instantaneous power cannot go negative.

The reactive voltamperes is denoted by the symbol Q and is the complement of average power P in (8.32)

$$Q = EI_1 \sin \psi_1 \quad (8.38)$$

Although there is no oscillation of energy between the load and the supply, the delay of current caused by SCR switching creates a definite power factor problem in that the average power is less than the apparent voltamperes. It should be noted that although the value of the reactive voltamperes is not associated with magnetic energy storage it is still possible to obtain some

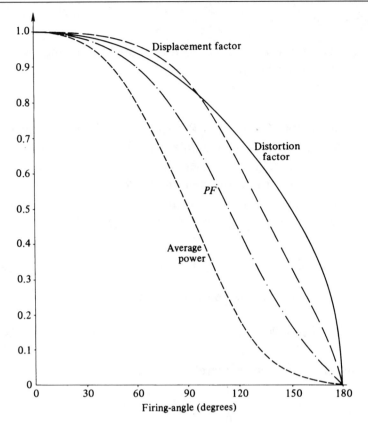

Fig. 8.7 Average power, power factor, distortion factor and displacement factor versus firing-angle. R load.

degree of power factor improvement by connecting a capacitor across the supply terminals as shown in Fig. 8.8. Substituting (8.8), (8.9) into (8.38) gives an expression for reactive voltamperes Q in terms of SCR firing-angle

$$Q = \frac{E^2}{2\pi R}(\cos 2\alpha - 1) \tag{8.39}$$

The per-unit variation of reactive voltamperes in the SCR controlled resistive circuit follows the characteristic $a_1 = c_3$ of Fig. 8.4.

In addition to the average power P and reactive voltamperes Q there is a distortion component of voltamperes D caused by combination of the fundamental frequency supply voltage with higher harmonic components of current.

$$D^2 = E^2(I_3^2 + I_5^2 + \ldots) \tag{8.40}$$

Apparent voltamperes S into the circuit can be thought of as the resultant of P, Q, D according to the relation

$$S^2 = E^2 I^2 = P^2 + Q^2 + D^2 \tag{8.41}$$

Although the components Q and D of the apparent voltamperes have no independent physical existence, it is sometimes useful to consider them individually when one is considering the problem of power factor compensation.

Some degree of power factor improvement may be obtained by connecting a capacitance across the circuit terminals, as in Fig. 8.8. The load current is unchanged but the supply current is now the sum of the load and capacitor currents

$$i_s(\omega t) = i_c(\omega t) + i(\omega t) \tag{8.42}$$

The capacitor current is continuous and is given by

$$i_c(\omega t) = \frac{E_m}{X_c} \sin(\omega t + 90°) \tag{8.43}$$

The instantaneous supply current is therefore given by

$$i_s(\omega t) = \frac{E_m}{X_c} \sin(\omega t + 90°) + \frac{E_m}{R} \sin \omega t \Big|_{\alpha, \pi+\alpha, \ldots}^{\pi, 2\pi, \ldots} \tag{8.44}$$

Since the supply voltage, the load current $i(\omega t)$ and the load power are unchanged by the presence of the capacitor, power factor improvement will occur if the r.m.s. supply current I_s is now smaller than the r.m.s. load current. The r.m.s. supply current is defined by

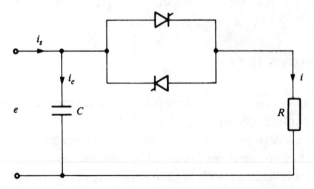

Fig. 8.8 Single-phase controller with terminal capacitance.

$$I_s = \sqrt{\frac{1}{2\pi}\int_0^{2\pi} i_s^2(\omega t)\, d\omega t} \tag{8.45}$$

Combining (8.44) and (8.45) is found to give the following expression

$$I_s^2 = E^2\omega^2 C^2 + I_1^2 - 2EI_1\omega C \sin\psi_1 + \sum_{n=3}^{\infty} I_n^2 \tag{8.46}$$

If (8.46) is differentiated with respect to C and equated to zero, the condition for minimum I_s is found to be when

$$C = \frac{I_1 \sin\psi_1}{\omega E} \tag{8.47}$$

Fig. 8.9 shows the time-phase relationships of the supply frequency components of (8.46). When $C = 0$ the supply current reduces to the load current. When the optimum compensation is used, then I_s is a minimum. Combining (8.46) and (8.47) gives

$$I_{s_{\min}}^2 = I_1^2 \cos^2\psi_1 + \sum_{n=3}^{\infty} I_n^2 \tag{8.48}$$

With optimum compensation the resultant current phasor I_{s_1} in Fig. 8.9 becomes in phase with phasor voltage E. Reactive voltamperes Q of the

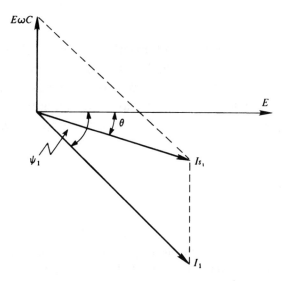

Fig. 8.9 Phasor diagram of the supply frequency components with terminal capacitance.

load (8.39) is then neutralised by the compensating leading reactive voltamperes of the capacitor $E\omega C$, as stated mathematically in (8.47).

In the presence of the capacitor the power factor of the circuit of Fig. 8.8 is given by

$$PF = \frac{P}{EI_s} \tag{8.49}$$

Combining (8.48) and (8.49) gives for the maximum power factor, where $I_s = I_{s_{\min}}$,

$$PF_{\max} = \frac{P}{E\sqrt{I_1^2 \cos^2 \psi_1 + \sum_{n=3}^{\infty} I_n^2}} \tag{8.50}$$

But $P = EI_1 \cos \psi_1$, from (8.32), and therefore

$$PF_{\max} = \frac{P}{E\sqrt{\frac{P^2}{E^2} + \sum_{n=3}^{\infty} I_n^2}} \tag{8.51}$$

If the total r.m.s. current I is known then the summation of the higher harmonic current components is given, from (8.24), by

$$\sum_{n=3}^{\infty} I_n^2 = I_3^2 + I_5^2 + I_7^2 = \ldots = I^2 - I_1^2 \tag{8.52}$$

Even with complete compensation of the load reactive voltamperes, the power factor is less than unity because of the higher harmonic components of the load (and supply) current I_n. In terms of components of the apparent voltamperes one can say that, in (8.45), although Q has been eliminated, the power factor is still reduced due to the effect of distortion voltamperes D.

8.1.4 Worked examples

Example 8.1
Derive an expression for the r.m.s. load current in a single-phase resistive circuit in which the load voltage is controlled by symmetrical phase-angle triggering of a pair of inverse-parallel connected SCRs. If the supply voltage is given by $e = 100 \sin \omega t$ and $R = 50 \Omega$, what is the r.m.s. load current at $\alpha = 30°, 60°, 90°, 120°$ and $150°$?

8.1 Resistive load with symmetrical phase-angle triggering

Solution. The instantaneous current in the circuit of Fig. 8.1 is given by (8.21)

$$i(\omega t) = \frac{E_m}{R} \sin \omega t \Big|_{\alpha, \pi+\alpha}^{\pi, 2\pi}$$

and has the same shape as instantaneous load voltage $e_L(\omega t)$ in Fig. 8.2. The r.m.s. value of this waveform is obtained from the defining integral

$$I = \sqrt{\frac{1}{2\pi} \int_0^{2\pi} i^2(\omega t)\, d\omega t}$$

Combining the above two equations, noting that the r.m.s. value of every half-wave is equal, gives

$$I^2 = \frac{1}{\pi} \int_\alpha^\pi \frac{E_m^2}{R^2} \sin^2 \omega t\, d\omega t$$

$$= \frac{E_m^2}{\pi R^2} \int_\alpha^\pi \frac{1 - \cos 2\omega t}{2}\, d\omega t$$

$$= \frac{E_m^2}{\pi R^2} \left[\frac{\omega t}{2} - \frac{\sin 2\omega t}{4} \right]_\alpha^\pi$$

$$= \frac{E_m^2}{2\pi R^2} \left[(\pi - \alpha) + \frac{\sin 2\alpha}{2} \right]$$

$$\therefore I = \frac{E_m}{R} \sqrt{\frac{1}{2\pi} \left[(\pi - \alpha) + \frac{\sin 2\alpha}{2} \right]}$$

$$= \frac{E}{R} \sqrt{\frac{1}{2\pi} [2(\pi - \alpha) + \sin 2\alpha]}$$

Substituting values of α into the equation gives the values of I shown in Table 8.1. The ratio E/R has the value $(100/\sqrt{2})(1/50) = \sqrt{2} = 1.4142\, \text{A}$.

Table 8.1 *Evaluation of I in Example 8.1.*

α	$2(\pi - \alpha)$	$\sin 2\alpha$	$\dfrac{1}{2\pi}[(\pi - \alpha) + \sin 2\alpha]$	$I(A)$
0	2π	0	1	1.414
30	$5\pi/3$	0.866	0.97	1.39
60	$4\pi/3$	0.866	0.804	1.27
90	π	0	0.5	1.0
120	$2\pi/3$	-0.866	0.196	0.625
150	$\pi/3$	-0.866	0.03	0.24
180	0	0	0	0

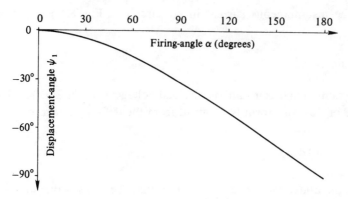

Fig. 8.10 Displacement-angle versus firing-angle for the single-phase controller. R load.

Example 8.2

An ideal single-phase source of instantaneous voltage $e = E_m \sin \omega t$ supplies power to a resistive load via a pair of inverse-parallel connected SCRs. Calculate and plot the displacement angle ψ_1 between the supply voltage and the fundamental component of the current at successive values of the firing-angle α.

Solution. The value of ψ_1 is given in (8.9) and is evaluated in Table 8.2:

$$\psi_1 = \tan^{-1}\left(\frac{a_1}{b_1}\right) = \tan^{-1}\left[\frac{\cos 2\alpha - 1}{\sin 2\alpha + 2(\pi - \alpha)}\right]$$

The variation of ψ_1 with α is given in Fig. 8.10. For firing-angles $\alpha \geq 90°$ the variation is seen to be virtually linear and is found to satisfy the relation $\psi_1 = 30° - 2\alpha/3$.

Example 8.3

For the resistively loaded SCR circuit of Example 8.1 calculate the r.m.s. value of the fundamental component of the current at thyristor firing-angles $\alpha = 0, 30°, 60°, 90°$ and $120°$.

Solution. The Fourier coefficients a_1 and b_1 for the load voltage with the inverse-parallel SCR controller of Fig. 8.1 are summarised in (8.8), which represents the peak value of the fundamental component. In this case $E_m = 100V$.

Utilising the data from Table 8.2 reduces the necessary manipulations. The required r.m.s. fundamental current values I_1 are given in Table 8.3.

8.1 Resistive load with symmetrical phase-angle triggering   299

Table 8.2 Evaluation of ψ_1 in Example 8.2.

α	$\cos 2\alpha$	$\cos 2\alpha - 1$	$\sin 2\alpha$	$2(\pi - \alpha)$	$\sin 2\alpha + 2(\pi - \alpha)$	$\dfrac{a_1}{b_1}$	ψ_1
0	1	0	0	2π	2π		
30	0.5	−0.5	0.866	$5\pi/3$	6.1	−0.082	−4.7°
60	−0.5	−1.5	0.866	$4\pi/3$	5.05	−0.3	−16.5°
90	−1	−2	0	π	3.142	−0.637	−32.5°
120	−0.5	−1.5	−0.866	$2\pi/3$	1.23	−1.22	−50.6°
150	0.5	−0.5	−0.866	$\pi/3$	0.181	−2.76	−70°
180	1.0	0	0	0	0		−90°

Table 8.3 Evaluation of I_1 in Example 8.3.

α	a_1^2 $(\cos 2\alpha - 1)^2$	b_1^2 $[2(\pi - \alpha) + \sin 2\alpha]^2$	$\sqrt{a_1^2 + b_1^2}$	c_1	I_1(A)
0	0	$4\pi^2$	2π	100	1.4142
30	0.25	37.2	6.12	97.4	1.37
60	2.25	25.5	5.27	83.9	1.186
90	4	9.87	3.72	59.2	0.837
120	2.25	1.51	1.94	30.9	0.437
150	0.25	0.032	0.532	8.47	0.12
180	0	0	0	0	0

Example 8.4
A resistive load of 50 Ω is supplied with power from an ideal single-phase source, of instantaneous e.m.f. $e = 100 \sin \omega t$ at 50 Hz, via a pair of ideal SCRs connected in inverse-parallel. Calculate and plot the variation of system power factor for a range of values of firing-angle α. If a variable capacitor C is connected across the supply terminals calculate the values that will give maximum power factor over the range of α. Calculate the maximum power factor realisable by pure capacitance compensation and plot this, for comparison, with the uncompensated power factor.

Solution. The power factor of the uncompensated load is given by (8.36), repeated here,

$$PF = \sqrt{\frac{1}{2\pi}[2(\pi - \alpha) + \sin 2\alpha]}$$

Values for the bracketed term in the PF expression may be obtained from the table in Example 8.2. Variation of the uncompensated PF is given in Fig. 8.11 with the appropriate values recorded in Table 8.4. The capacitance to give optimum compensation is given by (8.47), quoted here,

$$C = \frac{I_1 \sin \psi_1}{\omega E} = \frac{I_1 \sin \psi_1 \times \sqrt{2}}{2\pi \times 50 \times 100}$$

Utilising the values of ψ_1 in Table 8.2 and the values of I_1 in Table 8.3 enables $I_1 \sin \psi_1$ and C to be calculated, as given in Table 8.4.

The variation of the optimum capacitance is shown in Fig. 8.11. It was seen in (8.38) that the variation of load reactive voltamperes Q is proportional to the product $I_1 \sin \psi_1$. But, from (8.39), Q is also proportional to fundamental Fourier coefficient a_1, so that product $I_1 \sin \psi_1$ is proportional to a_1. It follows, from (8.47), that the variation of optimum capacitance C is also proportional to coefficient a_1. A comparison of Fig. 8.11 with Fig. 8.4 confirms that this is so.

In the presence of the capacitor the PF is given by (8.49). But the load power P may be written in terms of r.m.s. load current. Therefore

$$PF = \frac{P}{EI_s} = \frac{I^2 R}{EI_s}$$

When the optimum value of the capacitor is used and $PF = PF_{max}$ the equation forms (8.50) and (8.51) may be used. The values of I from Example 8.1 and I_1 from Example 8.3 are used to calculate $\Sigma I^2 3_n$ via

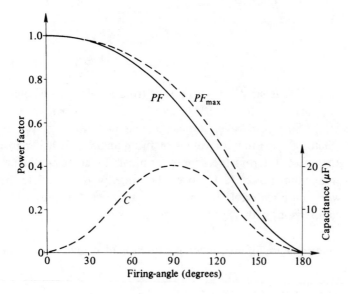

Fig. 8.11 Power factor, maximum power factor and optimum capacitance for the single-phase controller. R load.

Table 8.4 Evaluation of maximum power factor PF_{max} in Example 8.4.

α	$\dfrac{1}{2\pi}[2(\pi-\alpha)+\sin 2\alpha]$	PF	$I_1 \sin\psi_1$	C	$\sum_3^\infty I_n^2$	$I_1^2 \cos^2\psi_1$	$\sqrt{I_1^2 \cos^2\psi_1 + \sum I_n^2}$	$P(I^2R)$	PF_{max}
0	1	1	0	0	0	2	1.414	100	1.0
30	0.971	0.985	0.112	5.04	0.055	1.86	1.38	96.6	0.99
60	0.804	0.896	0.337	15.17	0.206	1.29	1.223	80.6	0.93
90	0.5	0.707	0.45	20.2	0.3	0.5	0.894	50	0.79
120	0.196	0.442	0.338	15.2	0.2	0.077	0.53	19.5	0.52
150	0.0288	0.17	0.112	5.04	0.043	0.0017	0.21	2.88	0.19
180	0	0	0	0	0	0	0	0	0

(8.52). The result is included in Table 8.4. The component of the r.m.s. fundamental current in-phase with the supply voltage $I_1 \cos \psi_1$ is obtained from the information of Tables 8.2 and 8.3. This is also included in Table 8.4 along with the load power dissipation $P = I^2 R$. Substituting values from Table 8.4 into (8.50) gives the corresponding values of $PF_{\max}$, which is plotted in Fig. 8.11. It is seen that the degree of power factor improvement realised by optimum capacitance compensation is small, being a maximum of about 10% improvement at $\alpha = 90°$.

Example 8.5

The power to a resistive load R from an ideal single-phase voltage supply $e = E_m \sin \omega t$ is to be regulated by a voltage controller consisting of a pair of SCRs connected in inverse-parallel. Each device is triggered at an identical point α with respect to its positive-going anode voltage zero to produce symmetrical phase-angle control. Load voltage is required at various levels from zero up to full supply voltage, with a particular requirement to supply 65% of full load power.

(a) Show that the load power $P_L(\alpha)$ at triggering angle α, compared with the load power $P(0)$ for sinusoidal operation, is given by

$$\frac{P(\alpha)}{P(0)} = \frac{1}{\pi}\left[(\pi - \alpha) + \frac{1}{2}\sin 2\alpha\right]$$

(b) Calculate the value of α to make the ratio $P_L(\alpha)/P_L(0)$ equal to 65%. What is the per-unit value of the r.m.s. load current at this value of α?

Solution. (a) The load power dissipation is given by (8.28)

$$P(\alpha) = \frac{E^2}{2\pi R}[2(\pi - \alpha) + \sin 2\alpha]$$

for an arbitrary firing-angle α.
At $\alpha = 0$, the power equation becomes

$$P(0) = \frac{E^2}{2\pi R} \times 2\pi = \frac{E^2}{R}$$

The ratio $P(\alpha)/P(0)$ is therefore

$$\frac{P(\alpha)}{P(0)} = \frac{1}{\pi}[(\pi - \alpha) + \tfrac{1}{2}\sin 2\alpha] \qquad \text{QED}$$

(b) Now let

$$\frac{1}{\pi}[(\pi - \alpha) + \tfrac{1}{2}2\alpha] = 0.65$$

Rearranging,

Table 8.5 *Iterative evaluation of α in Example 8.5.*

α degrees	rads	$-\frac{1}{2}\sin 2\alpha$	$\alpha - \frac{1}{2}\sin 2\alpha$
90	1.57	0	1.57
80	1.396	0.171	1.225
70	1.222	0.321	0.9
75	1.31	0.25	1.06
76	1.336	0.235	1.091

$$\alpha - \tfrac{1}{2}\sin 2\alpha = \pi(1 - 0.65) = 1.1$$

This equation in α is transcendental and has to be solved by iteration, as shown in Table 8.5. The value $\alpha = 76°$ is very close to the required value. When $P = 0.65$ p.u. the corresponding r.m.s. load current is given by

$$I = \sqrt{P(\text{p.u.})} = \sqrt{0.65} = 0.806 \,\text{p.u.}$$

8.2 SERIES *R–L* LOAD WITH SYMMETRICAL PHASE-ANGLE TRIGGERING

Consider the single-phase voltage controller circuit in which the load now consists of resistor R in series with inductor L. Any resistance associated with coil L is presumed to be included in the resistor R, Fig. 8.12. The instantaneous supply voltage, e, is now the sum of three components, consisting of the SCR voltage drop, e_T, the resistance drop, $e_R = iR$, and the voltage e_l across the inductor

$$\begin{aligned} e &= e_T + e_R + e_l \\ &= e_T + e_L \end{aligned} \quad (8.53)$$

In order to initiate and maintain conduction in a circuit with series inductance, it is necessary to sustain the triggering pulse or to use a train of short pulses. Typical pulse-train frequencies used in practice are of the order of 2.5 kHz with a mark–space ratio 1/10 for 50 Hz supply voltage. The pulse width is of necessity long compared with the turn-on time of an SCR.

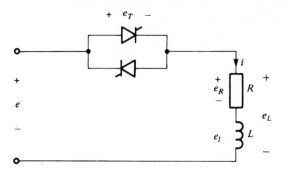

Fig. 8.12 Single-phase controller with series R–L load.

8.2.1 Analysis of the instantaneous current variation

When the switches are triggered by long pulses at a triggering angle α less than or equal to the sinusoidal phase angle, Φ, of the load impedance, sinusoidal operation results. The use of single short triggering pulses when $\alpha < \Phi$ could cause only one device to conduct because the continuation of conduction after the end of the voltage half cycle ensures that gating of the reverse device would have no effect. The SCR pair would then act as a rectifier.

When the triggering angle α is greater than the load phase-angle Φ the current occurs in discontinuous, nonsinusoidal, alternating pulses. Typical waveforms are shown in Fig. 8.13 for a load of phase angle $\Phi = 60°$ or power factor 0.5 lagging and a triggering angle $\alpha = 120°$. For such cases, when the triggering or firing-angle α is greater than the sinusoidal phase-angle Φ, the onset of the nonsinusoidal load current pulses always coincides with the triggering angle and the conduction of current is found to cease prior to the end of the sinusoidal current cycle. The range of possible load current waveforms varies from the dashed sinusoid of Fig. 8.13, when $\alpha < \Phi$, to pulses of current with their leading edges at $\alpha + n\pi$ radians of the applied voltage cycle and their conduction periods approaching zero as α approaches $180° + \Phi$.

The analytical components of the current waveform are demonstrated in Fig. 8.14. At the instant α of SCR triggering the load current is zero.

If the applied voltage $e(\omega t)$ is given by (8.2) then the steady-state component of current $i_{ss}(\omega t)$ in Fig. 8.14 is given by

$$i_{ss}(\omega t) = I_m \sin(\omega t - \Phi)$$
$$= \frac{E_m}{|Z|} \sin(\omega t - \Phi) \qquad (8.54)$$

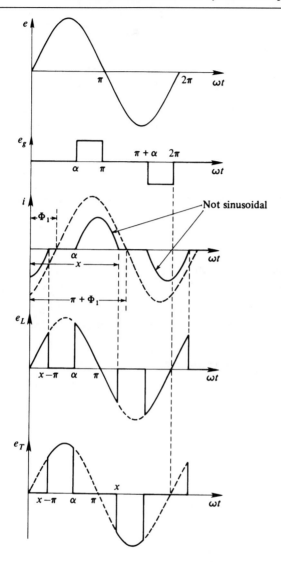

Fig. 8.13 Voltage and current waveforms for single-phase controller. Series R–L load, $\Phi = 60°$, $\alpha = 120°$.

where

$$|Z| = \sqrt{R^2 + \omega^2 L^2} \tag{8.55}$$

and

$$\Phi = \tan^{-1} \frac{\omega L}{R} \tag{8.56}$$

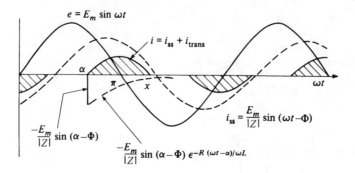

Fig. 8.14 Analytical components of current waveform for single-phase controller. Series R–L load, $\Phi = 60°$, $\alpha = 120°$.

At $\omega t = \alpha$, in Fig. 8.14, the steady-state current component has the value $i_{ss}(\alpha) = I_m \sin(\alpha - \Phi)$. The resultant load current is considered to consist of the steady-state sinusoid $i_{ss}(\omega t)$ plus a transient component $i_{trans}(\omega t)$ that decays exponentially with a time constant τ determined by the load impedance, where $\tau = L/R$.

At the instant $\omega t = \alpha$, the total current is zero. Therefore,

$$i(\omega t) = i_{ss}(\alpha) + i_{trans}(\alpha) = 0 \tag{8.57}$$

Combining (8.54) and (8.57) gives

$$i_{trans}(\alpha) = -i_{ss}(\alpha) = -\frac{E_m}{|Z|}\sin(\alpha - \Phi) \tag{8.58}$$

The transient component that pertains in the interval $\alpha \leq \omega t \leq x$, Fig. 8.14, is defined by

$$i_{trans}(\omega t) = -\frac{E_m}{|Z|}\sin(\alpha - \Phi)\varepsilon^{-(\omega t - \alpha)/\omega \tau} \tag{8.59}$$

Now

$$\frac{1}{\omega \tau} = \frac{R}{\omega L} = \cot \Phi = c \tag{8.60}$$

In the interval $\alpha \leq \omega t \leq x$, the total current is therefore described by the equation

$$i(\omega t)|_{\alpha \leq \omega t \leq x} = \frac{E_m}{|Z|}\sin(\omega t - \Phi) - \frac{E_m}{|Z|}\sin(\alpha - \Phi)\varepsilon^{-\cot\Phi(\omega t - \alpha)} \tag{8.61}$$

In Fig. 8.14 for the part of the current cycle such that $x \leq \omega t \leq \pi + \Phi$, the resultant current is mathematically negative. However, the conducting switch will not permit the flow of reverse current so that conduction ceases at point x, which is called the extinction angle or cut-off angle. Over the first supply voltage cycle, the current is described by the equation

$$i(\omega t) = \frac{E_m}{|Z|} \left[\sin(\omega t - \Phi) \big|_{0,\alpha,\alpha+\pi}^{x-\pi,x,2\pi} + \sin(\alpha - \Phi) e^{-\cot \Phi(\omega t + \pi - \alpha)} \big|_0^{x-\pi} \right.$$
$$\left. - \sin(\alpha - \Phi) e^{-\cot \Phi(\omega t - \alpha)} \big|_\alpha^x + \sin(\alpha - \Phi) e^{-\cot \Phi(\omega t - \pi - \alpha)} \big|_{\pi+\alpha}^{2\pi} \right]$$
(8.62)

The extinction angle x in Fig. 8.14 can be obtained from (8.61) by noting that $i(\omega t) = 0$ when $\omega t = x$. This results in the transcendental equation

$$\sin(x - \Phi) - \sin(\alpha - \Phi) e^{-\cot \Phi(x-\alpha)} = 0 \quad (8.63)$$

Only an iterative solution of (8.63) is possible. This yields the set of characteristics shown in Fig. 8.15. When $\alpha = \Phi$, which represents sinusoidal operation, (8.63) reduces to $\sin(x - \Phi) = 0$, which gives $x = \pi + \Phi$.

The values x for sinusoidal operation are seen to lie on the dashed linear characteristic of Fig. 8.15. For a purely inductive load, therefore, the variation of x with α is linear, as shown in the $\Phi = 90°$ characteristic of Fig. 8.15.

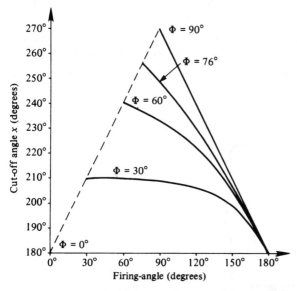

Fig. 8.15 Extinction angle versus firing-angle for single-phase controller. Series R–L load.

An approximate value of the extinction angle x can be found from the simple relationship

$$x = \pi + \Phi - \Delta \tag{8.64}$$

where Δ is $0°–5°$ for small phase-angles and $10°–25°$ for large phase-angles. Use of the rough value found from (8.64) is a helpful starting point in the iterative solution of (8.63). A numerical example comparing the approximate and iterative solutions is given in Example 8.6 below for the case $\alpha = 120°$, $\Phi = 60°$.

With a purely inductive load, analysis of the circuit is best approached in terms of classical differential equations. For forward conduction, in the circuit of Fig. 8.16, nonsinusoidal pulses of current occur in the interval $\alpha \leq \omega t \leq x$. Then

$$E_m \sin \omega t = \omega L \frac{di}{d\omega t} \tag{8.65}$$

Integrating both sides of (8.65) between the limits α and x gives

$$i(\omega t)|_{\alpha \leq \omega t \leq x} = \frac{E_m}{\omega L}(\cos \alpha - \cos \omega t) \tag{8.66}$$

For reverse conduction, integrating (8.65) in the range $\pi + \alpha \leq \omega t \leq \pi + x$ gives

$$i(\omega t)|_{\pi + \alpha \leq \omega t \leq \pi + x} = \frac{E_m}{\omega L}(-\cos \alpha - \cos \omega t) \tag{8.67}$$

The current with purely inductive load is therefore a sinusoid that is displaced vertically from the ωt axis by a value $\pm \cos \alpha$ for positive and negative conduction respectively. Specimen current waveforms are shown in Fig. 8.17 for

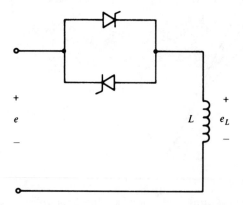

Fig. 8.16 Single-phase controller with purely inductive load.

8.2 Series R–L load with symmetrical phase-angle triggering

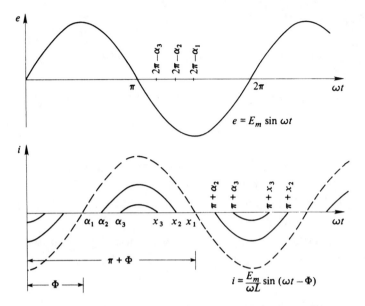

Fig. 8.17 Current waveforms for the single-phase controller with purely inductive load: $\alpha_1 = 90°$, $\alpha_2 = 120°$, $\alpha_3 = 150°$.

$\alpha = 90°$, $120°$ and $150°$. It can be seen that, for any firing-angle, the non-sinusoidal current pulses are symmetrical about $\omega t = 0$, π, 2π, etc.

8.2.2 Harmonic properties of the current

The Fourier harmonic components of the current function $i(\omega t)$, equation (8.61), are found to be

$$\frac{a_0}{2} = \frac{1}{2\pi}\int_0^{2\pi} i(\omega t)\, d\omega t = 0 \tag{8.68}$$

$$a_1 = \frac{1}{\pi}\int_0^{2\pi} i(\omega t)\cos\omega t\, d\omega t$$

$$= \frac{E_m}{2\pi|Z|}[\cos(2\alpha - \Phi) - \cos(2x - \Phi) - \sin\Phi(2x - 2\alpha)$$

$$+ 4\sin\Phi\sin(\alpha - \Phi)\{\cos(\Phi + x)\varepsilon^{-\cot\Phi(x-\alpha)} - \cos(\Phi + \alpha)\}] \tag{8.69}$$

$$b_1 = \frac{1}{\pi}\int_0^{2\pi} i(\omega t)\sin\omega t\, d\omega t$$

$$= \frac{E_m}{2\pi|Z|}[\sin(2\alpha - \Phi) - \sin(2x - \Phi) + \cos\Phi(2x - 2\alpha)$$

$$+ 4\sin\Phi\sin(\alpha - \Phi)\{\sin(\Phi + x)\varepsilon^{-\cot\Phi(x-\alpha)} - \sin(\Phi + \alpha)\}]$$

(8.70)

Alternative expressions for a_1, b_1 that do not contain exponential terms can be obtained by separately combining (8.69), (8.70) with (8.63).

The peak value $\hat{I}_1$ of the fundamental component of current is therefore obtained by substituting (8.69), (8.70) into (8.8).

Variation of the fundamental current with triggering angle is shown in Fig. 8.18 for several fixed values of phase-angle. Until the triggering angle is

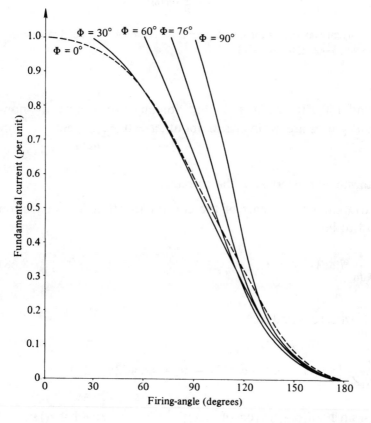

Fig. 8.18 Fundamental current versus firing-angle for single-phase controller. Series R–L load.

8.2 Series R–L load with symmetrical phase-angle triggering

greater than the load phase-angle, sinusoidal operation is maintained and the fundamental current retains unity per-unit value. At $\Phi = 60°$, for example, it is seen that controlled variation of the fundamental current is not possible until $\alpha > 60°$.

With purely resistive load, the exponential terms of the current equation (8.62) have zero value and the current wave consists of a sinusoid with symmetrical pieces chopped out, as discussed in Section 8.1 above, defined by (8.6) and (8.7).

With a highly inductive load the current waveform is symmetrical about $\omega t = 0$ and fundamental Fourier coefficient b_1 is therefore zero. This may be confirmed by substituting (8.66), (8.67) into (A.9) and (A.10), Appendix, or by putting $\Phi = 90°$ and $x = 2\pi - \alpha$ into (8.69) and (8.70). It is found that

$$a_1|_{\Phi=90°} = \frac{-E_m}{\pi \omega L}[(\pi - \alpha) + \sin 2\alpha] \tag{8.71}$$

$$b_1|_{\Phi=90°} = 0 \tag{8.72}$$

For the nth Fourier harmonic the coefficients a_n, b_n are given for general load phase-angle Φ, by

$$\begin{aligned}a_n &= \frac{1}{\pi}\int_0^{2\pi} i(\omega t)\cos n\omega t\, d\omega t \\ &= \frac{E_m}{2\pi|Z|}\Bigg[\frac{2}{n+1}\{\cos[(n+1)\alpha - \Phi] - \cos[(n+1)x - \Phi]\} \\ &\quad + \frac{2}{1-n}\{\cos[(1-n)\alpha - \Phi] - \cos[(1-n)x - \Phi]\} \\ &\quad + \frac{4\sin(\alpha - \Phi)}{n^2 + \cot^2\Phi}\{(\cot\Phi\cos nx - n\sin nx)\varepsilon^{-\cot\Phi(x-\alpha)} \\ &\quad - (\cot\Phi\cos n\alpha - n\sin n\alpha)\}\Bigg]\end{aligned} \tag{8.73}$$

$$b_n = \frac{1}{\pi}\int_0^{2\pi} i(\omega t)\sin n\omega t \, d\omega t$$

$$= \frac{E_m}{2\pi|Z|}\left[\frac{2}{n+1}\{\sin[(n+1)\alpha - \Phi] - \sin[(n+1)x - \Phi]\}\right.$$

$$- \frac{2}{1-n}\{\sin[(1-n)\alpha - \Phi] - \sin[(1-n)x - \Phi]\}$$

$$+ \frac{4\sin(\alpha - \Phi)}{n^2 + \cot^2\Phi}\{(\cot\Phi\sin nx - n\cos nx)\varepsilon^{-\cot\Phi(x-\alpha)}$$

$$\left. - (\cot\Phi\sin n\alpha - n\cos n\alpha)\}\right] \tag{8.74}$$

8.2.3 R.m.s. current

The dissipation in a series circuit is proportional to the square of the total r.m.s. current. If the magnitudes of the steady-state r.m.s. current harmonics are now denoted by I_n, the resultant r.m.s. current is given by

$$I = \sqrt{I_1^2 + I_3^2 + I_5^2 + \ldots}$$

$$= \sqrt{\sum_{n=1}^{\infty} I_n^2} \tag{8.75}$$

For *R–L* loads, the difference between the total r.m.s. current and the fundamental current diminishes as power factor decreases. Characteristics of r.m.s. current for different load power factors are similar in form to those of Fig. 8.18. The r.m.s. current can also be obtained from its defining integral

$$I = \sqrt{\frac{1}{2\pi}\int_0^{2\pi} i^2(\omega t)\, d\omega t} \tag{8.76}$$

Substituting (8.62) into (8.76) is found to give the result

$$I^2 = \frac{E_m^2}{2\pi|Z|^2}\left[(x-\alpha) - \frac{1}{2}\{\sin 2(x-\Phi) - \sin 2(\alpha - \Phi)\}\right.$$

$$+ \frac{\sin^2(\alpha - \Phi)}{\cot\Phi}\{1 - \varepsilon^{2\cot\Phi(\alpha-x)}\}$$

$$\left. + 4\sin\Phi\sin(\alpha - \Phi)\{\sin x\varepsilon^{\cot\Phi(\alpha-x)} - \sin\alpha\}\right] \tag{8.77}$$

When $\Phi = 0$, for resistive loads, $x = \pi$ and (8.77) reduces to (8.22).

8.2.4 Properties of the load voltage

It is seen in Fig. 8.13 that, in the presence of a series R–L load, the load voltage waveform has the shape of a sinusoid with a vertical segment chopped out. The 'missing' portions of the load voltage waveform form the voltage drop across the switches during the extinction periods. The general expression for time variation of the load voltage waveform for a typical steady-state cycle is

$$e_L(\omega t) = E_m \sin \omega t \Big|_{0,\alpha,\pi+\alpha}^{x-\pi,x,2\pi} \tag{8.78}$$

Oscillograms of circuit operation show waveforms very similar to the theoretical shapes of Fig. 8.13. Small spikes of load voltage occur at the current extinction points caused by recovery transients due to the dv/dt effect of the voltage suddenly reapplied to the SCRs. Fourier coefficients for the load voltage waveform are obtained by substituting (8.78) into (A.6), (A.7) and (A.8), Appendix, respectively. The average or d.c. component is

$$\frac{a_0}{2} = \frac{1}{2\pi} \int_0^{2\pi} e_L(\omega t) \, d\omega t = 0 \tag{8.79}$$

For the fundamental component

$$a_1 = \frac{1}{\pi} \int_0^{2\pi} e_L(\omega t) \cos \omega t \, d\omega t$$
$$= \frac{E_m}{2\pi} (\cos 2\alpha - \cos 2x) \tag{8.80}$$

$$b_1 = \frac{1}{\pi} \int_0^{2\pi} e_L(\omega t) \sin \omega t \, d\omega t$$
$$= \frac{E_m}{2\pi} [2(x - \alpha) - \sin 2x - \sin 2\alpha] \tag{8.81}$$

In the usual way the peak value of the fundamental load voltage is given by

$$\hat{E}_1 = c_1 = \sqrt{a_1^2 + b_1^2} \tag{8.82}$$

Now the fundamental component of the load voltage will be phase-displaced from the supply voltage by an angle ψ_{v_1}, which is different from the current displacement angle ψ_1,

$$\psi_{v_1} = \tan^{-1}\left(\frac{a_1}{b_1}\right) \tag{8.83}$$

where a_1, b_1 are now given by (8.80) and (8.81).

Variation of the fundamental load voltage with SCR firing-angle is found to be similar in form to the fundamental current characteristics of Fig. 8.18. For resistive load, $\Phi = 0$, $x = \pi$ and (8.80), (8.81) reduce to (8.6), (8.7) respectively. Fourier coefficients for the nth higher harmonic of the load voltage waveform are found to be

$$a_n = \frac{1}{\pi}\int_0^{2\pi} e_L(\omega t)\cos n\omega t\, d\omega t$$

$$= \frac{E_m}{2\pi}\left[\frac{2}{n+1}\{\cos(n-1)\alpha - \cos(n-1)x\} \right.$$
$$\left. - \frac{2}{n-1}\{\cos(n-1)\alpha - \cos(n-1)x\}\right] \tag{8.84}$$

$$b_n = \frac{1}{\pi}\int_0^{2\pi} e_L(\omega t)\sin n\omega t\, d\omega t$$

$$= \frac{E_m}{2\pi}\left[\frac{2}{n+1}\{\sin(n+1)\alpha - \sin(n+1)x\}\right.$$
$$\left. - \frac{2}{n-1}\{\sin(n-1)\alpha - \sin(n-1)x\}\right] \tag{8.85}$$

The peak value $\hat{E}_{L_n}$ of the nth higher harmonic component is given by coefficient c_n and the phase displacement is ψ_n, (A2) and (A3), Appendix. The nature of the load voltage waveform of Fig. 8.13 shows that only odd higher order harmonics ($n = 3, 5, 7, 9 \ldots$) are present.

Substituting (8.78) into (8.17) and applying the appropriate limits is found to give

$$E_L = E_m\sqrt{\frac{1}{2\pi}\left[(x-\alpha) + \frac{1}{2}\sin 2\alpha - \frac{1}{2}\sin 2x\right]} \tag{8.86}$$

With resistive load, $x = \pi$ and (8.86) reduces to (8.18).

8.2.5 Power and power factor

The instantaneous power $ei(\omega t)$ into the circuit of Fig. 8.12 involves the product of equations (8.2) and (8.62). This instantaneous power is a double supply frequency sinusoid of average value P that is negative for two short

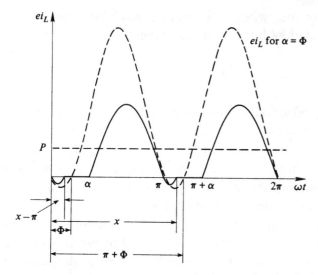

Fig. 8.19 Instantaneous terminal power for the single-phase controller. Series R–L load, $\Phi = 30°$, $\alpha = 60°$.

intervals in each supply voltage cycle. During the time intervals when ei is negative, energy is being transferred from the magnetic field of the load inductor back into the supply. With retarded triggering the time variation of the instantaneous power becomes discontinuous and the average value P reduces in value as shown in Fig. 8.19.

For a periodicity 2π the average power P is therefore defined by

$$P = \frac{1}{2\pi}\int_0^{2\pi} i(\omega t)E_m \sin \omega t \, d\omega t \tag{8.87}$$

For a series R–L load the per unit average power is given by

$$P = \frac{E^2}{2\pi|Z|}[\sin(2\alpha - \Phi) - \sin(2x - \Phi) + \cos\Phi \cdot (2x - 2\alpha)$$
$$+ 4\sin\Phi\sin(\alpha - \Phi)\{\sin(\Phi + x)^{\cot\Phi(\alpha-x)} - \sin(\Phi + \alpha)\}] \tag{8.88}$$

For sinusoidal operation $\alpha = \Phi$, $x = \pi + \Phi$ and (8.88) reduces to

$$P_{\alpha=0} = \frac{E^2}{|Z|}\cos\Phi = EI\cos\Phi \tag{8.89}$$

Equation (8.88) is tedious to evaluate and it may be more convenient to evaluate the r.m.s. current I and use the universal expression (8.30).

316 Single-phase voltage controllers

Since the dissipation in the switches is assumed negligible, the average power can also be associated with the load-side quantities as follows:

$$P = E_{L_1} I_1 \cos(\psi_{v_1} - \psi_1) + \sum_{n=3}^{\infty} I_n^2 R \tag{8.90}$$

The last term of (8.90) defines the power dissipation in load resistor R, Fig. 8.12, due to the higher harmonic components of current. Such a term does not occur explicitly in (8.32), which defines the total average power entering the circuit terminals. For sinusoidal operation, the angle $\psi_{v_1} - \psi_1$ between the fundamental components of the load voltage and current is equal to the load impedance angle Φ. Some degree of power factor improvement can be obtained by the connection of suitable parallel capacitance across the circuit terminals. The appropriate value of capacitance is found by determining the conditions for the minimum value of r.m.s. supply current, as described in Section 8.1.3(c), above.

8.2.6 Worked examples

Example 8.6
An ideal single-phase voltage source $e = E_m \sin \omega t$ supplies power to a series R–L load of phase-angle $\Phi = 60°$ via a pair of inverse-parallel connected SCRs. The SCR firing-angle $\alpha = 120°$. Calculate the current extinction angle x by an approximate method and use iteration to obtain the accurate value.

Solution. The circuit diagram is given in Fig. 8.12 and the nonsinusoidal current wave is shown in Fig. 8.13 and, in more detail in Fig. 8.14.
By the approximate method, (8.64), the extinction angle is

$$x = \pi + \Phi - \Delta$$

For $\Phi = 60°$, Δ has the estimated value of 15°. Therefore,

$$x \approx \pi + 60° - 15° = 220°$$

An accurate value of x is obtained by iteration from (8.63), from which it is seen that

$$\cot \Phi = \cot 60° = 0.58$$

$$\alpha = 120° = \frac{2\pi}{3} = 2.09 \text{ rad}$$

$$\sin(\alpha - \Phi) = \sin 60° = 0.866$$

8.2 Series R–L load with symmetrical phase-angle triggering

Table 8.6 *Iterative evaluation of x, in Example 8.6.*

x degrees	rads	$\dfrac{\sin(x-60°)}{}$ (LHS)	$\cot\Phi(x-\alpha)$	$\varepsilon^{-\cot\Phi(x-\alpha)}$	$\sin(\alpha-\Phi)\varepsilon^{-\cot\Phi(x-\alpha)}$ (RHS)
220	3.84	0.342	1.015	0.362	0.314
222	3.87	0.309	1.032	0.356	0.308

The iterated values are shown in Table 8.6, from which it is seen that the accurate value of extinction angle is

$$x = 222°$$

This value is seen to lie on the $\Phi = 60°$ characteristic of Fig. 8.15. The corresponding conduction angle is

$$\theta_c = x - \alpha = 222° - 120° = 102°/\text{half-cycle}$$

Example 8.7

Two SCRs are connected in inverse-parallel for the voltage control of a single-phase series R–L load of phase angle $\Phi = 45°$. These are each triggered at a firing-angle $\alpha = 90°$ with reference to their respective anode voltage waveforms. Estimate (do not calculate) the current extinction angle and use this to calculate the load voltage displacement factor.

Solution. From (8.64) the current extinction angle is given by

$$x = \pi + \Phi - \Delta$$

With $\Phi = 45°$, Δ is estimated to be $10°$. Therefore,

$$x = 180° + 45 - 10 = 215° = 3.75\,\text{rad}$$

Fourier coefficients a_1, b_1 for the fundamental component of the load voltage are given by (8.80) and (8.81) respectively,

$$a_1 = \frac{E_m}{2\pi}(\cos 2\alpha - \cos 2x)$$

$$= \frac{E_m}{2\pi}(-1 - 0.342) = \frac{-E_m}{2\pi}(1.342)$$

$$b_1 = \frac{E_m}{2\pi}[2(x-\alpha) - \sin 2x + \sin 2\alpha]$$

$$= \frac{E_m}{2\pi}[2(3.75 - 1.57) - 0.94 + 0]$$

$$= \frac{E_m}{2\pi}(3.42)$$

The load voltage displacement angle ψ_{v_1} is given, from (8.83), by

$$\psi_{v_1} = \tan^{-1}\frac{a_1}{b_1} = \tan^{-1}(-0.392) = -21.4°$$

The required displacement factor is therefore

$$\text{displacement factor} = \cos\psi_{v_1} = 0.93$$

Example 8.8

For the two-SCR series R–L circuit of Example 8.7, calculate the load voltage distortion factor when $\Phi = 45°$ and $\alpha = 90°$.

Solution. The load voltage distortion factor is given by

$$\text{distortion factor} = \frac{E_{L_1}}{E_L}$$

where E_{L_1} is the r.m.s. value of the fundamental component of the load voltage and E_L is the r.m.s. value of the total load voltage.

The Fourier coefficients a_1, b_1 of the fundamental load voltage are given by (8.80), (8.81) respectively and are calculated for the present case in Example 8.7:

$$a_1 = \frac{-E_m}{2\pi}(1.342)$$

$$b_1 = \frac{E_m}{2\pi}(3.42)$$

Therefore,

$$\hat{E}_{L_1} = c_1 = \sqrt{a_1^2 + b_1^2}$$

$$= \frac{E_m}{2\pi}\sqrt{(1.342)^2 + (3.42)^2}$$

$$= \frac{E_m}{2\pi}(3.674)$$

The r.m.s. value E_{L_1} of the fundamental component is therefore

$$E_{L_1} = \frac{c_1}{\sqrt{2}} = \frac{E_m}{2\pi}(2.598) = 0.413 E_m$$

8.2 Series R–L load with symmetrical phase-angle triggering

With $\Phi = 45°$ and $\alpha = 90°$, it was found in Example 8.7 that the estimated extinction angle is $x = 215° = 3.75$ rad. The r.m.s. load voltage E_L is found from (8.86)

$$E_L^2 = \frac{E_m^2}{2\pi}\left[(x-\alpha) + \frac{1}{2}\sin 2\alpha - \frac{1}{2}\sin 2x\right]$$

$$= \frac{E_m^2}{2\pi}\left[(3.75 - 1.57) + 0 - \frac{0.94}{2}\right]$$

$$= \frac{E_m^2}{2\pi}(2.18 - 0.47)$$

$$= \frac{E_m^2}{2\pi}(1.71)$$

Therefore,

$$E_L = E_m\sqrt{\frac{1.71}{2\pi}} = 0.522\,E_m$$

The distortion factor is therefore

$$\text{distortion factor} = \frac{0.413\,E_m}{0.522\,E_m} = 0.792$$

Note that the product of this load voltage distortion factor and the corresponding displacement factor, from Example 8.7, does not give a power factor. The power factor is the product of the distortion factor and displacement factor of the load current.

Example 8.9

A series R–L load with phase-angle $\Phi = 45°$ is supplied with power from an ideal single-phase source of instantaneous voltage $e = E_m \sin \omega t$ via a pair of inverse-parallel connected SCRs. If each SCR is triggered at an angle $\alpha = 90°$ with reference to its anode voltage, calculate the per-unit power and the power factor.

Solution. The per-unit power $P = I^2 R$, where I^2 is given by (8.77).

$\Phi = 45°, \alpha = 90° = \dfrac{\pi}{2} = 1.57\,\text{rad}$

$x = 215° = 3.75\,\text{rad}$ (from Example 8.7)

$x - \alpha° = 3.75 - 1.57 = 21.8\,\text{rad}$

$\sin 2(x - \Phi) = \sin 340° = -0.342$

$\sin 2(\alpha - \Phi) = \sin 90° = 1.0$

320 Single-phase voltage controllers

$$\sin(\alpha - \Phi) = \sin 45° = 0.707$$

$$\sin^2(\alpha - \Phi) = \sin^2 45° = 0.5$$

$$\cot \Phi = \cot 45° = 1.0$$

$$\varepsilon^{-\cot \Phi(x-\alpha)} = \varepsilon^{-2.18} = 0.113$$

$$\varepsilon^{-2\cot \Phi(x-\alpha)} = \varepsilon^{-4.36} = 0.0128$$

$$\sin x = \sin 215° = -0.574$$

$$\sin \alpha = \sin 90° = 1.0$$

In (8.77),

$$P = \frac{E_m^2}{2\pi|Z|^2} \left[2.18 - \frac{1}{2}\{-0.342 - 1.0\} \right.$$
$$\left. + \frac{0.5}{1.0}\{1 - 0.0128\} + 4.0707 \times 0.707\{(-0.574)(0.113) - 1.0\} \right]$$

$$= \frac{E_m^2}{2\pi|Z|^2} [2.18 + 0.671 + 0.4936 - 2(1.065)]$$

$$= \frac{E_m^2}{2\pi|Z|^2} (1.215)$$

Now the maximum power is transmitted with sinusoidal operation, when $\alpha = \Phi$ and $x = \pi + \Phi$. In (8.77) then

$$P = \frac{E_m^2}{2\pi|Z|^2} (\pi + \Phi - \alpha)$$

$$= \frac{E_m^2}{2\pi|Z|^2} (2.356)$$

At $\alpha = 90°$ the per-unit power is therefore

$$\frac{P_{\alpha=90°}}{P_{\alpha=45°}} = \frac{1.215}{2.356} = 0.515 \text{ p.u.}$$

This value is slightly greater than the value $P = 0.5$ p.u. that would be obtained with resistive load at $\alpha = 90°$.

The power factor, (8.33), is defined as

$$PF = \frac{P}{EI}$$

Using the relationship of (8.30)

8.2 Series R–L load with symmetrical phase-angle triggering

$$PF = \frac{I^2 R}{EI}$$

$$= \frac{IR}{E}$$

$$= \frac{IR}{E_m/\sqrt{2}}$$

Substituting for I from the expression for I^2 at $\alpha = 90°$ above gives

$$PF = \frac{E_m}{|Z|} \sqrt{\frac{1.215}{2\pi}} \times \frac{R\sqrt{2}}{E_m}$$

$$= \sqrt{\frac{1.215}{\pi}} \times \frac{R}{|Z|}$$

But $R/|Z| = \cos \Phi = 0.707$

$$\therefore PF = 0.622 \times 0.707 = 0.44$$

This compares with the value $PF = 0.707$ that is obtained with resistive load at $\alpha = 90°$.

Example 8.10

For the single-phase, series R–L, SCR controlled load of Example 8.9 calculate the distortion factor and the displacement factor of the current when $\alpha = 90°$.

Solution. Fourier components a_1 and b_1 for the fundamental component of the load current are given in (8.68) and (8.69) respectively.

$$\Phi = 45°, \; \alpha = 90° = \frac{\pi}{2} = 1.57 \, \text{rad}$$

$x = 215° = 3.75 \, \text{rad}$ (from Example 8.7)

$\cos(2\alpha - \Phi) = \cos 135° = -0.707$

$\cos(2x - \Phi) = \cos 385° = 0.906$

$\sin \Phi = \sin 45° = 0.707$

$2(x - \alpha) = 2(3.75 - 1.57) = 4.36 \, \text{rad}$

$\sin(\alpha - \Phi) = 45° = 0.707$

$\cos(x + \Phi) = \cos 260° = -0.174$

$\cos(\alpha + \Phi) = \cos 135° = -0.707$

$\cot \Phi = \cot 45° = 1.0$

$$(x - \alpha) = 3.75 - 1.57 = 2.18 \text{ rad}$$

$$\varepsilon^{-\cot \Phi (x-\alpha)} = \varepsilon^{-2.18} = 0.113$$

$$\begin{aligned}
a_1 &= \frac{E_m}{2\pi |Z|} [(-0.707 - 0.906 - 0.707 \times 4.36 + 4 \times 0.707 \times 0.707 \\
&\quad \times (-0.174 \times 0.113 + 0.707)] \\
&= \frac{E_m}{2\pi |Z|} [-0.707 - 0.906 - 3.082 + 4 \times \tfrac{1}{2}(-0.020 + 0.707)] \\
&= \frac{E_m}{2\pi |Z|} (-4.695 + 1.374)
\end{aligned}$$

$$a_1 = -\frac{E_m}{2\pi |Z|} (3.321)$$

$$\sin(2\alpha - \Phi) = \sin 135° = 0.707$$

$$\sin(2x - \Phi) = \sin 385° = 0.427$$

$$\cos \Phi = \cos 45° = 0.707$$

$$\sin(x + \Phi) = \sin 260° = -0.985$$

$$\sin(\alpha + \Phi) = \sin 135° = 0.707$$

$$\begin{aligned}
b_1 &= \frac{E_m}{2\pi |Z|} [0.707 - 0.427 + 0.707 \times 4.36 + 4 \times 0.707 \times 0.707 \\
&\quad \times (-0.985 \times 0.113 - 0.707)] \\
&= \frac{E_m}{2\pi |Z|} [0.707 - 0.427 + 3.082 + 4 \times \tfrac{1}{2}(-0.111 - 0.707)] \\
&= \frac{E_m}{2\pi |Z|} (3.789 - 0.427 - 1.636) \\
&= \frac{E_m}{2\pi |Z|} (1.726)
\end{aligned}$$

The current displacement angle ψ_1 is given by

$$\psi_1 = \tan^{-1} \frac{a_1}{b_1} = \tan^{-1} \frac{3.321}{1.726} = \tan^{-1} -1.924 = -62.5°$$

Displacement factor $= \cos \psi_1 = 0.461$

The peak value $\hat{I}_1$ of the fundamental component of the current is given by

$$\hat{I}_1 = c_1 = \sqrt{a_1^2 + b_1^2}$$

$$= \frac{E_m}{2\pi|Z|}\sqrt{(3.321)^2 + (1.726)^2}$$

$$= \frac{E_m}{2\pi|Z|}(3.74)$$

The r.m.s. value of the fundamental current is therefore

$$I_1 = \frac{c_1}{\sqrt{2}} = \frac{E_m}{2\pi|Z|}(2.65) = \frac{E_m}{|Z|}(0.422) \text{ A}$$

Now the r.m.s. value of the total current, from Example 8.9, is

$$I = \frac{E_m}{|Z|}\sqrt{\frac{1.215}{2\pi}} = \frac{E_m}{|Z|}(0.44) \text{ A}$$

The distortion factor is therefore

$$\text{distortion factor} = \frac{0.422}{0.44} = 0.959$$

Combining the distortion factor and the displacement factor shows that

$$PF = 0.959 \times 0.461 = 0.442$$

which agrees with the value in Example 8.9 that was obtained by another route.

8.3 RESISTIVE LOAD WITH INTEGRAL-CYCLE TRIGGERING

In the voltage controller circuit of Fig. 8.1 the SCRs may be gated, at $\alpha = 0°$, to permit complete cycles of supply voltage to be applied to the load. If the gating signal is withheld, in any cycle, then no conduction will occur. It is therefore possible to permit complete cycles of supply voltage to be applied to the load followed by complete cycles of extinction.

A typical waveform is given in Fig. 8.20, in which the number of conducting cycles $N = 2$ and the overall supply period (on + off cycles) $T = 3$. Since the repetition period of the waveform is over T supply voltage cycles it is mathematically convenient to express the instantaneous load voltage $e_L(\omega t)$ in terms of the period $T\omega t$

$$e_L(\omega t) = E_m \sin T\omega t \Big|_0^{2\pi N/T} + 0\Big|_{2\pi N/T}^{2\pi} \tag{8.91}$$

In a resistive circuit the current will also have the waveform of Fig. 8.20. Seen from the supply terminals, this is therefore a circuit where the application of a

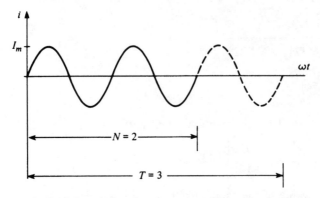

Fig. 8.20 Integral-cycle current waveform. $N = 2$, $T = 3$.

continuous voltage signal $e = E_m \sin \omega t$ results in a discontinuous current signal $i = e_L/R$, where e_L is given by (8.91).

8.3.1 Harmonic and subharmonic properties

Fourier coefficients a, b for an integral-cycle load voltage of N conducting cycles followed by an extinction period of $T - N$ cycles are given by

$$a_0 = \frac{1}{\pi} \int_0^{2\pi N/T} e_L(\omega t) \, d\omega t = 0 \tag{8.92}$$

$$\begin{aligned} a_n &= \frac{1}{\pi} \int_0^{2\pi N/T} e_L(\omega t) \cos n\omega t \, d\omega t \\ &= \frac{E_m T}{\pi(T^2 - n^2)} \left[1 - \cos(2\pi nN/T) \right] \end{aligned} \tag{8.93}$$

$$\begin{aligned} b_n &= \frac{1}{\pi} \int_0^{2\pi N/T} e_L(\omega t) \sin n\omega t \, d\omega t \\ &= \frac{E_m T}{\pi(T^2 - n^2)} \left[-\sin(2\pi nN/T) \right] \end{aligned} \tag{8.94}$$

The peak magnitude $\hat{E}_{L_n}$ of the nth harmonic load voltage, for $n \neq T$, is found to be

8.3 Resistive load with integral-cycle triggering

$$\hat{E}_{L_n} = c_n = \sqrt{a_n^2 + b_n^2}$$

$$= \frac{E_m T}{\pi(T^2 - n^2)} \sqrt{2[1 - \cos(2\pi nN/T)]} \qquad (8.95)$$

$$= \frac{2E_m T}{\pi(T^2 - n^2)} \sin(\pi nN/T)$$

Unlike the corresponding equation (8.15) for symmetric phase-angle waveforms, equation (8.95) shows that, in general, an integral-cycle waveform contains even order harmonics as well as odd order harmonics, depending on the values of N and T.

The phase-angle ψ_n between the supply voltage and the nth current harmonic is given by

$$\psi_n = \tan^{-1} \frac{a_n}{b_n}$$

$$= \tan^{-1} \left[\frac{1 - \cos(2\pi nN/T)}{-\sin(2\pi nN/T)} \right]$$

$$= \tan^{-1} \left[\frac{\sin(\pi nN/T)}{-\cos(\pi nN/T)} \right]$$

For $n < T$,

$$\psi_n = \pi - \frac{\pi nN}{T} \qquad (8.96)$$

For $n < T$,

$$\psi_n = \frac{\pi nN}{T} - \pi \qquad (8.97)$$

The condition $n = T$ is a special case, described below, in which $\psi_n = 0$.

In general, integral-cycle waveforms contain not only higher harmonic components of the supply frequency (when $n > T$) but subharmonic components also (when $n < T$). The subharmonic frequency components represent a serious disadvantage in most possible applications.

The case $n = 1$ represents the $1/T$th subharmonic of the supply frequency which is, by definition, the lowest order subharmonic that can occur. In Fig. 8.20, for example, the lowest order subharmonic voltage is one-third of the supply frequency since $T = 3$. The $1/T$th subharmonic is not necessarily the harmonic of the smallest magnitude. It is also possible for a subharmonic to exceed the value of the supply frequency component. A judicious choice of control period T may be necessary to avoid setting up subharmonic resonances in the power supply system or to avoid the natural frequencies of motor loads.

If $n < T$ the phase-angle of the nth harmonic current is negative (i.e. lagging) if $nN > T$. For the waveform of Fig. 8.20, for example, $T = 3$, $N = 2$ and the current harmonic components are negative for $n > \frac{3}{2}$ and $n < 3$. The $n = 1$ or $1/T$th subharmonic is seen, from (8.96), to have a phase-angle $\psi_n = \pi - 2\pi/3 = +\pi/3$ rad; the second or $n = 2$ or $2/T$th subharmonic has a phase angle $\psi_n = \pi - 4\pi/3 = -\pi/3$ rad.

The value $n = T$ represents the supply frequency component and forms a special case for which (8.93), (8.94) are indeterminate. If one proceeds from the basic integrals the following result is obtained:

$$a_{n=T} = \frac{1}{\pi} \int_0^{2\pi N/T} e_L(\omega t) \cos T\omega t \, d\omega t$$
$$= \frac{E_m}{4\pi T} [-\cos 2T\omega t]_0^{2\pi N/T} = 0 \tag{8.98}$$

$$b_{n=T} = \frac{1}{\pi} \int_0^{2\pi N/T} e_L(\omega t) \sin T\omega t \, d\omega t$$
$$= \frac{E_m}{2\pi} \left(\frac{2\pi N}{T}\right) \tag{8.99}$$
$$= E_m \frac{N}{T}$$

The peak magnitude $c_{n=T}$ of the supply frequency component of the load voltage is therefore given in terms of the r.m.s. supply voltage E by

$$\hat{E}_{L_{n=T}} = c_{n=T} = \sqrt{2} E \frac{N}{T} = E_m \frac{N}{T} \tag{8.100}$$

For the waveform of Fig. 8.20 the magnitude of the supply frequency component of current is, from (8.100), of value $\frac{2}{3}$ per-unit of the corresponding sinusoidal value at the same supply voltage. It is of interest that the magnitude of the supply frequency component of current is proportional to the number of conducting cycles N.

Note that with integral-cycle waveforms the supply frequency harmonic component (i.e. the $n = T$ component) is not the same as the $n = 1$ component. With integral-cycle waveforms it is wise to avoid the use of the term 'fundamental' component which applies, when $n = 1$, with phase-controlled waveforms.

The displacement angle ψ_T between the supply voltage and the supply frequency component of current is zero,

8.3 Resistive load with integral-cycle triggering

$$\psi_T = \tan^{-1}\left(\frac{a_{n=T}}{b_{n=T}}\right) = 0 \tag{8.101}$$

Equations (8.100), (8.101) can be interpreted to define the instantaneous value of the supply frequency component of the current

$$i_{n=T} = \frac{E_m}{R}\frac{N}{T}\sin\omega t \tag{8.102}$$

The higher (than the supply) harmonic components always have values smaller than the supply frequency component.

8.3.2 R.m.s. voltage and currrent

The r.m.s. value E_L of the load voltage function $e_L(\omega t)$ of Fig. 8.20 is given by combining (8.91) with the definite integral which corresponds to (8.17),

$$E_L = \sqrt{\frac{1}{2\pi}\int_0^{2\pi N/T} e_L^2(\omega t)\,d\omega t}$$

$$= \sqrt{\frac{1}{2\pi}\int_0^{2\pi N/T} E_m^2 \sin^2 T\omega t\,d\omega t} \tag{8.103}$$

$$= \frac{E_m}{\sqrt{2}}\sqrt{\frac{N}{T}} = E\sqrt{\frac{N}{T}}$$

With sinusoidal operation, $N = T$ and E_L is equal to the r.m.s. supply voltage E.

8.3.3 Power and power factor

In a circuit with series resistance R the average power is defined by (8.30) irrespective of the waveforms of voltage and current. Substituting E_L from (8.103) into (8.30) gives

$$P = \frac{E^2}{R}\frac{N}{T} = \frac{E_m^2}{2R}\frac{N}{T} = \frac{N}{T}\ \text{(p.u.)} \tag{8.104}$$

Like the r.m.s. load voltage and current the load power can only exist at the discrete levels defined by the values of N and T.

One can deliver (say) 50% power by an infinitely large number of values of N and T provided that $N/T = 0.5$. The choice of N and T, however, affects the harmonic content of the waveform and the frequency spectrum that defines it. In general, increasing the value of T causes the spectrum lines to be more closely clustered around the supply frequency or $n = T$ harmonic.

The distortion factor of the load voltage is defined by (8.35) as the ratio of the r.m.s. value of the supply frequency component to the total r.m.s. value. For an integral cycle waveform

$$\text{distortion factor} = \frac{E_{L_{n=T}}}{E_L} = \sqrt{\frac{N}{T}} \qquad (8.105)$$

Because displacement angle ψ_T is zero, from (8.101), the supply frequency component of the current is always in time-phase with the supply voltage. This does not mean that an integral-cycle circuit operates at unity power factor because, for part of the control period, the supply current is not in time-phase with the supply voltage. Indeed for part of the control cycle, there is no supply current at all.

The zero value of $a_{n=T}$ in (8.98) means that the displacement factor is unity:

$$\text{displacement factor} = \cos\psi_T = \cos\tan^{-1}\left[\frac{a_{n=T}}{b_{n=T}}\right] = 1 \qquad (8.106)$$

The power factor is therefore

$$PF = \text{distortion factor} \times \text{displacement factor} = \sqrt{\frac{N}{T}} \qquad (8.107)$$

It is seen from (8.107) that only when $N = T$, resulting in sinusoidal operation, can the power factor be unity.

8.3.4 Comparison between integral-cycle operation and phase-controlled operation

The choice between integral-cycle triggering and phase-angle triggering, in the circuit of Fig. 8.1, depends partly on whether the longer dwell-time ('off' time) with integral-cycle control is acceptable in the particular application.

8.3.4.1 Lighting control

It was pointed out in Section 8.3 above that even the minumum off-time of one supply cycle creates lamp flicker on a 50 Hz supply system. With one cycle on followed by one cycle off (i.e. $N = 1$, $T = 2$), the flicker occurs once every two cycles and has a frequency of 25 Hz. With two cycles on followed by one cycle off (i.e. $N = 2$, $T = 3$), Fig. 8.20, the flicker occurs once every three supply cycles and therefore has a frequency of $16\frac{2}{3}$ Hz.

When phase-angle triggering is used there is a dwell or off interval in every supply half-cycle. The fluctuation of current therefore has a frequency of

100 Hz on a 50 Hz supply. It is a feature of human vision that a flicker of 100 Hz is barely perceptible whereas a flicker of less than about 30 Hz is both perceptible and irritating. For this reason integral-cycle controlled SCRs are unsuitable as lamp dimmers and such equipment usually consists of phase-controlled SCRs.

8.3.4.2 Motor speed control

It is invariably desirable to maintain continuity of the armature current in a motor. Intermittent armature current results in pulsating electromagnetic torque and speed oscillation unless the system inertia is high. For motors with high inertia, large pulsations of torque create pulsations of torsional stress on the shaft and can cause failure due to shaft shearing.

The use of integral-cycle current waveforms, even with a minimum off time of one supply cycle, causes severe armature current variations in d.c. motors with rectified supply because of low armature inductance. Even in induction motors, which have much greater inductance, an off-time of one supply cycle causes severe current pulsations. For this reason the integral-cycle control of motors is not usually practicable. The much shorter off-times with phase-angle control, being usually some fraction of a half-cycle, permit the use of phase-angle control with a.c. motors. For the control of d.c. motors with rectified supply it is usually necessary to include additional inductance in the armature circuit, even with phase-angle control.

8.3.4.3 Heating loads

A discontinuous current waveform of integral-cycle nature is suitable only for loads with a large thermal time period, such as an electric furnace. Once the load element has reached the required temperature the intermittent nature of the current would not cause any significant drop in temperature unless the 'off' periods $T - N$ were very long. The use of an integral-cycle current in a thermal load of short time period, such as an ordinary incandescent electric lamp, is quite unsuitable as discussed in Section 8.3.4.1 above.

Where an application such as an electric heating load can be suitable for both integral-cycle control and phase-angle control, one has to decide which is the better alternative. One criterion of the choice is in terms of the distortion factor of the current and load voltage.

The displacement factor for a symmetrical phase-angle waveform such as that of Fig. 8.2 is obtained by substituting (8.8) and (8.20) into (8.35) to give

$$\text{distortion factor} = \frac{E_{L_1}}{E_L} = \frac{I_1}{I}$$

$$= \sqrt{\frac{[\cos 2\alpha - 1]^2 + [2(\pi - \alpha) + \sin 2\alpha]^2}{2\pi[2(\pi - \alpha) + \sin 2\alpha]}} \qquad (8.108)$$

In comparison, the distortion factor of the load current and voltage for integral-cycle control is, from (8.106),

$$\text{distortion factor} = \frac{E_T}{E_L} = \sqrt{\frac{N}{T}} \qquad (8.109)$$

If one wishes to dissipate, for example, 50% of the maximum possible sinusoidal power in the load resistor with phase-angle control it is necessary to use a firing-angle $\alpha = 90°$ and the distortion factor then has the value 0.84. To dissipate 50% load power with integral-cycle control requires that $N/T = 0.5$ so that the distortion factor has a value 0.707. It is found to be generally true for any value of power transfer that the relevant integral-cycle waveform contains a higher content of non-supply frequency harmonics and therefore has a lower (i.e. worse) distortion factor than the corresponding phase-angle control waveform. For both types of waveform the relationship $PF = \sqrt{P(\text{p.u.})}$ is true so that the power factor is identical when transferring the same amount of power.

8.3.4.4 Electromagnetic interference

An obvious advantage of integral-cycle current waveforms is that the di/dt value of the switch-on current is low, being of supply frequency sinusoidal value. The steep wavefront of current at switch-on, with its high di/dt value, that is characteristic of symmetrical phase-angle waveforms such as Fig. 8.2 is avoided. This means that the radio-frequency electromagnetic interference created by phase-angle switched SCR circuits is virtually eliminated by the use of integral-cycle waveforms.

8.3.4.5 Supply voltage dip

If an electricity supply system has a significant series impedance then the interruption of a high current load will cause a change in the supply point voltage. At the instant of switch-on there is likely to be a supply voltage dip while at the instant of switch-off there is likely to be a supply voltage rise, both of which are undesirable. With phase-angle control the result of high current switching is to produce a ripple voltage of 100 Hz superimposed on the supply voltage because current interruption occurs twice every supply

8.3 Resistive load with integral-cycle triggering

cycle. With integral-cycle control the harmonic ripple is of maximum frequency 25 Hz (when the switching pattern is alternate on and off cycles). For greater values of T the ripple frequency is of still lower subharmonic order. In general, subharmonic components of current and voltage are very undesirable in power supply systems. These interfere with metering and protection devices. A subharmonic voltage tends to cause a disproportionately large subharmonic current because of the low (inductive) impedance of the supply lines and transformers to currents of low harmonic order.

8.3.5 Worked examples

Example 8.11

The flow of power to a resistive load R from an ideal sinusoidal supply $e = E_m \sin \omega t$ is controlled by a pair of ideal inverse-parallel connected SCRs. The two switches are gated to produce bursts of load current consisting of two cycles of conduction followed by two cycles of extinction. What is the percentage power transfer compared with sinusoidal operation? What firing-angle would be required with phase-angle-controlled SCRs to produce the same load power?

Solution. In this case $N = 2$, $T = 4$. From (8.104)

$$P = \frac{N}{T} \text{ p.u.} = 50\% \text{ of the sinusoidal value}$$

To produce the same load power requires the same value of r.m.s. load current. By inference, from (8.103),

$$I = \frac{E_L}{R} = \frac{E}{R}\sqrt{\frac{N}{T}}$$

Therefore, with $N = 2$, $T = 4$

$$I = \sqrt{\frac{1}{2}} = 0.707 \text{ p.u.}$$

From (8.23) the per-unit r.m.s. current with phase-angle triggering is

$$I(\text{p.u.}) = \sqrt{\frac{1}{2\pi}[2(\pi - \alpha) + \sin 2\alpha]}$$

For the value $I(\text{p.u.}) = 0.707$ it is seen that the criterion for α is, once again,

$$0.5 = \frac{1}{\pi}[(\pi - \alpha) + \tfrac{1}{2}\sin 2\alpha]$$

Therefore,

$$\alpha - \tfrac{1}{2}\sin 2\alpha = \tfrac{1}{2}\pi$$

332 Single-phase voltage controllers

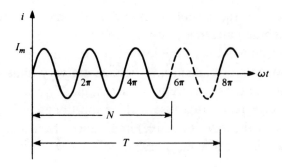

Fig. 8.21 Integral-cycle current waveform. $N = 3$, $T = 4$.

By inspection it is seen that the required value is $\alpha = \pi/2 = 90°$.

Example 8.12
A single-phase voltage source $e = E_m \sin \omega t$ supplies power to a resistive load via a pair of inverse-parallel connected SCRs. These are gated to provide a current waveform as shown in Fig. 8.21. What is the average value of the current? Explain what is the lowest order of harmonic associated with the waveform. Calculate the per-unit power transfer and the power factor. What are the values of the distortion factor and the displacement factor? Would the connection of a capacitor across the supply terminals improve the power factor of operation?

Solution. The average value of the current in Fig. 8.21 is seen to be zero, over any number of complete conduction cycles. For this waveform $N = 3$ and $T = 4$. The periodicity of the repetitive cycle is seen to be T times the periodicity of the supply voltage cycle. Therefore,

$$\text{lowest frequency harmonic} = \frac{1}{T} \times \text{supply frequency}$$

$$= \frac{1}{4} \times \text{supply frequency}$$

From (8.104) the per-unit power transfer is

$$P(\text{p.u.}) = \frac{N}{T} = \frac{3}{4} = 0.75$$

The power factor is found from (8.107),

$$PF = \sqrt{\frac{N}{T}} = \sqrt{\frac{3}{4}} = 0.866$$

The supply frequency or $n = T$ component of the current has a per-unit r.m.s. value, from (8.100), of

8.3 Resistive load with integral-cycle triggering

$$I_{n=T} = \frac{N}{T}$$

The total r.m.s. current is obtained from (8.103),

$$I = \frac{E_L}{R} = \frac{E}{R}\sqrt{\frac{N}{T}} = \sqrt{\frac{N}{T}} \text{(p.u.)}$$

The distortion factor of the current is therefore given by

$$\text{distortion factor} = \frac{I_{n=T}}{I} = \sqrt{\frac{N}{T}} = 0.866$$

The displacement factor is given by (8.106)

$$\text{displacement factor} = \cos\psi_T = \cos 0° = 1.0$$

Since the displacement angle ψ_T between the supply voltage and the supply frequency component of the current is zero the reactive voltamperes absorbed by the load, $Q = EI_{n=T}\sin\psi_T$, is also zero. It is therefore not possible to 'correct' the power factor by the connection of an inductor or a capacitor at the circuit terminals. In fact, the connection of an energy storage device across the supply terminals would introduce a phase-angle between the supply voltage and the supply frequency component of the supply current (not the load current) and thereby make the power factor worse.

Example 8.13

An ideal sinusoidal voltage supply $e = 300\sin\omega t$ provides power to a resistive load via a pair of inverse-parallel connected SCRs. The SCRs are triggered to provide a conduction pattern of one cycle on followed by three cycles off. Calculate the value of the supply frequency harmonic voltage and its immediate neighbours in the frequency spectrum.

Solution. In this case, $N = 1$, $T = 4$. The peak value of the supply frequency component is, from (8.100),

$$\hat{E}_{n=T} = E_m \frac{N}{T} = \frac{300}{4} = 75\,\text{V}$$

The general harmonic amplitude, if $n \neq T$, is given in (8.100). The lowest order harmonic or $n = 1$ value is the $1/T$ or $1/4$ of the supply frequency. From (8.95)

$$c_{n=1} = \frac{2 \times 300 \times 4}{\pi(16 - 1)} \sin\frac{\pi}{4} = 36\,\text{V}$$

$$c_{n=2} = \frac{2 \times 300 \times 4}{\pi(16 - 4)} \sin\frac{\pi}{2} = 63.7\,\text{V}$$

334 Single-phase voltage controllers

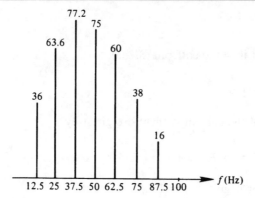

Fig. 8.22 Harmonic amplitude spectra (p.u.) for integral-cycle waveform with $N=1$, $T=4$, with 50 Hz supply.

$$c_{n=3} = \frac{2 \times 300 \times 4}{\pi(16-9)} \sin\frac{3\pi}{4} = 77.2\,\text{V}$$

Note that the $n=3$ value is greater than the $n=4$ (supply frequency) value. The harmonic spectrum is shown in Fig. 8.22 assuming that the supply frequency is 50 Hz. Because of the poor waveform and the significant size of the immediate harmonics, a firing pattern $N=1$, $T=4$ is characterised by the low distortion factor of 0.5.

Example 8.14
A load of 100 Ω resistance is supplied with power from an ideal single-phase supply of 240 V, 50 Hz. The load current is controlled by a pair of inverse-parallel connected SCRs triggered to produce an integral-cycle waveform of four on cycles followed by one off cycle. Calculate the maximum values of

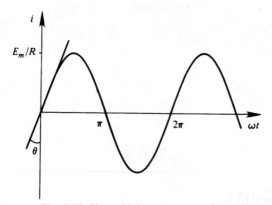

Fig. 8.23 Sinusoidal current waveform demonstrating initial slope.

8.3 Resistive load with integral-cycle triggering

di/dt and dv/dt generated by the switching of the thyristors. Compare these values with typical values for phase-angle switching.

Solution. The current waveform for the first cycle of conduction, Fig. 8.23, is given by $i(\omega t) = E_m R \sin \omega t$. The maximum current is given by $I_m = E_m/R = 240\sqrt{2}/100 = 3.394$ A. The time rate of change of current is therefore

$$\frac{di}{dt} = \frac{\omega E_m}{R} \cos \omega t$$

This has a maximum value when $\omega t = 0$ so that the initial slope θ in Fig. 8.23 is

$$\left(\frac{di}{dt}\right)_{max} = \frac{\omega E_m}{R}$$

$$= \frac{2 \times \pi \times 50 \times 140 \times \sqrt{2}}{100}$$

$$= 1066 \text{ A/s}$$

$$= 0.0011 \text{ A/}\mu\text{s}$$

The maximum value of the slope of the voltage waveform also occurs at $\omega t = 0$,

$$\frac{dv}{dt} = \omega E_m \cos \omega t$$

$$\left(\frac{dv}{dt}\right)_{max} = \omega E_m$$

$$= 2 \times \pi \times 50 \times 240 \times \sqrt{2}$$

$$= 106\,600 \text{ V/s}$$

$$= 0.106 \text{ V/}\mu\text{s}$$

In phase-angle control the voltage rises from zero to its instantaneous value (which depends on the firing-angle) in about 1 µs. For a voltage wave of 240 V r.m.s. value a typical working value might be of the order of 200 V. The value of dv/dt (and di/dt in a resistive circuit) is then 200 V/µs or about 2000 times the rise time of the supply voltage.

Example 8.15
A controller consisting of two SCRs connected in inverse-parallel is to be used to adjust the voltage to a single-phase resistive load. Load voltage is required over a range of levels from zero to fully supply voltage with a particular need to supply 73% of full power. The controller operates in the integral-cycle mode with a choice of overall repetition period ('on' + 'off' cycles) of 18, 20, 22, 24, or 26 supply cycles. What period would you

select to most nearly satisfy the 73% power demand? What is the actual per-unit power transfer with your selected firing pattern and what is the power factor? How would you tackle the problem of power factor improvement. Specify the lowest order of harmonic in your selected waveform.

Solution.

$$P \propto \frac{N}{T}\left(\frac{\text{'on' cycles}}{\text{'on'} + \text{'off' cycles}}\right)$$

Consider the necessary number of cycles 'on' to provide 0.73 p.u. power from the specified repetition periods

if $\dfrac{a}{18} = 0.73$, $a = 13.14$ \qquad if $\dfrac{d}{24} = 0.73$, $d = 17.52$

if $\dfrac{b}{20} = 0.73$, $b = 14.6$ \qquad if $\dfrac{e}{26} = 0.73$, $e = 18.98$

if $\dfrac{c}{22} = 0.73$, $c = 16.06$

A pattern $N = 19$ and $T = 26$ gives the nearest value to $P = 0.73$ p.u.

Actual power transfer is $P = \dfrac{19}{26} = 73.1\%$.

$$PF = \sqrt{P(\text{p.u.})} = \sqrt{0.731} = 0.855$$

The power factor does not have an associated lagging (or leading) phase-angle. The current displacement angle is zero, i.e. the fundamental current is in time-phase with the voltage.

Power factor reduction is due entirely to distortion effect, not displacement effect. No power factor correction is possible by the use of energy storage devices such as parallel-connected capacitors. The distorted supply current must be improved by (say) harmonic injection compensation or by-pass circuits must be used to swamp the distorted load current into becoming a small component of the total supply current.

Lowest order harmonic $= \dfrac{1}{T}$ th subharmonic

which, in this case $= \dfrac{1}{26}$ th subharmonic

8.4 PROBLEMS

Single-phase SCR voltage controllers with resistive load

8.1 For the circuit of Fig. 8.1 excited by an applied e.m.f. $e = E_m \sin \omega t$, show that the Fourier coefficients of the fundamental load voltage are given by (8.6) and (8.7). Calculate the p.u. magnitude and phase-angle of the fundamental load current when the firing-angle is $60°$.

8.2 Derive an expression for the r.m.s. load voltage in a single-phase resistive circuit where load voltage is varied by firing-angle adjustment of a pair of SCRs connected in inverse-parallel. If $e = 100 \sin \omega t$ and $R = 50\,\Omega$, what is the r.m.s. load voltage at $\alpha = 30°, 60°, 120°$?

8.3 An ideal voltage supply $e = E_m \sin \omega t$ provides a power to a single-phase load R by symmetrical phase-angle triggering of a pair of inverse-parallel connected SCRs in the supply lines. Sketch the load voltage waveform for firing-angle $\alpha = 90°$ and also sketch the corresponding fundamental component of the load current. Derive, from first principles, an expression for the r.m.s. load current I_L at any arbitrary angle α, in terms of E_m, R and α. Use I_L, or otherwise, to calculate the per-unit average power dissipated in the load at $\alpha = 80°$.

8.4 An ideal single-phase supply $e = E_m \sin \omega t$ provides power to a resistive load $R = 100\,\Omega$ using the circuit of Fig. 8.24. The SCRs of the inverse-parallel pair are gated to provide symmetrical phase-angle triggering.
(a) Sketch compatible waveforms, for two cycles, of the supply voltage, supply current and load current.
(b) Derive an expression for the load power dissipation in terms of E_m, R and α. Calculate the power at $\alpha = 60°$ if $E_m = 340\,\text{V}$.

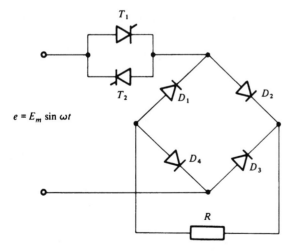

Fig. 8.24 Circuit diagram for Problem 8.4.

(c) If diode D_1 in Fig. 8.24 fails to an open circuit, sketch the waveform of the resulting load current. What effect does the failure of D_1 have on the load power dissipation?

(d) Suggest a modification to the circuit of Fig. 8.24 whereby the same load current waveform can be obtained using only one SCR. Would the SCR in your modified circuit be extinguished by natural commutation?

8.5 The power flow to a resistive load R from an ideal sinusoidal single-phase supply $e(\omega t) = E_m \sin \omega t$ is controlled by a pair of SCRs connected in inverse-parallel. These are gated to produce symmetrical angle triggering of the current waveform.

(a) Derive an expression for the load power dissipation P in terms of E_m, R and α. If the r.m.s. supply voltage is 240 V and $R = 25\,\Omega$, what is the power dissipation at $\alpha = 90°$?

(b) The fundamental component of the current at $\alpha = 90°$ can be described by the expression

$$i_1(\omega t) = 0.59 \frac{E_m}{R} \sin(\omega t - 32.5°)$$

Calculate the displacement factor, distortion factor and hence power factor of the circuit at $\alpha = 90°$.

8.6 A pair of SCRs connected in inverse-parallel, Fig. 8.1, is used to supply adjustable current to a resistive load. If the supply voltage is $e = E_m \sin \omega t$ and the switches are each triggered at an angle α after their respective anode voltage zeros, sketch the load current waveform for $\alpha = 60°$.

(a) Calculate the value of r.m.s. current I for $\alpha = 60°$ compared with the value for sinusoidal operation.

(b) If one of the SCRs is replaced by a diode, what effect would this have on the order of the harmonic components of the load current at $\alpha = 60°$?

8.7 A circuit consists of two SCRs, connected in inverse-parallel, supplying variable voltage to a resistive load. The single-phase supply voltage is given by $e = E_m \sin \omega t$. Each device is gated at an angle α of its respective anode voltage to result in a symmetrical load voltage waveform.

(a) State or derive mathematical definitions for the instantaneous current i and instantaneous power entering the circuit. Sketch a typical cycle of the instantaneous power at $\alpha = 60°$, over the supply period $0 < \omega t < 2\pi$.

(b) Use the standard integral definition to obtain an expression for average power dissipation in terms of α and calculate the per-unit value of this for $\alpha = 60°$.

8.8 In the rectifier circuit of Fig. 8.24 the current in the load resistor R is controlled by the switching of two ideal devices T_1, T_2. The ideal electrical

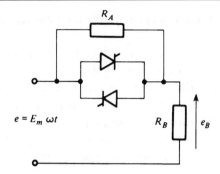

Fig. 8.25 Circuit diagram for Problems 8.9, 8.26 and 8.27.

supply has an instantaneous voltage $e(\omega t) = E_m \sin \omega t$. Each switch is gated at a firing-angle α with reference to its respective anode voltage waveform.
(a) Sketch compatible waveforms of the supply voltage, supply current and load current.
(b) Derive an expression for the average value of the load current in terms of E_m, R and α. Calculate the average load current at $\alpha = 60°$ if $R = 10\,\Omega$ and $E_m = 141$ V.
(c) If thyristor T_1 fails to an open circuit, sketch the load current waveform at $\alpha = 60°$.

8.9 The circuit of Fig. 8.25 is to be used to vary the current $i_B(\omega t)$ in resistor R_B. The two SCRs are gated at identical points on their respective anode voltage cycles to produce a periodic voltage $v_B(\omega t)$ that has symmetrical positive and negative alternations. The supply is ideal and is given by $e = E_m \sin \omega t$. $R_A = R_B = 10\,\Omega$.
(a) Sketch the waveform of $i_B(\omega t)$ if the SCR firing-angle $\alpha = 90°$. Write the equation that describes this waveform for an arbitrary angle α.
(b) Fourier coefficients a_1, b_1 for the fundamental component of the current $i_B(\omega t)$, with firing-angle α, are given by

$$a_1 = \frac{E_m}{2\pi}(\cos 2\alpha - 1)\frac{R_A}{R_B(R_A + R_B)}$$

$$b_1 = \frac{E_m}{2\pi}[2(\pi - \alpha) + \sin 2\alpha]\frac{R_A}{R_B(R_A + R_B)} + \frac{E_m}{R_A + R_B}$$

Use your equation for $i_B(\omega t)$ from part (a) above to prove the correctness of the equation for Fourier coefficient a_1.
(c) Calculate the displacement factor of current $i_B(\omega t)$ when $\alpha = 90°$.

8.10 At a firing-angle $\alpha = 90°$ in Fig. 8.1 the fundamental current lags the supply voltage by 32.5°. In a sinusoidal circuit a lagging phase-angle implies the existence of a magnetic field. But, in Fig. 8.1, no circuit element is capable of storing magnetic energy. Therefore, what happens to the fundamental lagging voltamperes?

8.11 Define a basic expression for the average electric power in any circuit in terms of instantaneous terminal properties. Use this to derive an expression for the average load power in a resistive circuit controlled by an inverse-parallel connected pair of SCRs with symmetrical phase-angle triggering. Sketch the waveform of the instantaneous voltampere input for $\alpha = 90°$.

8.12 Define the term 'power factor'. Show that the following relationship is true for a resistive circuit controlled by a pair of thyristor-type switches.

$$\text{power factor} = \sqrt{\text{power (p.u.)}}$$

8.13 The SCR voltage controller of Fig. 8.8 is compensated by a terminal capacitor of such value that $|X_c| = R$ at supply frequency. Sketch waveforms of the load current, capacitor current and supply current at $\alpha = 90°$. Do these waveforms indicate that any power factor improvement has occurred?

8.14 A resistive load $R = 50\,\Omega$ is supplied with power from a single-phase source $e = 100 \sin \omega t$. Load voltage variation is obtained by symmetrical phase-angle triggering of a pair of SCRs connected in inverse-parallel. Calculate the displacement angle ψ_1 of the fundamental component of current and hence calculate the real power P and reactive voltamperes Q into the circuit over the triggering range $0 < \alpha < \pi$.

8.15 For the circuit of Problem 8.14, calculate the apparent voltamperes at the circuit terminals over the whole range of SCR firing-angles. Utilise the values of P and Q from Problem 8.14 to obtain the variation of distortion voltamperes D.

8.16 The circuit of Problem 8.14 operates with an SCR firing-angle $\alpha = 75°$. Calculate the value of terminal capacitance that would minimise the r.m.s. supply current at 50 Hz. What is the compensated power factor, in the presence of optimal capacitance, compared with the uncompensated value?

8.17 The SCR voltage controller of Fig. 8.8 has a capacitance C connected across the terminals for the purpose of power factor correction. Show that power factor improvement is realised at firing-angle α if

$$|X_C| > \frac{\pi R}{\cos 2\alpha - 1}$$

8.18 Power is delivered to a resistive load R from an ideal single-phase supply of instantaneous voltage $e = E_m \sin \omega t$. Smooth variation of the load power is achieved by the symmetrical phase-angle triggering of a triac connected in the supply line.
(a) Sketch the load voltage for an arbitrary value of firing-angle α and derive an expression for the load average power in terms of E_m, R and α. If the r.m.s. supply voltage is 240 V and $R = 25\,\Omega$, calculate the power dissipation at $\alpha = 75°$.

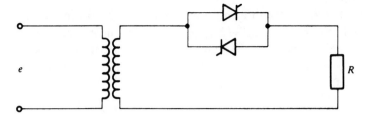

Fig. 8.26 Circuit diagram for Problem 8.19.

(b) Calculate the per-unit power dissipation at $\alpha = 75°$ and use this value to obtain the corresponding input power factor of the circuit.

(c) The connection of a suitable capacitor across the circuit input terminals is found to improve the overall circuit power factor. If the fundamental component of the load current at $\alpha = 75°$ is 0.72 p.u., lagging by 24.1° calculate the value of capacitance that will give the maximum power factor.

8.19 A resistive load is supplied through an SCR voltage controller connected in the secondary circuit of a single-phase transformer with low leakage reactance, Fig. 8.26. Sketch the forms of the load current and supply current.

8.20 An a.c. voltage controller incorporating SCR switches controls the load voltage $e_L(\omega t)$ of a resistive load. A typical load voltage waveform is shown in Fig. 8.2. The single-phase supply voltage is given by $e = E_m \sin \omega t$. For an arbitrary SCR firing-angle α the Laplace transforms $E(s)$ and $E_L(s)$ of the input and load voltages $e(\omega t)$ and $e_L(\omega t)$, respectively, are given by

$$E(s) = E_m \frac{\omega}{s^2 + \omega^2}$$

$$E_L(s) = E_m \frac{\omega \cos \alpha + s \sin \alpha + \omega \varepsilon^{-(\pi - \alpha/\omega)s}}{(s^2 + \omega^2)(1 + \varepsilon^{-\pi s/\omega})} \varepsilon^{s\alpha/\omega}$$

where s is the Laplacian operator.

(a) Show that the expression for the transfer function $(E_L/E)(j\omega)$ obtained from $E_L(s)$ and $E(s)$ above is indeterminate, having zeros in both the numerator and denominator.

(b) Show that, when the numerator and denominator of $(E_L/E)(s)$ are separately differentiated, with respect to s, then the following determinate function is obtained:

$$\frac{E_L}{E}(j\omega) = \frac{\sin \alpha + (\pi - \alpha)\varepsilon^{j\alpha}}{\pi \varepsilon^{j\alpha}}$$

(c) Why is the function $(E_L/E)(j\omega)$ independent of ω? Would it be appropriate to use the servomechanism term 'describing function' to describe the function given in part (b)?

(d) Obtain, in cartesian form, expressions for the magnitude $|(E_L/E)(j\omega)|$ and phase-angle $\angle(E_L/E)(j\omega)$ of the frequency response function $(E_L/E)(j\omega)$ given in (b) above. Sketch a complex plane (Nyquist) diagram of this function for different values of α, identifying the points $\alpha = 0°, 90°, 180°$.

Single-phase SCR voltage controllers with series R–L load and symmetrical phase-angle control

8.21 For the circuit of Fig. 8.12 show that current extinction is defined by the transcendental equation
$$\sin(x - \Phi)\varepsilon^{\cot\Phi(x-\alpha)} = \sin(\alpha - \Phi)$$

8.22 In the circuit of Fig. 8.12, $\Phi = 45°$ and $\alpha = 90°$. Calculate an accurate value for extinction angle x by iteration and compare this with the estimated value used in Example 8.7.

8.23 For the circuit of Fig. 8.12 sketch characteristics of the conduction angle versus firing-angle for the cases $\Phi = 0$, $\Phi = 90°$ and an intermediate value.

8.24 A pair of inverse-parallel connected SCRs controls the flow of current from a single-phase supply $e = E_m \sin \omega t$ to a highly inductive load. Deduce an expression for the instantaneous current $i(\omega t)$ and show that the current waveforms are sinusoids, vertically displaced from the ωt axis, that are centred at $\omega t = 0, \pi, 2\pi$ etc.

8.25 A series R–L load with phase-angle $\Phi = 60°$ is supplied with power from an ideal single-phase source of instantaneous voltage $e = E_m \sin \omega t$ via a pair of inverse-parallel connected SCRs. If each device is triggered at a firing-angle $\alpha = 120°$ calculate the per-unit power and the power factor.

8.26 For the circuit of Problem 8.25 calculate the displacement factor and distortion factor of the load current. Hence calculate the power factor and show that this agrees with the value obtained, more directly, from the r.m.s. current.

8.27 For the circuit of Problem 8.25 calculate the per-unit values of the load voltage Fourier coefficients a_1, b_1 and hence calculate the load voltage displacement factor.

8.28 For the circuit of Fig. 8.12 show that the r.m.s. value of the load voltage E_L is given by the expression (8.86).
If $\Phi = 60°$ and $\alpha = 120°$ calculate the per-unit value of E_L.

Single-phase SCR voltage controllers with resistive load and integral-cycle control

8.29 One feature of the harmonic properties of integral-cycle controlled SCR circuits is the presence of subharmonics of the supply frequency. With a control period T (on-time plus off-time) the lowest subharmonic is $1/T$ of

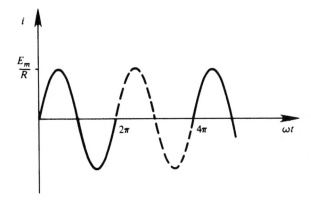

Fig. 8.27 Current waveform for Problem 8.34.

the supply frequency. What are the disadvantages caused by subharmonic currents flowing in the supply system?

8.30 Give reasons why you would consider a current waveform such as that of Fig. 8.20 to be suitable or unsuitable for (i) single-phase a.c. motor control, (ii) incandescent lighting control and (iii) electric heater control.

8.31 In an integral-cycle controlled SCR circuit with resistive load, switching at voltage zero, the current (during the periods that it flows) is always in time-phase with the supply voltage. Does this mean that the circuit operates at unity power factor?

8.32 In a resistive circuit the current waveform consists alternately of bursts of N complete cycles of conduction followed by $T - N$ complete cycles of extinction. Show that the ratio between the r.m.s. value of the harmonic load voltage E_h (i.e. the non-supply frequency components) and the r.m.s. load voltage E_L is given by

$$\frac{E_h}{E_L} = \sqrt{1 - \frac{N}{T}}$$

8.33 For the circuit of Problem 8.30 show that the r.m.s. value E_h of the harmonic load voltage (i.e. the non-supply frequency components) is given by

$$E_h = \frac{1}{T}\sqrt{N(T-N)}$$

8.34 An integral-cycle waveform consists of alternate conduction cycles and extinction cycles, as in Fig. 8.27. If the sinusoidal supply has a frequency of 50 Hz show that the 100 Hz harmonic component has a value of zero and the 75 Hz component has a value (very nearly) equal to one-half of the 50 Hz value.

8.35 Show that for an integral-cycle waveform pattern of N conducting cycles followed by $T - N$ extinction cycles the harmonic voltages of order n are zero if n is an integer of value TK/N, where $K = 1, 2, 3, \ldots$

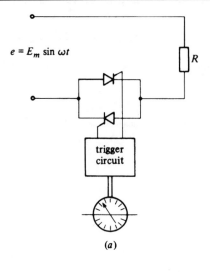

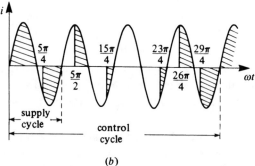

(b)

Fig. 8.28 (a) Circuit and (b) current waveform for Problem 8.38.

8.36 In integral-cycle operation the waveforms $N=1$, $T=2$; $N=2$, $T=4$; $N=3$, $T=6$ and $N=4$, $T=8$ all cause the same power transfer and have the same distortion factor. What would be the basis on which you would choose between these options for the purpose of supplying power to a resistive load from a 50 Hz supply?

8.37 Calculate the supply frequency harmonic voltage and the amplitudes of its immediate harmonic neighbours for an integral-cycle waveform $N=1$, $T=3$, if the peak supply voltage is 300 V.

8.38 Power is supplied from an ideal single-phase supply $e = E_m \sin \omega t$ to a purely resistive load R, Fig. 8.28(a), via a pair of inverse-parallel connected SCRs. The onset of SCR firing in any half-cycle is proportional to the value of the trigger signal pertaining at the beginning of the half-cycle. This trigger signal is, in turn, proportional to the angular displacement θ of a control potentiometer which varies from zero at its null setting to a maximum value θ_m.

This angular displacement is made time-dependent so as to produce the load current waveform given in Fig. 8.28(b). Deduce an expression for $\theta(t)$ in terms of θ_m and ωt and sketch the form of this on an ωt axis, indicating the intervals 2π, 4π etc. What is the d.c. value of the load current?

Comment on the possible harmonic components of the load current waveform.

8.39 It is required to transfer power from a standard single-phase, 50 Hz supply to a resistive load, R. A range of discrete load voltages is required between zero and full supply voltage, with a particular need to supply 71% of full sinusoidal power. The load power is controlled by a triac operating in integral-cycle mode and controllers are available that will deliver current with an overall repetition period ('on' + 'off' cycles) of either 18, 20 or 24 supply cycles.

(a) What particular triac firing pattern would you choose to achieve most nearly the desired transfer of 71% power?
What is the actual per-unit transfer with your selected firing pattern?

(b) What is the frequency and the per-unit magnitude of the lowest order of harmonic component in your selected waveform?

(c) Calculate values for the distortion factor, displacement factor and power factor of your selected waveform.
Calculate the new value of the displacement factor, if a capacitor C, $X_c = R$, is connected across the supply terminals.

9

Three-phase induction motor with constant frequency supply

9.1 THREE-PHASE INDUCTION MOTOR WITH SINUSOIDAL SUPPLY VOLTAGES

A three-phase induction motor contains a three-phase distributed winding that is housed in slots on the stationary part of the motor, usually called the stator. The rotating part of the machine, or rotor, also contains either a distributed three-phase winding or a cage of interconnected copper bars that serve as rotor winding conductors. When the rotor contains a distributed winding the three phases of this winding are connected to three slip rings on the motor shaft and the motor is known as a wound-rotor machine or slip-ring machine. When a cage of copper bars is used these bars are electrically connected by end rings inside the rotor, no electrical connection can be made to them and the motor is known as a squirrel-cage motor or, more simply, a cage motor.

One set of three-phase windings is connected to a three-phase voltage supply and this set becomes the primary or excitation (field) windings. With a slip-ring motor either the stator or the rotor windings may act as primary windings, although invariably the stator is used. With a cage motor only the stator windings can be used as primary windings. The other set of motor windings, known as secondary windings, is not connected to the electrical supply but is closed on itself. There is no electrical connection between the primary windings and the secondary windings but these are linked magnetically, as in a transformer. It is because the secondary e.m.f.s and currents are produced by electromagnetic induction that the motor is known as an induction motor. As with any form of electric motor the force on the rotating conductors, and hence the motor torque, is proportional to the product of the armature (in this case the secondary) current and the mutual flux in the air-gap. Note that, unlike the d.c. machine, the same motor (primary) wind-

Table 9.1 *Synchronous speeds of three-phase motors.*

Number of pole pairs, p	Synchronous speed N_1 (r.p.m.)	
	$f_1 = 50$ Hz	$f_1 = 60$ Hz
1	3000	3600
2	1500	1800
3	1000	1200
4	750	900
5	600	720
6	500	600
⋮	⋮	⋮

ings are used both to provide excitation current to set up the air-gap flux and also to provide the power component of the secondary current that creates torque and output power.

When a set of balanced three-phase voltages is applied to the distributed three-phase primary windings the resulting three-phase currents establish a rotating m.m.f. wave that results in a flux wave of constant amplitude rotating at a constant speed known as the synchronous speed. The value of the synchronous speed is fixed by two parameters:

(a) the supply frequency, f_1 cycles/second or hertz,
(b) the number of pair of poles p for which the primary is wound.

The synchronous angular speed N_1 of the rotating magnetic field is given by

$$N_1 = \frac{f_1}{p} \text{ rev/s} \quad (9.1(a))$$

$$= \frac{2\pi f_1}{p} = \frac{\omega_1}{p} \text{ rad/s} \quad (9.1(b))$$

$$= \frac{60 f_1}{p} \text{ r.p.m.} \quad (9.1(c))$$

For a two-pole winding, $p = 1$ and the synchronous speed N_1 in SI units (i.e. rad/s) is equal to the angular frequency ω_1 of the supply voltages.

If f_1 is fixed in value, which is customary in electricity supply systems, then N_1 has the values given in Table 9.1.

An induction motor runs at a shaft speed N that is less than the synchronous speed N_1. The speed difference $N_1 - N$ is called the slip speed. The ratio of slip speed to synchronous speed is the most important variable in induction motor operation and is called the per-unit slip S

$$S = \frac{N_1 - N}{N_1} \tag{9.2}$$

When the motor is running at synchronous speed, $N = N_1$ and $S = 0$. At standstill $N = 0$ and $S = 1$. If the motor is rotating at synchronous speed in the reverse direction, which sometimes happens in adjustable speed drives, then $N = -N_1$ and $S = 2$. At full load the per-unit slip is usually about 0.05 for a small motor because $N \simeq 0.95 N_1$.

If the rotor (secondary) conductors rotate at speed N and cut the constant rotating stator flux (which rotates at speed N_1) at a speed $N_1 - N$ the induced e.m.f. and current in the rotor are of frequency f_2 where

$$f_2 = \frac{N_1 - N}{N_1} f_1 = S f_1 \tag{9.3}$$

Also, since the flux is constant, the magnitude of the secondary e.m.f. $|E_2|$ is proportional to the time rate of flux cutting. In terms of the primary e.m.f. E_1, for a transformation ratio of unity,

$$|E_2| = S|E_1| \tag{9.4}$$

From (9.3) and (9.4) it is seen that, at high speeds, both the magnitude and frequency of the secondary induced e.m.f. are a small fraction of the supply-side values.

9.1.1 Equivalent circuits

An induction motor is similar in action to a transformer with rotating secondary windings. For a three-phase motor with balanced three-phase applied voltages a per-phase equivalent circuit, referred to primary turns, is shown in Fig. 9.1. The use of 'j' notation for the reactances implies that the equivalent circuit is valid for sinusoidal currents only.

The application of primary voltage V_1 causes magnetising current I_m to flow in magnetising inductance X_m which sets up a flux Φ_m mutual to both primary and secondary circuits. The time varying mutual flux induces a secondary e.m.f. which causes secondary current I_2 to flow. The total input current I_1 is the phasor sum of I_2 and no-load current I_0 (which consists mainly of the magnetising current I_m)

$$I_1 = I_2 + I_0 \tag{9.5}$$

In an induction motor the magnitude of the no-load current is of the order $|I_0| \simeq 0.2\text{--}0.3|I_1|$ at full load. In a small transformer the corresponding figures are $|I_0| \simeq 0.02\text{--}0.03|I_1|$ at full load. The relatively high value of I_0 for an

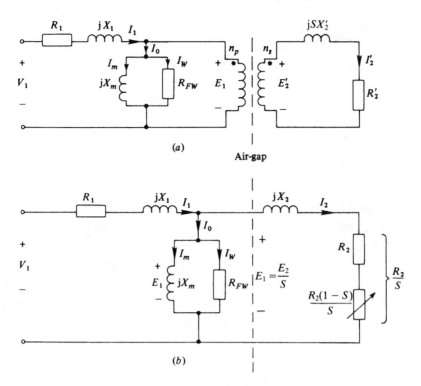

Fig. 9.1 Per-phase equivalent circuits of a three-phase induction motor: (a) 'transformer' circuit, (b) referred to primary turns. R_1 = primary winding resistance, R_2 = secondary (referred) winding resistance, X_1 = primary leakage reactance, X_2 = secondary (referred) leakage reactance, X_m = magnetising reactance, R_{FW} = friction and windage resistance.

induction motor is because of the large magnetising m.m.f. required to establish mutual flux in the high reluctance magnetic circuit containing the motor air-gap. The mutual flux path is subject to magnetic saturation which makes X_m variable and directly dependent on applied voltage V_1. Some of the magnetic flux created by the primary m.m.f. fails to link with the secondary windings and this is accounted for by the primary leakage reactance X_1. Similarly, not all of the magnetic flux associated with the secondary m.m.f. links the primary windings and the difference is accounted for in terms of the secondary leakage reactance X_2 in Fig. 9.1. The values of the components R_2, X_2 and the parameters E_2, I_2 in the 'referred' secondary circuit of Fig. 9.1 are related to the respective actual physical quantities on the secondary side by a transformation ratio similar, in effect, to a transformer turns ratio. If n_p, n_s are the respective turns per phase on the primary and secondary windings

$$E_2 = \left(\frac{n_p}{n_s}\right) \times \text{actual secondary e.m.f. } E_2' \tag{9.6}$$

$$I_2 = \left(\frac{n_s}{n_p}\right) \times \text{actual secondary current } I_2' \tag{9.7}$$

$$R_2 = \left(\frac{n_p}{n_s}\right)^2 \times \text{actual secondary resistance } R_2'$$

$$X_2 = \left(\frac{n_p}{n_s}\right)^2 \times \text{actual secondary reactance } X_2' \tag{9.8}$$

The transformation ratio (n_p/n_s) is usually greater than unity and is typically in the range 1.1–1.3.

The effect of motor speed is reflected by the presence of an impedance parameter R_2/S in the equivalent circuit. When $S = 0$ the motor shaft speed N is equal to the synchronous speed N_1 of the rotating flux field. The rotor (secondary) conductors do not then cut the rotating flux and hence the secondary induced e.m.f. is zero and so, therefore, is the motor secondary current. In the equivalent circuit of Fig. 9.1 it is seen that $I_2 = 0$ when $S = 0$.

In general

$$I_2 = \frac{E_1}{R_2/S + jX_2} = \frac{E_2}{R_2 + jSX_2} \tag{9.9}$$

The magnitude of the referred secondary current is therefore given by

$$|I_2| = \frac{|E_1|}{\sqrt{(R_2/S)^2 + X_2^2}} \tag{9.10}$$

9.1.2 Power and torque

The mechanical output power P_{out} from the motor is the product of the delivered torque T and the shaft speed N. In SI units there is no multiplying constant and

$$P_{\text{out}} = TN \tag{9.11}$$

where P_{out} is in watts, T in newton metres and N is radians/s. Combining (9.2) and (9.11) gives the output power in terms of per-unit slip S.

$$P_{\text{out}} = TN_1(1 - S) \tag{9.12}$$

The output power has to overcome the retarding forces due to bearing and brush friction (usually small) and due to ventilation windage (which may be considerable) as well as driving the load. If it is necessary to identify the load torque T_L and the windage torque T_{FW} separately, (9.11) may be written

$$P_{out} = (T_L + T_{FW})N \tag{9.13}$$

Since the motor secondary windings have a resistance R_2 there is an amount of 'winding loss' or 'copper loss' power $|I_2|^2 R_2$ dissipated as heat in each phase. But, in the secondary part of the equivalent circuit,

$$\frac{R_2}{S} - R_2 = \frac{R_2}{S}(1-S) \tag{9.14}$$

The portion of the internal electrical power per-phase that is converted to mechanical output power is therefore

$$P_{out} = |I_2|^2 \frac{R_2}{S}(1-S) \tag{9.15}$$

The total power per-phase delivered to the motor secondary windings across the air-gap is therefore

$$\begin{aligned} P_g &= P_{out} + |I_2|^2 R_2 \\ &= |I_2|^2 \frac{R_2}{S} = TN_1 \end{aligned} \tag{9.16}$$

The input power per-phase entering the machine terminals is given by

$$P_{in} = V_1 I_1 \cos\phi \tag{9.17}$$

where ϕ is the phase-angle between V_1 and I_1 and is a function of slip.

Input power P_{in} supplies the primary copper loss $|I_1|^2 R_1$ and the motor core loss, represented as $|I_W|^2 R_{FW}$, as well as the air-gap power. A diagram of the motor power flow is shown in Fig. 9.2.

It is significant to note that a fraction $(1-S)$ of the air-gap power is delivered by the motor as mechanical output power. The ratio of the output power P_{out} to the air-gap power P_g is therefore

$$\frac{P_{out}}{P_g} = 1 - S \tag{9.18}$$

Power ratio is shown versus speed in Fig. 9.3. The line $1-S$ represents the ratio P_{out}/P_g. Now motor efficiency η is defined as

$$\eta = \frac{P_{out}}{P_{in}} = \frac{P_{out}}{P_{out} + losses} = \frac{P_{in} + losses}{P_{in}} \tag{9.19}$$

Since $P_{in} > P_g$ it is obvious that the efficiency characteristic is smaller than the $1-S$ characteristic and must lie within the lower triangle of Fig. 9.3. A

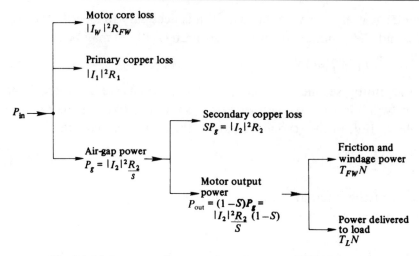

Fig. 9.2 Motor power flow per-phase, in terms of Fig. 9.1.

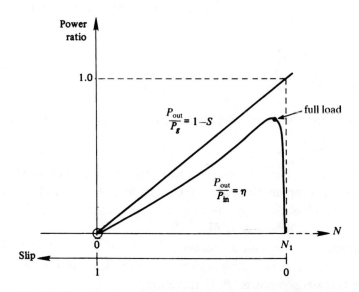

Fig. 9.3 Power ratio versus speed for a three-phase induction motor.

typical efficiency–speed characteristic for a small induction motor is shown. Maximum efficiency occurs at, or near to, the design rated speed. It is of interest to note that the per-unit efficiency is always lower than the per-unit speed at which it occurs. At 50% speed, for example, $N = 0.5$ p.u. but the efficiency is then about 0.4 p.u., in Fig. 9.3. The inherently poor efficiency of an induction motor, at reduced speed operation, is one of its great limitations.

The variation of delivered torque T per-phase with speed can be deduced from (9.10)–(9.16)

$$T = \frac{P_{\text{out}}}{N}$$

$$= \frac{P_{\text{out}}}{N_1(1-S)}$$

$$= \frac{P_g}{N_1}$$

$$= \frac{|I_2|^2}{N_1} \frac{R_2}{S}$$

$$= \frac{|I_2|^2}{N} \frac{R_2}{S}(1-S) \tag{9.20}$$

9.1.3 Approximate equivalent circuit

In the equivalent circuit of Fig. 9.1 the internal e.m.f. E_1 of the motor primary windings is related to the applied terminal voltage per-phase by the phasor equation

$$E_1 = V_1 - I_1(R_1 + jX_1) \tag{9.21}$$

In practice, the voltage drop across $R_1 + jX_1$ is relatively small at rated frequency so that $|E_1| \simeq |V_1|$.

Because of this it is possible to slightly modify the circuit of Fig. 9.1 into the approximate equivalent circuit of Fig. 9.4. From this it is seen that the primary voltage V_1 now falls across the magnetising and no-load branches so that

$$I_m = \frac{V_1}{jX_m} \tag{9.22}$$

$$I_W = \frac{V_1}{R_{FW}} \tag{9.23}$$

The secondary current I_2 is now given by

$$I_2 = \frac{V_1}{(R_1 + R_2/S) + j(X_1 + X_2)} \tag{9.24}$$

which has the magnitude

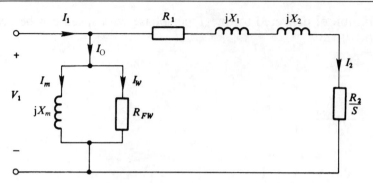

Fig. 9.4 Approximate per-phase equivalent circuit of the three-phase induction motor.

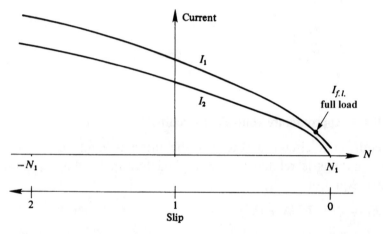

Fig. 9.5 Variation of primary and secondary current magnitudes with speed.

$$|I_2| = \frac{|V_1|}{\sqrt{(R_1 + R_2/S)^2 + (X_1 + X_2)^2}} \qquad (9.25)$$

At any fixed slip, the difference between $|I_2|$ in equations (9.10) and (9.25) is small, even at full-load current. The variation of $|I_2|$ with slip is shown in Fig. 9.5, for fixed supply voltage V_1. Since I_o is presumed to be constant, $|I_1|$ can be easily calculated from (9.5). A point $I_{f.l}$ shows a typical working value at full-load rated speed. As speed reduces, the current increases so that, at standstill, the current with full applied voltage may be three times (or more) the rated value.

The torque equations (9.20) are valid for both the full equivalent circuit, Fig. 9.1, and the approximate equivalent circuit, Fig. 9.4. Substituting (9.25) into (9.20) gives

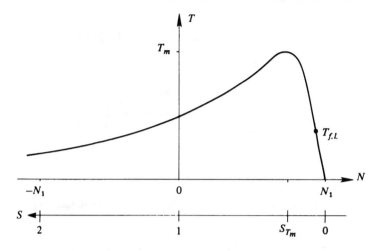

Fig. 9.6 Torque–speed characteristic of a three-phase induction motor.

$$T = \frac{\frac{|V_1|^2}{N_1} \frac{R_2}{S}}{\left(R_1 + \frac{R_2}{S}\right)^2 + (X_1 + X_2)^2} \tag{9.26}$$

In (9.26) slip S is the only variable. A typical form of torque–speed characteristic is shown in Fig. 9.6. The full-load operating point $T_{f.l}$ has a torque value roughly equal to the value of the starting torque $T(N=0)$. Peak torque T_m is usually two to three times the value of the full-load torque. The slip S_{T_m} at which peak torque occurs may be obtained by differentiating (9.26) with respect to S and equating to zero to give the result.

$$S_{T_m} = \frac{R_2}{\sqrt{R_1^2 + (X_1 + X_2)^2}} \simeq \frac{R_2}{X_1 + X_2} \tag{9.27}$$

If secondary resistance R_2 is increased the peak torque will occur at a lower speed in Figs. 9.6 and 9.8 (below). This will reduce the maximum possible working speed but will also slightly increase the range of speed control.

Substituting (9.27) into (9.26) gives an expression for the peak torque (sometimes called the maximum torque or pull-out torque)

$$T_m = \frac{\frac{|V_1|^2}{2N_1}}{R_1 + \sqrt{R_1^2 + (X_1 + X_2)^2}} \simeq \frac{|V_1|^2}{2N_1(X_1 + X_2)} \tag{9.28}$$

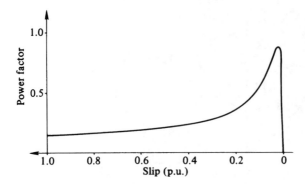

Fig. 9.7 Variation of power factor versus speed for a three-phase induction motor.

It is significant that expression (9.28) for T_m is independent of secondary resistance R_2, illustrated in Fig. 9.39 (below).

Since the motor is a balanced three-phase load and since the supply voltages are sinusoidal, the power factor is given by the cosine of the phase-angle between the supply (phase) voltage and the supply current

$$PF = \cos \psi_1 \qquad (9.29)$$

Variation of the power factor with slip is shown in Fig. 9.7 for the motor of Figs. 9.5, 9.6. It is an inherent feature of induction motor design that the power factor is usually low (i.e. poor) at low speeds.

9.1.4 Effect of voltage variation on motor performance

It is seen from (9.26) that at any fixed value of speed the developed torque T is proportional to the square of the applied voltage $|V_1|^2$. The slip at which peak torque occurs, equation (9.27), is not affected by change of V_1. Peak torque T_m is also proportional to $|V_1|^2$, as seen in (9.28).

When a motor is supplied by balanced three-phase sinusoidal voltages of fixed frequency, the torque–speed characteristics of the motor output therefore have the form shown in Fig. 9.8. If the supply voltage is reduced by one-half, for example, the peak torque reduces to one-quarter of the original value.

The point of operation of an induction motor, in the torque–speed plane, is defined by the point of intersection between the motor T/N characteristic and the load $(T_L + T_{FW})/N$ characteristic. In Fig. 9.6, for example, the point $T_{f.l}$ defines such an intersection although the load torque–speed characteristic is not shown.

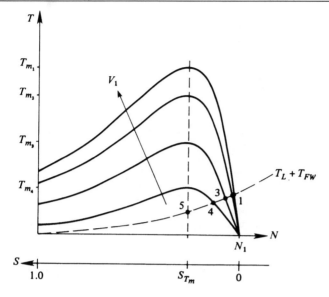

Fig. 9.8 The effect on the motor torque–speed characteristics of changing the applied voltage.

Consider the load characteristic in Fig. 9.8, which is typical of certain fan or pump loads.

Change of the motor voltage between the four states shown causes operation to vary between the four points of intersection. For the case of Fig. 9.8 it is seen that operation at the lowest voltage takes place at intersection 4 and the speed of operation has then reduced from $N = 0.97N_1$, at intersection 1, to $N = 0.9N_1$. It is a property of voltage-controlled induction motors, with torque–speed curves of the form of Fig. 9.8, that the range of possible speed variation is small. The maximum possible reduction of speed (i.e. about 20% reduction) would be to make $S = S_{T_m}$ by reducing the voltage so as to make the motor T/N characteristic pass through point 5 in Fig. 9.8.

Now the internal e.m.f. E_1 of the motor is related to the peak mutual flux $\hat{\Phi}_m$ by a relationship

$$E_1 = 4.44 \hat{\Phi}_m f_1 n \tag{9.30}$$

where n is a design constant. Since $|V_1| \simeq |E_1|$, as discussed above, one can say

$$V_1 \simeq K \hat{\Phi}_m f_1 \tag{9.31}$$

where K is a constant.

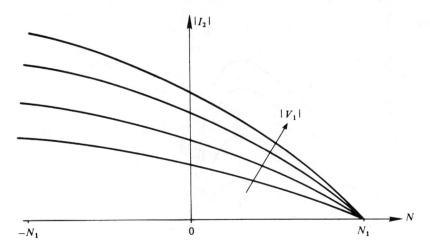

Fig. 9.9 The effect on the motor secondary current of changing the applied voltage.

If the supply frequency f_1 is constant, the motor flux is proportional (nearly) to the terminal voltage. Reduction of the terminal voltage therefore results in operation at reduced flux, which is very wasteful because the laminated iron core of the motor is then operating below its rated level. Also, because the torque is proportional to the product of flux and secondary current, any operation at reduced flux involves the use of excessive current to maintain high torque.

It is seen from (9.25) that, at fixed slip, the secondary current is directly proportional to the applied voltage. Voltage adjustment therefore results in secondary current characteristics of the form shown in Fig. 9.9. If the duty cycle requires prolonged operation at reduced speed, it is necessary to reduce the applied voltage in order to restrain the current to an acceptable level. Since the motor copper losses are proportional to the square of the current any prolonged operation with excessive current will cause a rapid temperature rise which could permanently damage the machine insulation.

9.1.5 M.m.f. space harmonics due to fundamental current

With an ideally distributed stator winding, consisting of an infinitely large number of slots, only the fundamental component m.m.f. space wave would be present. The need for a finite number of slots results in an m.m.f. waveform that is not sinusoidal but a periodic nonsinusoidal wave in which the fundamental component is dominant. The space harmonic m.m.f. waves produced by a symmetrical integral-slot winding have the orders $k = 5, 7, 11, 13 \ldots$. The 5th harmonic m.m.f. space wave is rotating backwards, i.e. in the opposite direction to the fundamental m.m.f. wave. All of the m.m.f. space harmonic waves created by the fundamental frequency current rotate at fundamental (supply) frequency but have a pole number corresponding to the space harmonic order. For example, the 5th space harmonic m.m.f. has 5 times the number of fundamental pole pairs so that it rotates backwards at a synchronous speed 1/5th of the supply frequency synchronous speed. On the other hand the $k = 7$ space harmonic m.m.f. due to supply frequency currents rotates forwards at a synchronous speed 1/7th of the supply frequency synchronous speed. In a well-designed a.c. motor the harmonic winding factor is much smaller than the fundamental winding factor and space harmonic m.m.f. waves have amplitudes so small that they are usually neglected.

9.2 THREE-PHASE INDUCTION MOTOR WITH PERIODIC NONSINUSOIDAL SUPPLY VOLTAGES

The application of symmetrical, nonsinusoidal three-phase voltages of constant periodicity to the motor terminals results in symmetrical, nonsinusoidal three-phase motor currents. These currents may be thought to consist of a fundamental component plus higher time harmonics.

9.2.1 Fundamental spatial m.m.f. distribution due to time harmonics of current

If the supply currents in the motor windings are nonsinusoidal functions of time each of the time harmonics contributes to the resultant air-gap m.m.f. wave. For a time harmonic order n each phase winding sets up a standing m.m.f. wave with the same spatial distribution as the fundamental (supply) frequency wave but pulsating at n times the supply frequency. The resultant fundamental space m.m.f. wave due to the nth time harmonic of current is given by

$$F(\omega_1 t) = \hat{F}_{k_n} \sin(n\omega_1 t - k\delta) \tag{9.32}$$

where space harmonic coefficient $k = 1$ for the fundamental component and δ is the fundamental spatial displacement angle. The space fundamental m.m.f. wave represented in (9.32) has two poles and travels at n times the synchronous speed of the fundamental frequency synchronous speed. Current time harmonics of order $n = 3h + 1$ for h integer, result in forward rotating or positive sequence m.m.f. waves while time harmonics of order $n = 3h + 2$ produce backward rotating or negative sequence m.m.f. waves. Triplen (i.e. multiples of three) order time harmonics are usually suppressed by the circuit connections in three-phase supply systems where the time harmonics present are invariably odd higher harmonics. In any event, when harmonic currents of order $n = 3h + 3$ flow in a set of (say) delta-connected windings they are in time-phase in each winding and are therefore of zero sequence nature – they do not combine to produce fundamental space m.m.f. waves.

Substituting $h = 1, 3, 5, \ldots$ into the above relationships for n shows that current time harmonics of order $1, 7, 13, \ldots$ result in forward rotating m.m.f. waves while current time harmonics of order $5, 11, 17, \ldots$ result in backward rotating m.m.f. waves, all related to the fundamental space harmonic $k = 1$.

9.2.2 Simultaneous effect of space and time harmonics

Each time harmonic present in the winding currents not only produces a fundamental component of spatial m.m.f. but also contributes towards certain higher harmonic components of the spatial m.m.f. The net result is that the air-gap m.m.f. wave contains space harmonic and time harmonic effects simultaneously. If a time harmonic of order n results in a space harmonic of order k the particular m.m.f. wave may be denoted by (9.32).

The amplitude $\hat{F}_{k_n}$ of a harmonic travelling m.m.f. wave varies inversely with the order k of the space harmonic but directly with the magnitude of the relevant time harmonic current. A combination of space and time harmonics of the same order $(n = k)$ results in a fundamental frequency synchronous speed. When the time harmonic order is greater than the space harmonic order $(n > k)$ the synchronous speed of the combination is greater than 1 p.u. When, on the other hand, the time harmonic order is smaller than the space harmonic order $(n < k)$ the synchronous speed of the combination is less than 1 p.u.

In any consideration of m.m.f. distribution in electrical machines it is well to remember that although m.m.f. systems can be accurately designed and calculated the resulting fluxes are not always so amenable to design or analysis. The flux patterns arising from a given m.m.f. source may take unexpected paths that can only be determined by search coils on the actual apparatus.

9.2.3 Equivalent circuits for nonsinusoidal voltages

For steady-state motoring operation the physical nonlinearities of an induction motor, such as magnetic saturation, and the resistance and inductance variation with current, can be neglected. If the motor can properly be regarded as a linear system a periodic nonsinusoidal supply voltage can be resolved into Fourier components and each sinusoidal, higher harmonic component applied separately to an appropriate equivalent circuit, the total effect being obtained by applying the *Principle of Superposition*.

Compared with the fundamental frequency equivalent circuit the following criteria apply to the nth time harmonic equivalent circuit:
(i) all reactances have a value n times the fundamental frequency value,
(ii) the operating slip is the harmonic slip S_n,
(iii) skin effect should be taken into account in calculating the primary resistance and the secondary resistance and reactance of cage rotors for high frequency harmonics.

For a synchronous angular speed N_1 at fundamental frequency a time harmonic of order n results in a harmonic synchronous speed nN_1. If the machine is rotating at angular velocity N the nth harmonic slip S_n is given by

$$S_n = \frac{nN_1 \mp N}{nN_1} \tag{9.33}$$

The negative sign in equation (9.33) refers to forward rotating fields, obtained with harmonic orders 1, 7, 13, etc., while the positive sign refers to backward rotating fields, obtained with harmonic orders 5, 11, 17, etc. In terms of fundamental frequency slip S the time harmonic slip S_n is found to be

$$S_n = \frac{n \mp (1-S)}{n} \tag{9.34}$$

At fundamental frequency, $n = 1$ so that $S_n = S$ for a positive sequence or forward rotating field and $S_n = 2 - S$ for a negative sequence or backward rotating field. The frequency f_{2_n} of the secondary (rotor) e.m.f.s and currents, in terms of fundamental frequency f_1, for the nth primary time harmonic is

$$\begin{aligned} f_{2_n} &= nS_n f_1 \\ &= [n \mp (1-S)]f_1 \end{aligned} \tag{9.35}$$

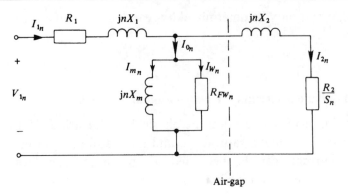

Fig. 9.10 Equivalent circuit for steady-state time harmonic currents of order n.

A general equivalent circuit for operation at the nth time harmonic frequency is shown in Fig. 9.10. In the forward motoring quadrant of the torque–speed plane, Fig. 9.3, the slip varies over the range $0 < S < 1$. For realistic values of harmonic order $n = 5, 7, 11, 13$, etc. it is seen from (9.34) that S_n is approximately constant at the value unity. At standstill $S = 1$ and therefore $S_n = 1$ for all harmonics. For the 7th time harmonic the lowest value of S_n is $6/7 = 0.857$, which occurs at speeds near to fundamental synchronous speed. It is of interest to note that when S is small $S_n \simeq 1$ and therefore $f_{2_n} \simeq nf_1$.

Now in Fig. 9.10 it is found that $nX_1 \gg R_1$ and $nX_2 \gg R_2/S_n$ so that, for the purpose of current calculations, one may use the approximate circuits of Fig. 9.11 noting that a harmonic current is roughly constant for all motoring speeds.

9.3 THREE-PHASE INDUCTION MOTOR WITH VOLTAGE CONTROL BY ELECTRONIC SWITCHING

Various connections of three-phase voltage controllers incorporating solid-state switches can be used to provide stepless voltage variation at the terminals of a three-phase load, Fig. 9.12. Most induction motors use star-connected primary windings and usually only one end of each winding is available. As with the single-phase voltage controller, described in Chapter 8, the conducting switches are extinguished by natural commutation. Gate turn-off switches are not needed.

The most common form of three-phase voltage control for small motors is the use of three pairs of inverse-parallel SCRs or three triacs, as shown in Fig.

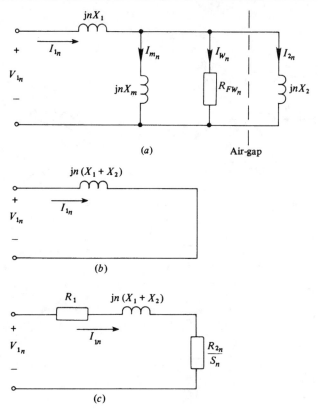

Fig. 9.11 Approximate equivalent circuits for steady-state time harmonic currents of order n: (a) including magnetising current and core losses, (b) neglecting magnetising current and core losses, (c) for the calculation of torque and power.

9.13. If reverse speed operation is required the phase sequence of the motor voltages must be changed and this can be achieved by the use of two additional pairs of SCRs, as, for example, in Fig. 9.14.

By appropriate triggering of the five SCR pairs of Fig. 9.14 unbalanced voltages can be applied to the motor. The fundamental harmonic components of these unbalanced motor voltages can be resolved into positive sequence and negative sequence symmetrical components by the use of standard techniques. The effect of unbalanced primary-side voltages is to create a resultant motor torque–speed characteristic that can be considered as the sum of two separate characteristics, due to the positive sequence and negative sequence motor voltages, respectively. Positive sequence motor voltages create a conventional (positive sequence) rotation. If the motor voltages are balanced then $T_{PS} = 1.0$ p.u. and $T_{NS} = 0$, so that operation occurs at

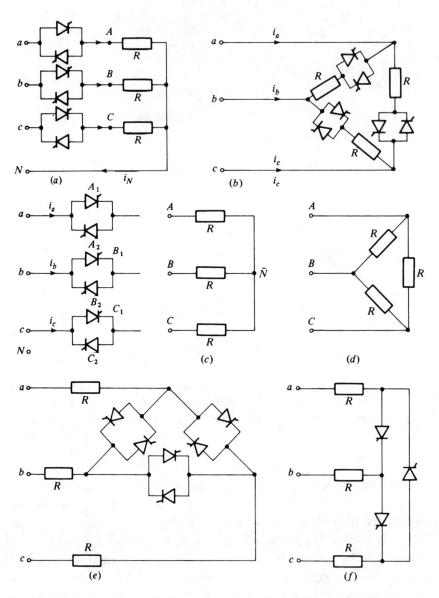

Fig. 9.12 Connections for symmetrical, three-phase voltage control of passive load: (*a*) four-wire star connection, (*b*) branch-delta connection, (*c*) three-wire star connection, (*d*) line controlled, delta connection, (*e*) neutral point control, (*f*) neutral point control.

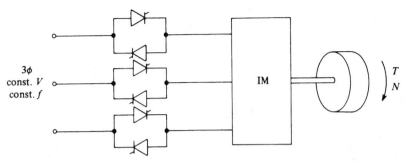

Fig. 9.13 Symmetrical three-phase voltage controller.

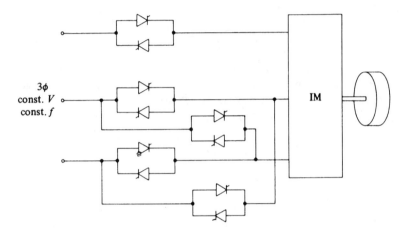

Fig. 9.14 Reversible symmetrical three-phase voltage controller.

point 1 in Fig. 9.15. With balanced voltages of the reverse phase sequence, $T_{PS} = 0$ and $T_{NS} = 1$ p.u. and operation occurs at point 5 in Fig. 9.15. The application of unbalanced voltages such that $T_{PS} = T_{NS} = 1.0$ p.u. results in a motor torque–speed characteristic that passes through the origin, Fig. 9.15, intersecting the load line T_L at point 3 in the overhauling quadrant II. When the primary voltage unbalance is such that $T_{NS} = T_{PS}/2$ the T/T_L intersection occurs at point 2.

By the use of an appropriate delay angle in the symmetrical firing of the six SCRs the motor in Fig. 9.13 can be 'soft started', whereby the motor starting currents are restricted by motor voltage reduction. The use of reduced voltage starting also reduces the starting torque and hence increases the motor run-up time.

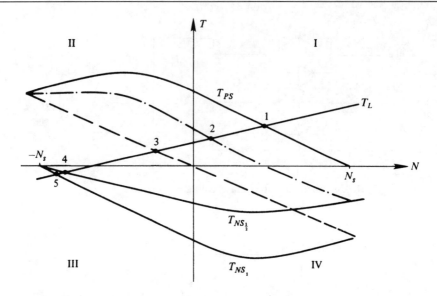

Fig. 9.15 Torque–speed characteristics for three-phase induction motor with high secondary resistance and unbalanced primary voltages.
$-\cdot-T = T_{PS} + T_{NS_1}$, $---T = T_{PS} + T_{NS_{\frac{1}{2}}}$

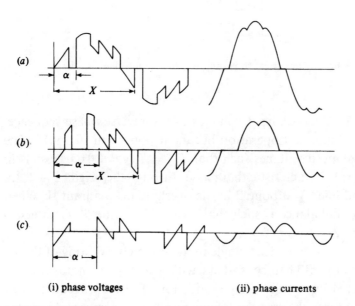

(i) phase voltages   (ii) phase currents

Fig. 9.16 Phase voltage and phase (line) current waveforms for three-wire star-connected series R–L load: (a) $\alpha = 60°$, $\phi = 45°$, (b) $\alpha = 120°$, $\phi = 45°$, (c) $\alpha = 120°$, highly inductive.

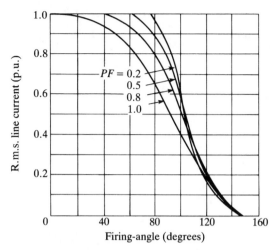

Fig. 9.17 R.m.s. line current versus firing-angle for three-wire, star-connected load.

Waveforms of the motor currents for the connection of Fig. 9.13 are very similar to corresponding waveforms for passive, series R–L loads, such as those of Fig. 9.16. At a fixed value of motor speed (i.e. fixed input phase-angle ϕ) the waveforms of the motor currents vary in three separate modes over the range of SCR firing-angle from 150° to α (i.e. 150° − ϕ). In the motor application, the phase voltages are also similar to those of Fig. 9.16 except that the zero value gaps are then partly 'filled' by supply frequency e.m.f.s induced in the open windings by coupling from conducting windings on both the primary and secondary sides. The voltage waveforms applied to the motor as actuating signals have the form of those in Fig. 9.16, since the induced e.m.f.s are not part of the forcing function, and the current amplitude characteristics of Fig. 9.17 apply directly to the motor fundamental equivalent circuit.

If the SCR firing-angle α is kept constant the motor current and voltage waveforms, together with their fundamental and r.m.s. values, vary with load phase-angle ϕ (i.e. motor speed) and the current extinction angle is a complicated function of ϕ and α.

A detailed analysis of the SCR-motor controlled drive would be very complex because of the interaction between the motor and its controller. The controller output voltage, which is also the motor input voltage, is simultaneously dependent on both the state of the controller and the state of the load. It is not possible, in general, to perform independent analyses of

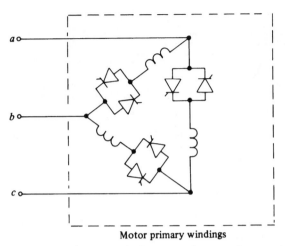

Fig. 9.18 SCR control of motor primary voltages using the branch-delta connection.

the motor and the SCR controller and to join these two analyses at their interface.

An accurate analysis of the induction motor to take account of instantaneous variations of the speed, electromagnetic torque and electrical variables would require a minimum of four loop voltage equations plus a torque equation. These differential equations are nonlinear and a general solution is only possible by the use of simulation methods and digital computation. Even with the case of steady-state operation at constant speed one has to contend with the interaction between the motor and controller, which renders the analysis intractable.

It can be shown that the three-phase, branch-delta connection, Fig. 9.12(b), results in a superior performance to that of the three-phase line controller, Fig. 9.12(c), with resistive load. This superior performance is also true with series R–L and motor loads. If both ends of the motor primary windings are available there is possible advantage to be gained in considering the connection of Fig. 9.18. The oscillogram of Fig. 9.19(a) shows the induced voltages at various speeds filling the gaps in the load phase voltages, compared with the waveform in Fig. 9.16 for passive R–L load. Correspondingly, the oscillogram of Fig. 9.19(b) shows the similarity of the phase current waveform to that of Fig. 8.13 and the line current waveform to that of Fig. 9.16(a). The waveforms of Fig. 9.19 suggest that the per-phase amplitude variations of the motor voltage and current, with the branch-delta connection, will be closely represented by the single-phase characteristics of Figs. 8.22–8.25.

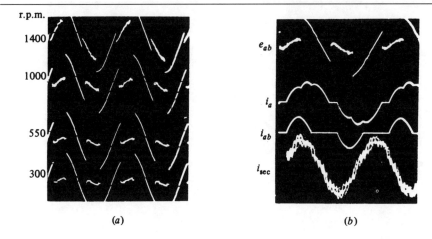

Fig. 9.19 Performance oscillograms for a 50 Hz, four-pole, branch-delta connected motor. $\alpha = 120°$: (a) primary phase voltage, (b) phase voltage e_{ab}, phase current i_{ab}, line current i_{ab}, at 1100 r.p.m.

9.3.1 Approximate method of solution for steady-state operation

9.3.1.1 Theory of operation

For steady-state operation over a range of average speeds it is usually necessary to determine the corresponding range of SCR firing-angles. This can be done by use of the motor fundamental equivalent circuit, for the particular speed, together with the curves of per-unit current versus firing-angle for the particular load phase-angle.

Suppose the problem is to find the necessary range of firing-angles to permit a motor of known rating to deliver the necessary torque to a fan load in a 2:1 speed range, as shown in Fig. 9.20. The most economical operation is that point P_1 should represent rated torque at rated speed. Torque T_2 at one-half rated speed is specified by the load line. From a knowledge of the torque at a given slip, one can calculate the corresponding secondary current I_2 from either of the last two statements in (9.20). The per unit primary current is then obtained corresponding to an equivalent circuit of known impedance and phase-angle ϕ (or power factor $\cos\phi$). From a knowledge of per-unit current and circuit phase-angle, the necessary firing-angle can be read from the data characteristics for passive circuit operation. The method is illustrated in Examples 9.1 and 9.2, below.

For star-connected motors of large phase-angle ϕ the current versus firing-angle curves of Fig. 9.17 can be further approximated by the straight-line relationship of Fig. 9.21, from which it can be deduced that

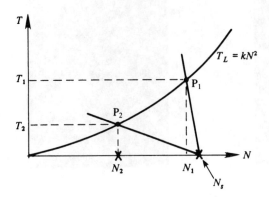

Fig. 9.20 Load torque $T_L = kN^2$ supplied by a voltage-controlled induction motor.

$$I(\text{p.u.}) = \frac{150 - \alpha}{150 - \phi} \qquad (9.36)$$

For branch-delta connected motors, of large phase-angle, the single-phase current versus firing-angle curves of Fig. 8.18 are appropriate. These may be represented, roughly, by the relationship

$$I(\text{p.u.}) = \frac{180 - \alpha}{180 - \phi} \qquad (9.37)$$

The use of (9.36) or (9.37) gives a useful rough check on the more accurate method using data curves.

9.3.1.2 Worked examples

Example 9.1
A pump has a torque–speed curve given by $T_L = (1.4/10^3)N^2$ N m. It is proposed to use a 240 V, 50 Hz, four-pole, star-connected induction motor with the equivalent circuit parameters (referred to stator turns) $R_1 = 0.2\,\Omega$, $X_1 = 0.36\,\Omega$, $R_2 = 0.65\,\Omega$, $X_2 = 0.36\,\Omega$, $X_m = 17.3\,\Omega$. The pump speed N is to vary from full speed 1250 r.p.m. to 750 r.p.m. by voltage control using pairs of inverse-parallel connected SCRs in the lines. Calculate the range of firing-angles required.

Solution. In Fig. 9.20, $N_1 = 1250$ r.p.m. and $N_2 = 750$ r.p.m. For a four-pole motor, at 50 Hz, from (9.1(a)),

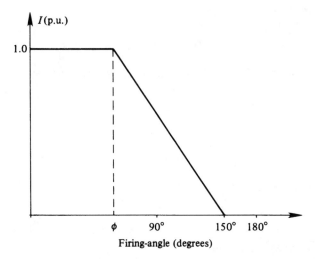

Fig. 9.21 Straight-line approximation of current (p.u.) versus firing-angle for three-wire, star-connected motors of large phase-angle.

synch. speed $N_s = 1500$ r.p.m.

$$S_1 = \frac{1500 - 1250}{1500} = 0.167$$

$$S_2 = \frac{1500 - 750}{1500} = 0.5$$

$$\frac{R_2}{S_1} = \frac{0.65}{0.167} = 3.9\,\Omega$$

$$\frac{R_2}{S_2} = \frac{0.65}{0.5} = 1.3\,\Omega$$

For the equivalent circuit of Fig. 9.4, neglecting core losses,

$$Z_{in_1} = j17.3 \parallel (4.15 + j0.72) = 3.65 + j1.54$$
$$= 3.96 \angle 22.9°$$
$$Z_{in_2} = j17.3 \parallel (1.55 + j0.72)$$
$$= 1.42 + j0.815 = 1.64 \angle 29.9°$$

Neglecting friction, $T = T_L$ and the total power converted to mechanical work is, from (9.15),

$$P_{out} = 3|I_2|^2 \frac{R_2}{S}(1 - S)$$

The torque may be obtained from (9.20),

$$T = \frac{3}{N} |I_2|^2 \frac{R_2}{S} (1-S)$$
$$= \frac{3}{N_1} |I_2|^2 \frac{R_2}{S} \qquad (a)$$

For the specified load, T_L is in Nm and therefore N must be given in radians/seconds,

$$T_L = \frac{1.4}{10^3} N^2$$

$$N_{750} = 750 \text{ r.p.m.} = 750 \times \frac{2\pi}{60} = 78.5 \text{ rad/s}$$

$$N_{1250} = 1250 \text{ r.p.m.} = 1250 \times \frac{2\pi}{60} = 157 \text{ rad/s}$$

Therefore,

$$T_2 = T_{750} = \frac{1.4}{10^3} (78.5)^2 = 8.63 \text{ N m}$$

$$T_1 = T_{1250} = \frac{1.4}{10^3} (130.9)^2 = 24 \text{ N m}$$

The secondary currents can be obtained by substituting into equation (a) above.

At 1250 r.p.m.,

$$24 = \frac{3}{130.9} I_2^2 \frac{0.65}{0.167} (1 - 0.167)$$
$$\therefore I_2 = 17.97 \text{ A}$$

At 750 r.p.m.,

$$8.63 = \frac{3}{78.5} I_2^2 \frac{0.65}{0.5} (1 - 0.5)$$
$$\therefore I_2 = 18.64 \text{ A}$$

In the equivalent circuit of Fig. 9.4, neglecting R_W, the primary current I_1 divides between the secondary and magnetising branches in the inverse ratio of the branch impedances.

$$I_2 = \left| \frac{jX_m}{(R_1 + R_2/S) + j(X_1 + X_2 + X_m)} \right| I_1$$

At 1250 r.p.m.,

$$I_1 = \left| \frac{4.15 + j18.02}{j17.3} \right| (17.97) = \frac{18.47 \times 17.97}{17.3} = 19.17 \text{ A}$$

At 750 r.p.m.,

$$I_1 = \left| \frac{1.42 + j18.02}{j17.3} \right| (18.64) = \frac{18.16 \times 18.64}{17.3} = 19.57 \text{ A}$$

9.3 Voltage control by electronic switching

Base currents. For a three-wire, star-connection, the primary base current at a given speed is the per-phase voltage divided by the per-phase impedance.
At 1250 r.p.m.,

$$I_{base} = \frac{240}{\sqrt{3}} \times \frac{1}{3.96} = 35 \text{ A}$$

At 750 r.p.m.,

$$I_{base} = \frac{240}{\sqrt{3}} \times \frac{1}{1.64} = 84.5 \text{ A}$$

Per-unit currents. The per-unit primary current is defined as the actual current divided by the base current.
At 1250 r.p.m.,

$$I = \frac{19.17}{35} = 0.55 \text{ p.u.}$$

At 750 r.p.m.,

$$I = \frac{19.57}{84.5} = 0.23 \text{ p.u.}$$

Motor power factor,
At 1250 r.p.m., $PF = \cos 22.9° = 0.920$
At 750 r.p.m., $PF = \cos 29.9° = 0.866$

Calculation of firing-angles. From the characteristics of Fig. 9.17, for series R–L loads, it is seen that:
  (i) For 0.55 p.u. r.m.s. current at $\phi = 22.9°$ (i.e. operation at 1250 r.p.m.) the necessary firing-angle is $\alpha = 95°$,
  (ii) for 0.23 p.u. r.m.s. current at $\phi = 29.9°$ (i.e. operation at 750 r.p.m.) the firing-angle is $\alpha = 118°$.

The necessary range of firing-angles is therefore

$$95° \leq \alpha \leq 118°$$

If one uses characteristics of fundamental current rather than r.m.s. current for a three-wire, star-connection, the range of necessary firing-angles is found to be similar to those shown.

If one uses the characteristics of r.m.s. current versus firing-angle for a single-phase, R–L load, the corresponding firing-angle range turns out to be $90° \leq \alpha \leq 115°$.

Since this motor has relatively low phase-angles the very approximate method of (9.36) gives the inaccurate result $80° \leq \alpha \leq 122.4°$.

Example 9.2

A three-phase, 140 V, four-pole, 50 Hz, delta-connected squirrel-cage induction motor is to be used to provide stepless speed control for a load represented by the relation

$$T_L = \frac{N^2}{500} - \frac{N^3}{10^6} \text{ SI units}$$

The per-phase equivalent circuit parameters of the motor, referred to primary turns, are $R_1 = 0.32\,\Omega$, $R_2 = 0.18\,\Omega$, $X_1 = X_2 = 1.65\,\Omega$, $X_m =$ very large. The motor terminal voltages are to be varied by the symmetrical triggering of triacs connected in series with each primary phase winding. If speed control is required in the range 1200–1450 r.p.m., calculate the necessary range of triac firing-angles. Calculate, approximately, the efficiencies of the two speeds of operation.

Solution. The load characteristic is similar to that of Fig. 9.17 where

$$N_1 = 1450 \text{ r.p.m.} = 1450 \times \frac{2\pi}{60} = 151.8 \text{ rad/s}$$

$$N_2 = 1200 \text{ r.p.m.} = 1200 \times \frac{2\pi}{60} = 125.7 \text{ rad/s}$$

Therefore, the delivered torques and output power are, at 1450 r.p.m.,

$$T_{1450} = \frac{(151.8)^2}{500} - \frac{(151.8)^3}{10^6} = 46 - 3.5 = 42.5 \text{ N m}$$

$$P_{1450} = TN = 42.5 \times 151.8 = 6451.8 \text{ W}$$

At 1200 r.p.m.,

$$T_{1200} = \frac{(125.7)^2}{500} - \frac{(125.7)^3}{10^6} = 31.6 - 1.98 = 29.62 \text{ N m}$$

$$P_{1200} = TN = 29.62 \times 125.7 = 3723 \text{ W}$$

Since X_m is very large the induction motor is now represented by a series R–L circuit of the form of Fig. 9.22.

At 1450 r.p.m.,

$$S = \frac{1500 - 1450}{1500} = 0.0333$$

$$\frac{R_2}{S} = \frac{0.18}{0.0333} = 5.4\,\Omega$$

$$Z_{\text{in}} = (0.32 + 5.4) + j3.3 = 6.6\angle 30°.$$

$$PF = \cos\phi_{\text{in}} = 0.87$$

At 1200 r.p.m.,

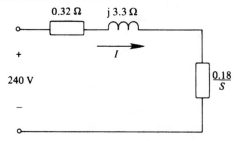

Fig. 9.22 Per-phase equivalent circuit for the induction motor of Example 9.2.

$$S = \frac{1500 - 1200}{1500} = 0.2$$

$$\frac{R_2}{S} = \frac{0.18}{0.2} = 0.9\,\Omega$$

$$Z_{in} = (0.32 + 0.9) + j3.3 = 3.52\angle 69.7°$$

$$PF = \cos 69.7° = 0.347$$

Rearranging equation (*a*), from Example 9.1, gives

$$I = \sqrt{\frac{TNS}{3R_2(1-S)}} \tag{b}$$

At 1450 r.p.m.,

$$I = \sqrt{\frac{42.5 \times 151.8 \times 0.0333}{3 \times 0.18 \times (1 - 0.0333)}}$$

$$= \sqrt{411.8} = 20.3\,\text{A}$$

At 1200 r.p.m.,

$$I = \sqrt{\frac{29.62 \times 125.7 \times 0.2}{3 \times 0.18 \times (1 - 0.2)}}$$

$$= \sqrt{1724} = 41.52\,\text{A}$$

The base and per-unit currents at the upper speed are found to be

$$I_{base} = \frac{140}{6.6} = 21.21\,\text{A}$$

$$I = \frac{20.3}{21.21} = 0.96\,\text{p.u.}$$

Note that in some motor applications the per-unit current demanded at full load is greater than unity. This defines an overload condition for which $\alpha < \phi$ is needed to give maximum conduction.

At 1200 r.p.m., in this example,

$$I_{\text{base}} = \frac{140}{3.52} = 39.8 \text{ A}$$

$$\therefore I = \frac{41.52}{39.8} = 1.043 \text{ p.u.}$$

From the characteristics of Fig. 9.17, a current of 0.96 p.u. at a *PF* of 0.87 requires $\alpha = 45°$ (to give 1450 r.p.m.) while a current of 1.043 p.u. at $PF = 0.35$ requires $\alpha = 65°$ (for 1200 r.p.m.). The required solution is therefore

$$45° \leq \alpha \leq 65°$$

The rough approximation of (9.36) gives an erroneous corresponding result $36° \leq \alpha \leq 107°$.

The motor copper losses at the two speeds are

$$P_{\text{loss}} = I^2(R_1 + R_2) = (20.3)^2 0.5 = 206 \text{ W, at } 1450 \text{ r.p.m.}$$

$$= (41.52)^2 0.5 = 862 \text{ W, at } 1200 \text{ r.p.m.}$$

Neglecting core losses and rotational effects, the approximate values of motor efficiency are given below.

At 1450 r.p.m.,

$$\eta = \frac{P_{\text{out}}}{P_{\text{out}} + P_{\text{loss}}} = \frac{6451.8}{6451.8 + 206} = 96.9\%$$

at 1200 r.p.m.,

$$\eta = \frac{3723}{3723 + 862} = 81.2\%$$

Example 9.3

A three-phase, six-pole, star-connected squirrel-cage induction motor is to be controlled by terminal voltage variation using pairs of inverse-parallel connected SCRs in each supply line. Sketch a diagram of this arrangement and list the advantages and disadvantages of SCR control using symmetrical phase-angle triggering, compared with sinusoidal voltage variation.

The motor is rated at 20 kW, 240 V, 50 Hz. If it operates at a full-load efficiency of 0.85 p.u. and power factor 0.8 lagging, calculate the r.m.s. current rating and maximum voltage rating required of the SCRs. The motor drives a load characterised by the relation $T_L = kN$ where T_L is the shaft torque and N is the motor speed. Operation is required to give rated torque at rated speed (which corresponds to 4% slip) and also to supply the appropriate load at two-thirds of rated speed. If the motor operates at full voltage at its upper speed, explain, and roughly estimate, the change of SCR firing-angle necessary to realise satisfactory operation at the lower speed.

Table 9.2 *SCR control of motor voltage (compared with sinusoidal control).*

Advantages	Disadvantages
Less expensive	Motor current nonsinusoidal (harmonic copper loss)
Fast transient response	Motor voltage nonsinusoidal (harmonic iron loss)
Easily adaptable to closed-loop control	Supply current distorted
Less bulky	Limited overload capacity

Solution. The circuit diagram of Fig. 9.13 describes the connection. The advantages and disadvantages of this form of control are summarised in Table 9.2.

motor output power = 20 000 W

$$\text{motor input power} = \frac{20\,000}{0.85} \text{ W}$$

$$\text{motor input VA} = \frac{20\,000}{0.85 \times 0.8} = 29.41 \text{ kVA}$$

$$\text{kVA} = \sqrt{3} I_L V_L$$

$$I_L = \frac{29\,410}{\sqrt{3} \times 240} = 70.75 \text{ A/phase}$$

Since the line SCRs share the phase current equally,

$$\text{SCR r.m.s. current rating} = \frac{70.75}{\sqrt{2}} = 50 \text{ A}$$

$$\text{peak voltage for normal working} = \frac{240\sqrt{2}}{\sqrt{3}} = 196 \text{ V/phase}$$

The torque–speed operation is described in Fig. 9.23. For a six-pole, 50 Hz motor, the synchronous speed is 1000 r.p.m. At 4% slip the full-load speed is $0.96 \times 1000 = 960$ r.p.m. The lower speed is therefore $0.667 \times 960 = 640$ r.p.m.

At rated power, from (9.11),

$$T_{960} = \frac{20\,000}{960 \times \frac{2\pi}{60}} = 199 \text{ N m}$$

Therefore,

$$T_{640} = \frac{640}{960} \times 199 = 132.7 \text{ N m}$$

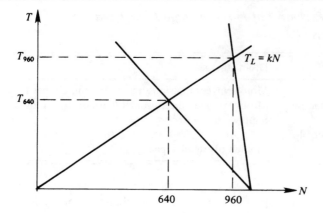

Fig. 9.23 Torque–speed characteristics for the motor of Example 9.3.

If

$$P_{\text{out}_{960}} = 1.0 \, \text{p.u.}$$

then

$$P_{\text{out}_{640}} = \frac{132.7}{199} \times \frac{640}{960} = 0.445 \, \text{p.u.}$$

Therefore, if the change of motor impedance is small,

$$I_{960} = 1.0 \, \text{p.u.}$$

then

$$I_{640} = \sqrt{0.445} = 0.667 \, \text{p.u.}$$

The current is therefore required to reduce from 1.0 p.u. to 0.667 p.u. by firing-angle retardation. At full load the power factor is 0.8 so that $\phi = \cos^{-1} 0.8 = 36.9°$. Assume that the power factor falls to 0.6 at 640 r.p.m. so that $\phi = \cos^{-1} 0.6 = 53°$. From the curves of Fig. 9.17 it is found that $37° \leq \alpha \simeq 90°$.

9.3.2 Control system aspects

9.3.2.1 Representation of the motor

For speed control purposes a three-phase induction motor can be considered to be an electrical system driving a mechanical system. The electrical system is a torque generator with an output variable T, which is the electromagnetic torque developed by the motor at a speed N with applied voltage V_1. This electrical system contains physical nonlinearities such as saturation and also

9.3 Voltage control by electronic switching   379

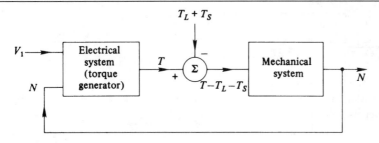

Fig. 9.24 Basic block diagram for the time control function of a three-phase induction motor.

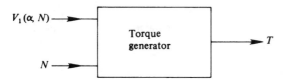

Fig. 9.25 Control nature of the 'electrical system' of the induction motor with SCR voltage control.

mathematical nonlinearities because the output variable depends on product functions of the two input variables V_1 and N. All of the variables N, T and V_1 are functions of the independent variable time. If the speed is constant the steady-state torque depends on the square of the steady-state voltage magnitude, as shown in (9.26). If the r.m.s. supply voltage is constant the relationship between steady-state speed and torque is defined by the motor physical parameters, (9.26). With SCR control a further analytical difficulty arises because the waveform and r.m.s. value of the applied voltage (with fixed firing-angle) depends on the motor impedance and therefore on the speed. The torque generator of Fig. 9.24 may be more accurately described by the relation of Fig. 9.25 in which the terminology $V_1(\alpha, N)$ means that the r.m.s. (or fundamental) value of the driving voltage is a function of both α and N. For small changes of V_1 or N one might consider the respective responses of T separately and add these in the form

$$dT = \left(\frac{\partial T}{\partial V_1}\right)_{N\mathrm{const}} \cdot dV_1 + \left(\frac{\partial T}{\partial N}\right)_{V_1\mathrm{const}} \cdot dN \tag{9.38}$$

The time response of torque T to time changes of applied voltage can be approximated by a first-order single time constant response, Fig. 9.26. This is consistent with representing the motor electrical system as a series R–L

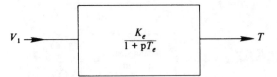

Fig. 9.26 Simplified representation of the induction motor transfer relationship, torque versus voltage.

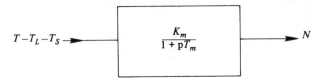

Fig. 9.27 Simplified representation of the induction motor mechanical transfer relationship, speed versus torque.

circuit in the manner of Fig. 9.22. The electrical gain constant K_e has the dimension of N m/volt and the motor electrical time constant T_e is given by

$$T_e = \frac{L_e}{R_e} \simeq \frac{X_1 + X_2}{2\pi f_1 R_2} \tag{9.39}$$

where L_e, R_e represent the effective series inductance and resistance, respectively. In reality the motor electrical system is of at least second (and probably third) order so that the first-order representation gives an optimistic expectation with regard to transient response and to closed-loop stability. The value of T_e generally varies over the range 0.004–0.05 seconds, tending to increase with the size of motor. For very large motors of the order of several thousand horsepower, $T_e \simeq 0.1$ second.

The mechanical system of Fig. 9.24 can be interpreted in terms of equation (4.5) and is shown in Fig. 9.27. The differential operator $p = d/dt$ is used and the mechanical time constant is given by

$$T_m = \frac{J}{B} \tag{9.40}$$

The mechanical gain constant K_m has the dimension r.p.m./N m. Time constant T_m is usually 10–100 times the electrical time constant T_e. If the motor is unloaded and has zero static friction then $T_L = T_S = 0$. Also, if the effect of the intrinsic speed loop in Fig. 9.24 is initially ignored, the electrical and mechanical 'boxes' can be combined in cascade to give a relationship of speed to motor voltage,

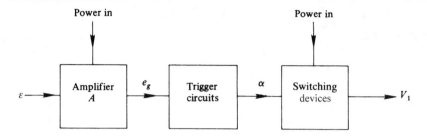

Fig. 9.28 Block diagram of the SCR voltage controller.

$$\frac{N}{V_1} = \frac{N}{T}\frac{T}{V_1} = \frac{K_m K_e}{(1+pT_m)(1+pT_e)} \tag{9.41}$$

Equation (9.41) is an operational form of expressing the relationship between the speed–time response caused by some specified voltage–time signal. It may be rearranged, in classical form, as

$$T_m T_e \frac{d^2 N}{dt^2} + (T_m + T_e)\frac{dN}{dt} + N = K_m K_e V_1 \tag{9.42}$$

In the steady state, both dN/dt and d^2N/dt^2 are zero. The remaining relationship $N = K_m K_e V_1$ in (9.42) means that any slow time change of V_1 is mapped by a corresponding slow time change of no-load speed N.

9.3.2.2 Representation of the SCR controller

The main components of the SCR voltage controller are shown in Fig. 9.28. The actuating signal ε (usually a d.c. voltage) is amplified by a linear electronic amplifier of gain A so that

$$e_g = A\varepsilon \tag{9.43}$$

Signal e_g creates an appropriate gating signal, usually a pulse train, to fire the SCRs at the predetermined value of phase delay angle α. The relationship between α and e_g depends on the particular design of the pulse circuitry, as discussed in Chapter 2, but is usually of the inverse-slope form of Fig. 9.29. Trigger circuit input signal e_g usually operates in the range $0 \leq e_g \leq 2.5\,\text{V}$. For operation with a three-phase motor load the commercial trigger circuit of one manufacturer (Ref. TP2) was found to satisfy the relation

$$\alpha = 226\left(1 - \frac{e_g}{2.1}\right) \tag{9.44}$$

A more general definition of the characteristics of Fig. 9.29 is

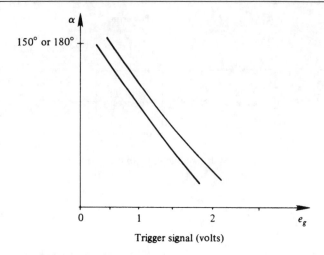

Fig. 9.29 Form of the trigger circuit calibration curve.

$$\alpha = \alpha_0 - K_g e_g \qquad (9.45)$$

where α_0 is the α-axis intercept and K_g is the slope of the characteristic. A correlation with (9.45) shows that the particular characteristics of (9.44) give $\alpha_0 = 226°$ and $K_g = 226/2.1 = 107.6$ degrees/volt.

The relationship between V_1 and α depends on the load phase-angle and is extensively discussed earlier in this section and in Chapter 8. For low values of load phase-angle ϕ the characteristics of Fig. 9.17 and the corresponding voltage characteristics (not shown) might be approximated by the relationship

$$V = V_{\max} \cos \alpha$$

or (9.46)

$$I = I_{\max} \cos \alpha$$

For high values of load phase-angle the approximation of (9.36) or (9.37) may be appropriate. But for more accurate calculations, particularly in the mid-firing-angle range, it is necessary to use the precise relationships of V (or I) in terms of α, such as those of (8.77), (8.82), (8.86), or corresponding relations for three-wire, star-connected loads (reference 32).

The overall response of V_1 to changes of ε in Fig. 9.28 cannot usually be accurately obtained by the linear system technique of multiplying the transfer relationships of the individual boxes, except for incremental variations about a fixed operating point.

$$\frac{V_1}{\varepsilon} = \frac{V_1}{\alpha} \frac{\alpha}{e_g} \frac{e_g}{\varepsilon} \tag{9.47}$$

The transient time response of each of the component systems in Fig. 9.28 is virtually instantaneous. This means that if the relationship V/ε is represented by only one box, the corresponding time constant is zero but the gain factor would depend on the particular firing circuit and the particular SCR configuration, the load and the operating point.

9.3.2.3 Closed-loop operation using tachometric negative feedback

A tachogenerator or encoder connected to the motor shaft in the circuit of Fig. 9.13 will give a signal proportional to the instantaneous shaft speed. For signal processing it is most convenient to have this speed signal in the form of a direct voltage by rectifying and smoothing (if necessary) the output from an a.c. tachogenerator. If the signal processing is to be conducted in digital form, rather than the analogue form implicit in the use of a tachogenerator, the shaft speed signal can be obtained as a pulse train by the use of a shaft encoder.

With a linear tachogenerator

$$e_N = K_{TG} N \tag{9.48}$$

Voltage e_N, representing the actual speed condition, is fed back, negatively, to an electronic signal discriminator (usually a summing circuit) and added algebraically to a reference voltage V_{ref} which represents the desired speed. The difference between V_{ref} and e_N is the error signal ε that actuates the drive

$$\varepsilon = V_{ref} - e_N \tag{9.49}$$

If (say) loss of load causes the drive speed to transiently rise above the desired steady-state speed defined by V_{ref}, then e_N is increased and ε reduces. This causes an increase of SCR firing-angle α which, in turn, causes reduction of the motor voltage. The effect of motor voltage reduction is to cause speed reduction. The negative feedback operation is therefore to cause self-correcting response of the speed to changes of load torque and also to cause fast response of the speed to changes of the drive command signal V_{ref}.

A more sophisticated form of drive might also contain a current feedback loop, whereby the line current at a given speed is detected and limited by adjustment of the SCR firing-angle.

The various components of the closed-loop drive of Fig. 9.30 are represented in the more detailed signal flow diagram of Fig. 9.31. A 3 h.p. motor, using the branch-delta connection of Fig. 9.18, resulted in the experimental

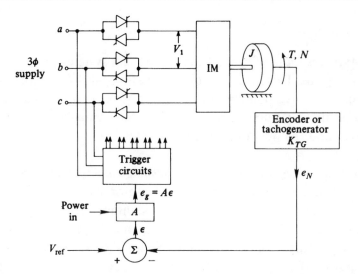

Fig. 9.30 Schematic diagram of an SCR controlled induction motor with tachometric feedback.

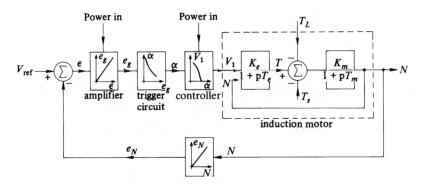

Fig. 9.31 Signal flow representation of a closed-loop SCR controlled induction motor.

characteristics of Fig. 9.32. These are seen to be a good realisation, in quadrants I and II, of the ideal torque–speed characteristics specified for an adjustable speed drive in Fig. 4.6. By change of the motor voltage, phase sequence characteristics can also be obtained in quadrants III and IV of the torque–speed plane. For the drive of Fig. 9.32 there was found to be a linear relationship between the reference voltage and the no-load speed for both positive and negative speeds.

The drive of Fig. 9.31 can be represented in the simplified form of Fig. 9.33 for ease of distinguishing between the open-loop and closed-loop responses.

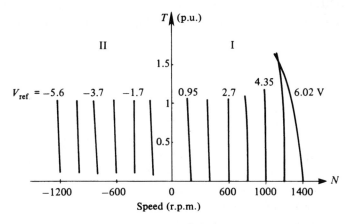

Fig. 9.32 Measured torque–speed performance of a 200 V, four-pole, 50 Hz, 3 h.p. induction motor with tachometric feedback, $V_{\text{line}} = 150$ V.

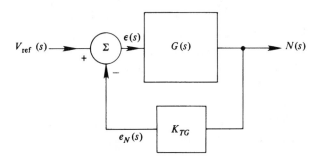

Fig. 9.33 Generalised control system representation of a closed-loop drive.

In the frequency domain, the forward loop transfer function in terms of the Laplace variable s is

$$\frac{N(s)}{\varepsilon(s)} = G(s) \tag{9.50}$$

With the feedback loop closed through a tachogenerator of gain K_{TG} the overall transfer function becomes

$$\frac{N(s)}{V_{\text{ref}}(s)} = \frac{G(s)}{1 + K_{TG}G(s)} \tag{9.51}$$

A plot of the function $K_{TG}G(j\omega)$ in the complex plane is usually called a Nyquist diagram and gives useful information about the transient response and closed-loop stability of the drive.

9.3.2.4 Worked examples

Example 9.4
A three-phase, squirrel-cage induction motor is to be controlled by terminal voltage variation using pairs of inverse-parallel SCRs in the supply lines with symmetrical phase-angle triggering. Sketch a diagram of this arrangement for open-loop control and point out the advantages and disadvantages compared with (say) auto-transformer control.

The control loop is now closed using a tachogenerator to give negative feedback. Sketch the system in block diagram form, defining transfer functions or transfer characteristics for each block. If the system is operating at low speed, what would be the effect of suddenly increasing the reference signal on (i) the error signal, (ii) the SCR firing-angles, (iii) the motor voltage and (iv) the speed? Sketch the type of speed response versus time that you would consider acceptable and reasonable in such a system. How would increase of the motor inertia affect the transient speed response? Sketch this effect on the speed response diagram.

Solution. An SCR controlled induction motor on open loop is shown in Fig. 9.13 or Fig. 9.18. The advantages and disadvantages of SCR control compared with auto-transformer (i.e. sinusoidal) voltage variation are listed in Table 9.2, as part of Example 9.3. A representation of the SCR controlled motor drive with negative tachometric feedback is given in Fig. 9.31, showing some detail of the transfer characteristics of individual parts of the system.

Consider operation after a step increase of reference signal. The speed cannot change instantaneously so that the speed signal e_N remains constant. From (9.49), if V_{ref} increases then $\varepsilon (= V_{ref} - e_N)$ also increases. Let increase in the value of V_{ref} be denoted by the terminology $V_{ref} \uparrow$.

If $V_{ref} \uparrow$,
then $\varepsilon (= V_{ref} - e_N) \uparrow$
then $e_g (= A\varepsilon) \uparrow$
then $\alpha \downarrow$
then $V_1 \uparrow$
then $N \uparrow$, from (9.41).

An increase of V_{ref} is therefore followed by a corresponding increase of speed, the nature of the speed–time transition being determined by the physical nature of the drive components. The dominant physical parameter

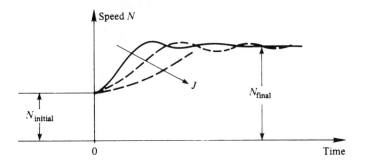

Fig. 9.34 Speed–time response of an SCR controlled induction motor drive.
——— acceptable response, — — — sluggish response.

affecting speed–time changes is known to be the inertia J of the motor and load. Increase of the drive inertia causes a more sluggish response and increases the mechanical drive constant T_m. An acceptable speed–time response together with the response modified by increase of J is shown in Fig. 9.34.

Example 9.5
A three-phase induction motor with voltage control by SCR switching is used as a drive motor in a closed-loop system with tachometric feedback. Sketch such a system in block diagram form. If the system is operating at low speed what would be the effect of suddenly increasing the reference signal on the error signal, the SCR firing-angles, the motor voltage and the speed?

Over the low-speed range, at no-load, the forward loop transfer function is represented by

$$G(s) = \frac{K}{s(1+Ts)}$$

If $T = 0.15$ s and $K = 10$, sketch, roughly to scale, a diagram of attenuation (in dB) versus angular frequency ω (in logarithmic scale) for $G(s)$. At what frequency does $|G(j\omega)|$ in dB cross the frequency axis? What is the physical significance of this?

Solution. A diagram in partial block diagram form is given in Fig. 9.30 with a more detailed diagram in Fig. 9.31. The effect of a sudden increase of the reference voltage is described in Example 9.4 above.

The forward loop transfer function is given as

$$G(s) = \frac{K}{s(1+sT)}$$

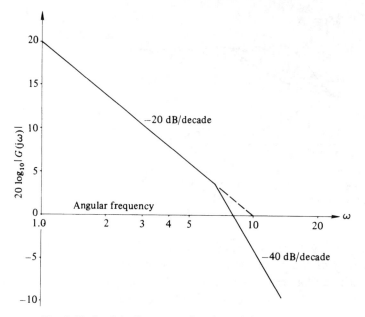

Fig. 9.35 Straight-line approximation of the gain versus frequency (i.e. Bode diagram) for $|G(j\omega)|$ in Example 9.5.

This has the corresponding frequency response function

$$G(j\omega) = \frac{K}{j\omega(1+j\omega T)}$$

where ω is the angular frequency of the actuating signal. Function $G(j\omega)$ has a break point at $1/T = 1/0.15 = 6.67$ rad/s. On the Bode diagram of Fig. 9.35 the attenuation characteristic at $\omega = 1$ has the magnitude

$$|G(j\omega)| = \frac{K}{\omega} = 10$$

The gain in dB at $\omega = 1$ is therefore

$$|G|_{dB} = 20\log_{10}|G(j\omega)| = 20\,dB$$

The straight-line approximation of the gain versus frequency reduces at the rate of 20 dB/decade, Fig. 9.35, until it reaches the break point at $\omega = 6.67$ rad/s. For frequencies greater than 6.67 rad/s both of the denominator terms of $G(j\omega)$ are operational and the gain thereafter reduces at the rate of 40 dB/decade.

It is seen that $|G|_{dB} = 0$ at $\omega = 8$ rad/s.

At this point,

9.3 Voltage control by electronic switching

Table 9.3 *Values for G(jω).*

ω	Real part $\dfrac{-1.5}{1+0.0225\omega^2}$	Imaginary part $\dfrac{-j10}{1+0.0225\omega^2}$	$\|G\|$ $\dfrac{K}{\omega\sqrt{1+\omega^2 T^2}}$	$\angle G$ $\tan^{-1}\left(\dfrac{1}{\omega T}\right) - 90°$
0	−1.5	−∞	∞	−90°
1	−1.467	−9.78	9.89	−98.5°
5	−0.96	−1.28	1.6	−116.9°
7	−0.714	−0.68	0.986	−136.4°
8	−0.615	−0.512	0.8	−140.2°
10	−0.46	−0.31	0.55	−146°
20	−0.15	−0.05	0.16	−161.5°
∞	−0	−0	0	−180°

$$20 \log_{10} |G(j\omega)| = 0$$
$$\therefore \log_{10} |G(j\omega)| = 0$$
$$|G(j\omega)| = 1.0$$

If the expression $G(j\omega)$ is rationalised, it may then be written in the cartesian form

$$G(j\omega) = \frac{-KT}{1+\omega^2 T^2} + \frac{jK}{\omega(1+\omega^2 T^2)}$$
$$= \frac{-1.5}{1+0.0225\omega^2} + \frac{j10}{\omega(1+0.0225\omega^2)}$$

Values for $G(j\omega)$ are presented in Table 9.3.

It is seen that with accurate calculation, the value $|G(j\omega)| = 1.0$ occurs at $\omega = 7$ rad/s rather than at $\omega = 8$ rad/s as suggested in the straight-line approximation of Fig. 9.35.

The function $G(j\omega)$ is plotted in the complex (Nyquist) plane in Fig. 9.36. If the Nyquist diagram of an open-loop frequency response function $G(j\omega)$ does not encompass the point $(-1, j0)$ then the system on closed-loop, with direct feedback, will be stable. If $G(j\omega)$ in the Nyquist plane does not approach the point $(-1, j0)$ then the system on closed loop with direct feedback is amply stable and is likely to have little oscillation in its transient response. The second-order function $G(s)$ above specifies that the present drive will be intrinsically stable on closed-loop with direct feedback because $G(j\omega)$ cannot cross the negative real axis. If the gain constant K is increased then $G(j\omega)$ in Fig. 9.36 is moved to the left and approaches more nearly the point $(-1, j0)$. The system transient response then becomes more oscillatory.

In the present problem there is no direct physical significance in the crossover frequency of the Bode diagram in Fig. 9.35.

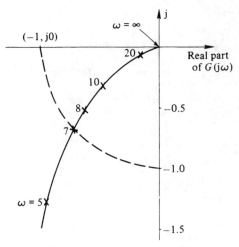

Fig. 9.36 Complex-plane plot of $G(\mathrm{j}\omega)$ from Example 9.5.

Example 9.6

An induction motor with SCR voltage control of the form shown in Fig. 9.13 has its open-loop performance approximately represented by the transfer function

$$G(s) = \frac{N}{\varepsilon}(s) = \frac{K_c K_e K_m}{(1 + sT_c)(1 + sT_e)(1 + sT_m)}$$

where

$K_e = 2.4$, $K_m = 12$, $K_c =$ adjustable

$T_c = 0.003$ s, $T_e = 0.04$ s, $T_m = 0.14$ s

Gain constant K_c represents the overall effect of the amplifier, trigger circuit and SCR control configuration of Fig. 9.31. The system is to be operated on closed loop with tachometric feedback where the tachogenerator constant $K_{TG} = 1$ V/rad/s. What is the maximum permitted value for K_c?

Solution. The Nyquist diagram for a transfer function of the third-order type above has the form shown in Fig. 9.37. Such a diagram can encompass the critical point $(-1, \mathrm{j}0)$ on the negative real axis and hence the corresponding closed-loop system may be unstable, even if the open-loop system is stable.

Rewriting N/ε as a frequency response function gives

$$\frac{N}{\varepsilon}(\mathrm{j}\omega) = \frac{K_c K_e K_m}{(1 + \mathrm{j}\omega T_c)(1 + \mathrm{j}\omega T_m)(1 + \mathrm{j}\omega T_e)}$$

9.3 Voltage control by electronic switching

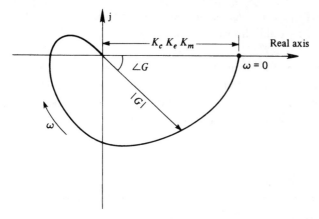

Fig. 9.37 Form of Nyquist plot $G(j\omega)$ for a third-order transfer function in Example 9.6.

Rationalising gives

$$\frac{N}{\varepsilon}(j\omega)$$
$$= \frac{K_c K_e K_m}{[1 - \omega^2(T_c T_m + T_c T_e + T_e T_m)] - j\omega(\omega^2 T_c T_e T_m - T_c - T_e - T_m)}$$

$$= K_c K_e K_m \left\{ \frac{[1 - \omega^2(T_c T_m + T_c T_e + T_e T_m)] + j\omega(\omega^2 T_c T_e T_m - T_c - T_e - T_m)}{[1 - \omega^2(T_c T_m + T_c T_e + T_e T_m)]^2 + \omega^2(\omega^2 T_c T_e T_m - T_c - T_e - T_m)^2} \right\}$$

At the point where $(N/\varepsilon)(j\omega)$ crosses the negative real axis its j term has zero value. Therefore, if

$$K_c K_e K_m \omega(\omega^2 T_c T_e T_m - T_c - T_e - T_m) = 0$$

then

$$\omega^2 = \frac{T_c + T_e + T_m}{T_c T_e T_m}$$

At this frequency the gain of $(N/\varepsilon)(j\omega)$ is

$$\left| \frac{N}{\varepsilon}(j\omega) \right| = \frac{K_c K_e K_m}{1 - \omega^2(T_c T_m + T_c T_e + T_e T_m)}$$

$$= \frac{K_c K_e K_m}{1 + \frac{(T_e + T_c + T_m)(T_c T_e + T_c T_m + T_e T_m)}{T_c T_e T_m}}$$

$$= \frac{12 \times 2.4 K_c}{1 + \frac{(0.183)(0.006\,14)}{0.000\,017}}$$

$$= \frac{28.8 K_c}{1 + 66.1} = 0.429 K_c$$

392 Three-phase induction motor with constant frequency supply

To make $|(N/\varepsilon)(j\omega)| = 1$, K_c would have the the value $K_c = 1/0.429 = 2.33$. In practice, one would choose a value for K_c of 0.5–0.7 times this critical value.

If this induction motor is incorporated in a negative feedback, closed-loop system with non-unity feedback constant K_{TG}, the closed-loop stability can be tested by plotting $K_{TG}G(\omega)$ in the Nyquist plane. Increase in the value of K_{TG} would increase the possibility of instability.

Example 9.7

For the induction motor drive of Example 9.6 calculate the gain of the amplifier if the trigger circuit and the SCR configuration are represented by the approximations of (9.44) and (9.37) respectively.

Solution. The amplifier, trigger circuit and SCR configuration are in cascade as shown in Fig. 9.28 and Fig. 9.31. If the transfer characteristics of the three 'boxes' were dependent only on the system parameters, independently of signal level, then the system would be linear and the overall transfer function V/ε could be obtained by multiplication of the individual transfer functions, as in (9.47).

The transfer characteristics of the SCR controller box in Fig. 9.31 can be represented approximately by (9.37) if the branch-delta connection, Fig. 9.18, is used

$$\frac{V}{\alpha} = -\left(\frac{1}{180-\phi}\right) + \left(\frac{180}{180-\phi}\right)\frac{1}{\alpha} \text{ V/degree} \qquad (a)$$

Transfer charactersitic V/α depends not only on system parameters but on the signal variable α, which is a feature of nonlinear systems. The transfer characteristic of the trigger circuits in Fig. 9.31 is obtained from (9.44):

$$\frac{\alpha}{e_g} = -107.6 + \frac{226}{e_g} \text{ degrees/V} \qquad (b)$$

Similarly to (a), the transfer characteristic depends on the signal level. The transfer characteristic of the amplifier in Fig. 9.31 is its gain A, which is independent of signal level.

Substituting from (a) and (b) into (9.47) gives

$$\frac{V}{\varepsilon} = \left[-\left(\frac{1}{180-\phi}\right) + \left(\frac{180}{180-\phi}\right)\frac{1}{\alpha}\right]\left(-107.6 + \frac{226}{e_g}\right)(A) \qquad (A)$$

The magnitude of V/ε corresponds to the gain constant K_c in Example 9.6, for no-load operation of the motor.

One must conclude that the SCR controlled drive cannot be readily analysed in terms of linear control theory. The 'constant' K_c in Example 9.6 is,

Table 9.4 Evaluation of gain A.

	High speed	Low speed
ϕ	25°	45°
α	45°	90°
e_g	2 V	0.65 V
A	22.3	1.3

in reality, a parameter that undergoes wide variation and is a function of ϕ, α and e_g, all of which depend on the drive speed.

Choosing typical values gives the results for A (with $K_c = 2.33$) shown in Table 9.4.

9.4 THREE-PHASE INDUCTION MOTOR WITH FIXED SUPPLY VOLTAGES AND ADJUSTABLE SECONDARY RESISTANCES

9.4.1 Theory of operation

One of the most common ways of achieving speed control in a three-phase induction motor is to vary the secondary circuit resistance R_2. This can easily be done for wound-rotor motors, by connecting a set of ganged resistors to the secondary brushes, as shown in Fig. 9.38. It is an advantage that the heat developed in the secondary resistors is dissipated outside the machine frame. In very large motors, of several thousand horsepower rating, wire-wound resistors of suitable current range would be very large and expensive.

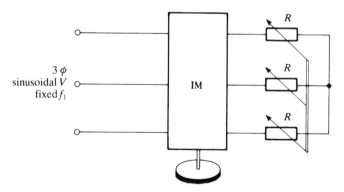

Fig. 9.38 Three-phase slip-ring induction motor with equal secondary circuit resistors.

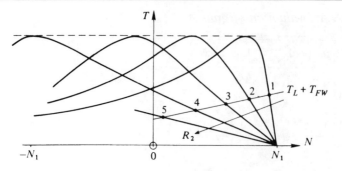

Fig. 9.39 Torque–speed characteristics of three-phase induction motor with adjustable secondary resistance.

Satisfactory control can be obtained by the use of columns of water with plungers that vary the column length.

Typical torque–speed characteristics due to constant primary voltage, adjustable secondary resistance are shown in Fig. 9.39. The peak torque T_m is not changed, (9.28). The slip S_{T_m} at which peak torque occurs is directly proportional to the total secondary circuit resistance per-phase, (9.27), and hence the torque characteristics are displaced progressively to the left as R_2 increases. A wide range of speed control is available, Fig. 9.39, but increase of R_2 causes reduction of the motor current and hence reduction of motor torque at low speeds.

Effective variation of the secondary current can be obtained by the use of inverse-parallel connected pair of SCRs in the secondary lines, Fig. 9.40. It is seen that the external resistance per-phase R_{ext} varies between two extremes,

$$R_{\text{ext}} = R_B, \text{ when } \alpha = 0° \\ R_{\text{ext}} = R_A + R_B, \text{ when } \alpha = 180° \tag{9.52}$$

If $R_B = 0$ the motor secondary windings can be short circuited to give maximum operating speed. The use of secondary side devices greatly reduces the primary current distortion present in supply-side control.

Although the system of Fig. 9.40 is conceptually straightforward it suffers from a number of serious disadvantages. These arise because the anode voltages of the SCRs are now the motor secondary voltages, rather than the supply voltages. The e.m.fs that occur in the secondary windings are of slip frequency, (9.3), and have magnitudes proportional to slip, (9.4). At high speeds the secondary e.m.f. E_2' and its corresponding value referred to primary turns E_2 are small. Near to synchronous speed, voltage E_2' may become too small to reliably ensure SCR switch-on after gating.

9.4 Secondary resistance control

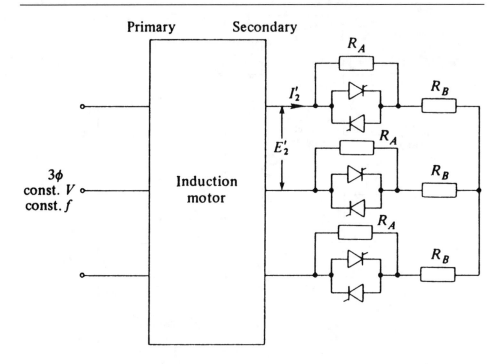

Fig. 9.40 SCR control of secondary current in a three-phase induction motor.

An SCR firing-angle α is normally measured relative to its respective anode voltage waveform. With primary voltage control the anode voltage datum is fixed by the supply. In the secondary circuit of an induction motor the induced e.m.f. varies in time-phase with both the motor speed and current relative to the fixed datum of the primary voltages. The anode voltage datum with secondary side SCRs is therefore not fixed. A given gating signal in the circuit of Fig. 9.40 will produce a different firing-angle for each value of motor speed at a fixed current and for each value of current (i.e. torque) at a fixed speed. The control problem arising here is complex and secondary SCR control has not, so far, proved to be a commercially attractive proposition.

An alternative form of secondary SCR control is illustrated in Fig. 9.41. A unidirectional voltage V_s is produced at the output of a full-wave diode bridge circuit and smoothed by inductor L. Current I_R through the resistor R is adjusted by the chopping rate of the parallel connected d.c. chopper circuit, typically of the order of 1 kHz. The external resistance is zero during chopper conduction and R during chopper extinction with a time variation corresponding to the voltage variation of Fig. 5.7(b). Therefore,

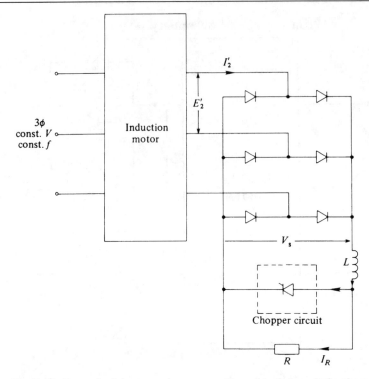

Fig. 9.41 Control of the secondary current in a three-phase induction motor by means of a secondary side chopper circuit (commutation circuit not shown).

$$R_{\text{ext}} = \frac{T_{\text{off}}}{T_{\text{on}} + T_{\text{off}}} R \qquad (9.53)$$

The chopper circuit requires to be force commutated and the commutation circuit results in high losses at high chopping frequencies. At high motor speeds the secondary e.m.f. is low and may be inadequate as a commutating voltage. Only a small range of speed control is possible. The use of a rotor chopper raises an interesting technical challenge but has not yet proved to be a commercially viable system.

9.4.2 Worked example

Example 9.8
A three-phase, wound-rotor induction motor is connected directly to a three-phase supply of constant voltage and frequency. Pairs of inverse-parallel SCRs are incorporated into the secondary windings using the connction of Fig. 9.42. In values referred to primary turns, $R_A = 10R_B$ and R_A is five

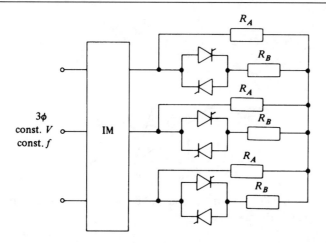

Fig. 9.42 Motor secondary configuration in the circuit of Example 9.8.

times the secondary winding resistance R_2. Sketch approximate torque–speed characteristics for the two conditions (i) $\alpha = 0°$ and (ii) $\alpha = 180°$. What approximate range of speed control is possible?

Solution. At $\alpha = 0°$,

$$R_{ext} = R_A \| R_B = \frac{R_A R_B}{R_A + R_B}$$

$$R_{sec} = R_2 + \frac{R_A R_B}{R_A + R_B}$$

At $\alpha = 180°$,

$$R_{ext} = R_A$$
$$R_{sec} = R_2 + R_A$$

Now $R_A = 10 R_B$ and $R_A = 5 R_2$. Therefore, at $\alpha = 0°$,

$$R_{sec} = \frac{R_A}{5} + \frac{R_A}{11} = \frac{16}{55} R_2 = \frac{16}{11} R_2$$

At $\alpha = 180°$,

$$R_{sec} = R_A + \frac{R_A}{5} = \frac{6}{5} R_A = 6 R_2$$

From (9.28) the peak torque T_m is not affected by a change of R_{sec} but the slip S_{T_m} at which peak torque occurs, (9.27), is proportional to R_{sec}/phase. In this case

$$\frac{S_{T_m} \text{ at } \alpha = 0°}{S_{T_m} \text{ at } \alpha = 180°} = \frac{\frac{16}{11} R_2}{6 R_2} = \frac{8}{33}$$

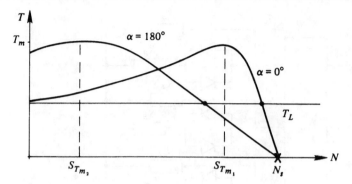

Fig. 9.43 Torque–speed characteristics obtained by SCR control in the circuit of Fig. 9.42, referring to Example 9.8.

Corresponding torque–speed characteristics for $\alpha = 0°$ and $\alpha = 180°$ are shown in Fig. 9.43. With a constant torque load of about rated value the speed range available for control is from full speed to about two-thirds of full speed.

9.5 PROBLEMS

Three-phase induction motor with primary voltage variation at constant frequency

9.1 A three-phase induction motor operating from a power supply of constant frequency can be used to supply a range of speed by variation of
  (a) the magnitudes of the terminal voltages, or
  (b) the magnitudes of the secondary circuit resistances.
  Sketch motor torque–speed characteristics to demonstrate each of these control methods, showing intersections with a load line representing a constant torque requirement of about rated torque. Point out the relative features of the two schemes for speed control purposes in this case.
  Show, by diagrams, how pairs of SCRs connected in the inverse-parallel arrangements may be used to obtain speed control by each of the two methods. Explain the scheme of SCR triggering that you would recommend. What particular difficulties would you anticipate in correctly triggering the SCRs to obtain secondary resistance control?

9.2 A three-phase squirrel-cage induction motor is supplied from a source of three-phase sinusoidal voltages of constant frequency but variable magnitude. Explain, using an equivalent circuit and corresponding equations, the effect of terminal voltage reduction on
  (a) motor current–speed
  (b) torque–speed

(c) efficiency–speed

How is the motor torque per ampere affected if the supply voltage is halved?

Terminal voltage variation can be achieved by the symmetrical phase-angle triggering of pairs of SCRs in the supply lines. Sketch this proposed arrangement and explain the advantages and disadvantages of using SCRs compared with the use of a variable voltage auto-transformer.

9.3 The torque–speed curve of a water pump is described by the equation $T = N^2/200$ SI units, where N is the speed of the pump motor. This pump is to be controlled by a three-phase, 240 V, six-pole, 50 Hz, star-connected induction motor with pairs of inverse-parallel connected SCRs in each supply line. The per-phase equivalent circuit parameters of the motor, referred to primary turns, are $R_1 = 0.3\,\Omega$, $R_2 = 0.2\,\Omega$, $X_1 = X_2 = 0.6\,\Omega$, $X_m = $ infinity. The required speed range is 600–975 r.p.m. Use performance curves of current versus firing-angle to calculate, approximately, the necessary range of firing-angles.

9.4 A three-phase, 240 V, four-pole, 50 Hz, star-connected, squirrel-cage induction motor is to be used for the speed control of a fuel pump. In the required control range the load line can be represented by

$$T_L = \frac{N^2}{500} - \frac{N^3}{10^6} \text{ SI units}$$

where $N = $ motor speed.

Per-phase equivalent circuit parameters of the motor, referred to primary turns, are $R_1 = 0.3\,\Omega$, $R_2 = 0.2\,\Omega$, $X_1 = X_2 = 0.72\,\Omega$, $X_m = $ very large. The motor terminal voltage are to be controlled by pairs of inverse-parallel connected SCRs in the supply lines. If steady-state speed control is required in the range 750–1450 r.p.m., calculate the necessary range of firing-angles.

9.5 A three-phase, six-pole, squirrel-cage induction motor is to be controlled by terminal voltage variation using pairs of inverse-parallel SCRs in the supply lines with symmetrical phase-angle triggering. Sketch a diagram of this arrangement and point out the advantages and disadvantages compared with, say, auto-transformer control.

The motor is rated at 50 kW, 240 V, 50 Hz. If it operates at a full-load efficiency of 0.9 p.u. and power factor 0.85, calculate the r.m.s. current rating and maximum voltage rating required of the SCRs.

The motor drives a fan load characterised by the relation $T_L = kN^2$, where T_L is the shaft torque and N is the motor speed. Operation is required at full speed, which corresponds to 5% slip, and at 750 r.p.m. What approximate reduction of motor power is required (compared with full-load, full-speed operation) in order to realise operation at 750 r.p.m.?

If the motor operates at full voltage at the upper speed, calculate, approximately, the voltage required at the lower speed. Explain and estimate the change of SCR firing-angle necessary to obtain the required change of voltage.

9.6 A three-phase, six-pole, star-connected squirrel-cage induction motor is to be controlled by terminal voltage variation using pairs of inverse-parallel SCRs in the supply lines with symmetrical phase-angle triggering. The motor is rated at 10 kW, 240 V, 50 Hz, and its equivalent circuit parameters, referred to stator (primary) turns, are $R_1 = 0.25\,\Omega$, $R_2 = 0.2\,\Omega$, $X_1 = X_2 = 0.8\,\Omega$, $X_m =$ infinity. Core losses and friction and windage may be neglected. The motor is to be used to drive a water pump which has the load requirement $T_L = 0.01 N^2$, where T_L is the load torque and N the load speed in SI units. Smooth speed control is required from rated speed (955 r.p.m.) to 500 r.p.m. Calculate, approximately, the necessary range of SCR firing-angles.

9.7 The voltages to the terminals of a three-phase, 50 kW, 240 V induction motor are to be controlled by pairs of inverse-parallel connected SCRs in the supply lines. If the motor full-load efficiency is 0.9 p.u. and the full-load power factor is 0.85, calculate the r.m.s. current, mean current and maximum voltage ratings of the SCR switches.

9.8 A three-phase, delta-connected, 240 V, six-pole, 50 Hz induction motor is to be used for the speed control of a water pump by the connection of pairs of inverse-parallel SCRs in series with each load leg. The required speed range is from 500–950 r.p.m. and the load characteristic is defined by $T = kN^2$, where $k = 7 \times 10^{-3}$ SI units. Standard tests on the motor gave the following data for the equivalent circuit parameters (referred to primary turns): $R_1 = 0.294\,\Omega$, $R_2 = 0.144\,\Omega$, $X_1 = 0.503\,\Omega$, $X_2 = 0.209\,\Omega$, $X_m = 13.25\,\Omega$. No-load loss = 1209 W. If the SCRs are triggered to give symmetrical phase-angle firing, calculate the limits of firing-angle to be used.

9.9 A three-phase, star-connected induction motor has its terminal voltages controlled by the symmetrical triggering of pairs of inverse-parallel connected SCRs in the supply lines. At a certain fixed speed the motor equivalent circuit has an input phase-angle ϕ. Show that the firing-angle α necessary to give a per-unit current I is given by the approximate expression

$$I = \frac{150° - \alpha}{150° - \phi}$$

9.10 Show that the transfer characteristic I versus α obtained from the SCR controller of Problem 9.9 has the form

$$\frac{I}{\alpha} = -K_1 + \frac{K_2}{\alpha}$$

What implication does this have for the development of a control system transfer function relating I and α?

9.11 A three-phase induction motor with voltage control by SCR switching is used as a drive motor in a closed-loop system with tachometric feedback. Sketch such a system in a 'black box' form. If the system is initially operating at high speed, what would be the effect of suddenly decreasing the reference signal on the error signal, the SCR firing-angles, the motor voltage and the speed?

Sketch the type of speed response versus time that you would consider acceptable and reasonable in such a system.

What are the advantages and disadvantages of using SCRs for primary voltage control of an induction motor?

9.12 A three-phase induction motor with terminal voltage variation controlled by the triggering of pairs of SCRs in the supply lines is used as the drive motor in a closed-loop system with tachometric feedback. Sketch the likely circuit arrangement and give block diagrams of the various system components and their interconnection for negative feedback.

Over the low-speed range, at no-load, the forward loop transfer function is represented by

$$G(s) = \frac{K}{s(1+Ts)}$$

If $T = 0.12$ s and $K = 8$, sketch, roughly to scale, a diagram of attenuation (in dB) versus angular frequency ω (in logarithmic scale) for $G(j\omega)$. At what frequency does $|G(j\omega)|$ in dB cross the frequency axis? How would you modify the system so that the frequency of the $|G(j\omega)|$ versus ω crossover is increased?

9.13 Sketch the drive torque–speed characteristics necessary for ideal steady-state control of a constant torque load. Also sketch and briefly comment on the possible dynamic step responses to sudden decrease of (a) load torque, (b) control (speed) signal.

A three-phase induction motor has its r.m.s. terminal voltages V controlled by a.c. voltage controllers in the supply lines. This forms part of a closed-loop control system with negative feedback provided via a shaft encoder. Sketch the arrangement diagrammatically.

Describe the likely response of the drive variables to (a) sudden increase, (b) sudden decrease, of the reference voltage, between the same extreme values.

9.14 The steady-state torque–speed characteristics of an ideal variable speed drive are given in Fig. 9.44 together with the characteristics of a certain load, at no-load, full load torque and twice full load torque.

Sketch the form of speed–time transitions that you consider acceptable in the cases (i) $A \to B$, (ii) $B \to C$, (iii) $C \to D$, (iv) $D \to A$.

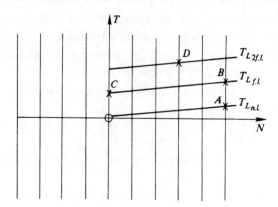

Fig. 9.44 Load and drive torque–speed characteristics in Problem 9.14.

9.15 A three-phase, squirrel-cage induction motor is to be controlled by terminal voltage variation using pairs of inverse-parallel SCRs in the supply lines with symmetrical phase-angle triggering.

The control loop is now closed using a tachogenerator to give negative feedback. Sketch the system in block diagram form, defining transfer functions or transfer characteristics for each block. Discuss the inherent difficulties of analysing such a system by linear control theory. How would you approach the analysis of such a system?

The response of motor shaft speed to a step increase of control signal in a typical case is given in Fig.9.45. What do you deduce from this about the nature of the system?

9.16 A three-phase induction motor drive with SCR voltage control is incorporated in a closed-loop speed-control system with tachometric negative feedback of gain constant K_{TG}. The performance of the forward loop can be represented by a three-time-constant transfer function of the form

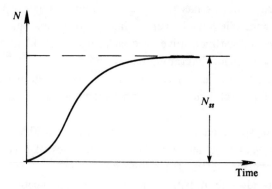

Fig. 9.45 Motor speed transient response in Problem 9.15.

$$G(s) = \frac{N}{\varepsilon}(s) = \frac{K_c K_e K_m}{(1+sT_c)(1+sT_e)(1+sT_m)}$$

where

$K_e = 2, \ K_m = 10, \ K_c = 1.5$

$T_c = 0.0025 \text{ s}, \ T_e = 0.05 \text{ s}, \ T_m = 0.15 \text{ s}$

Sketch the Nyquist diagram for this function and calculate the magnitude and frequency for the point where the locus crosses the negative real axis of the complex plane. What is the maximum value of feedback constant K_{TG} that you would advise?

9.17 For the closed-loop drive of Problem 9.16 derive an expression for the closed-loop frequency response function $(N/V_{\text{ref}})(j\omega)$.

What is $|(N/V_{\text{ref}})(j\omega)|$ when $\omega = 1 \text{ rad/s}$?

9.18 Derive an overall expression for the transfer characteristic between the speed N and the speed (reference) signal V_{ref} for a closed-loop induction motor drive of the form shown in Figs. 9.30, 9.31. Show that this characteristic is an implicit function of several interrelated variables and cannot be expressed in terms of physical parameters only.

9.19 A three-phase induction motor has its voltages controlled by the symmetrical triggering of pairs of SCRs in the supply lines, as illustrated in Fig. 9.30. What would you estimate to be the order of the transfer characteristic relating motor shaft speed N to the trigger circuit control signal e_g? Sketch the form of torque–speed curves obtained on closed-loop with corresponding characteristics obtained for open-loop operation.

Three-phase induction motor with fixed supply voltages and adjustable secondary resistances

9.20 A three-phase, wound-rotor induction motor has its speed controlled by a secondary side chopper connection of the form of Fig. 9.41. If $R = 5R_2$ sketch, approximately, the comparative forms of torque–speed characteristics obtained with (i) full conduction of the chopper, (ii) $T_{\text{on}}/T_{\text{off}} = 4/5$ and (iii) $T_{\text{on}}/T_{\text{off}} = 1/5$.

9.21 A three-phase, wound-rotor induction motor is controlled by a.c. chopping using pairs of inverse-parallel connected SCRs in the secondary windings, as shown in Fig. 9.40. Referred to primary turns, the resistance values are $R_A = 6R_B, \ R_B = R_2$.

Sketch curves of torque–speed for (1) $\alpha = 0°$, (ii) $\alpha = 180°$, compared with operation with a directly short-circuited secondary. Comment on the comparative performance with a constant torque load of about rated value.

10

Induction motor slip-energy recovery

10.1 THREE-PHASE INDUCTION MOTOR WITH INJECTED SECONDARY VOLTAGE

10.1.1 Theory of operation

It is possible to open the secondary windings of a wound-rotor induction motor and replace the resistors of Fig. 9.38 by a voltage source of adjustable magnitude and phase-angle. Such an external injected voltage must operate at slip frequency for all motor speeds. The approximate equivalent circuit of Fig. 9.4 becomes modified to the form of Fig. 10.1. The magnitude of the secondary current is now given by

$$I_2 = \frac{|V_1 \pm E_s|}{\sqrt{\left(R_1 + \frac{R_2}{S}\right)^2 + (X_1 + X_2)^2}} \qquad (10.1)$$

The phasor magnitude of the two voltages in the numerator of (10.1) can vary from zero, when E_s is equal to but in antiphase with V_1, to a value much larger than V_1 if E_s has a component in phase with it. The torque–speed performance is greatly modified if voltage E_s is a function of motor slip, which is the usual case. If separate control can be established of the in-phase and quadrature components of E_s then independent control of both speed and power factor can be realised.

Many methods have been described for achieving a secondary injected e.m.f. These include induction machines with elaborate and expensive auxiliary windings and multi-machine systems involving one or more rotating machines additional to the induction motor.

The principle of secondary e.m.f. injection can be most effectively implemented by the use of a slip-energy recovery (SER) scheme, as described in the

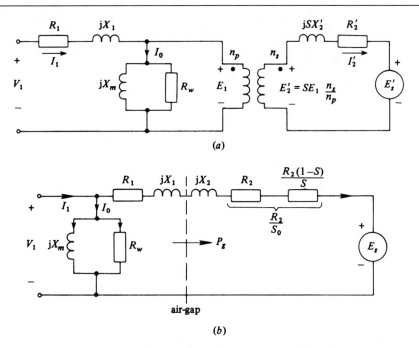

Fig. 10.1 Per-phase equivalent circuits of a three-phase induction motor with injected secondary e.m.f.: (*a*) transformer circuit, (*b*) approximate circuit (referred to primary turns).

following section. This arrangement uses a solid-state bridge rectifier and a controlled inverter to act as a static frequency changer.

10.1.2 Worked example

Example 10.1
A three-phase, wound-rotor induction motor has a slip frequency e.m.f. E'_s injected into the slip rings. This e.m.f. is adjustable by external means and is in time-phase or antiphase with the secondary induced e.m.f. Deduce the shape of the torque–speed characteristic for a range of values of E'_s.

Solution. The equivalent circuit of Fig. 10.1(*a*) is appropriate here, in which the secondary current I'_2 is given by

$$I'_2 = \frac{\bar{E}'_2 - \bar{E}'_s}{R'_2 + jSX'_2}$$

If the voltage drops of the primary series components are neglected the net driving voltage in the secondary circuit at no-load ($I'_2 \to 0$) is

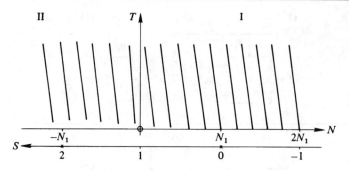

Fig. 10.2 Torque–speed characteristics of induction motor with injected secondary e.m.f.

$$SV_1 \frac{n_s}{n_p} - E'_s = 0$$

Given that E'_s can only be in phase or antiphase with V_1, the no-load slip S_0 is therefore

$$S_0 = \pm \left| \frac{n_p}{n_s} \cdot \frac{E'_s}{V_1} \right|$$

The no-load speed N_0 may then be expressed in terms of the synchronous speed N_1 as

$$N_0 = (1 - S_0)N_1 = \left(1 \pm \left|\frac{n_p}{n_s} \cdot \frac{E'_s}{V_1}\right| N_1 \right)$$

In the working range of a polyphase induction motor the torque is proportional to the slip and the speed regulation is small. The speed equation above describes the characteristics shown in Fig. 10.2. If E'_s in Fig. 10.1(a) is reversed in polarity the positive sign becomes applicable in the speed equation above and supersynchronous speeds are realised.

10.2 INDUCTION MOTOR SLIP-ENERGY RECOVERY (SER) SYSTEM

Speed control in the subsynchronous region, from zero to rated motor speed, can be realised by the system shown in Fig. 10.3. The slip frequency e.m.f.s of the motor secondary windings are rectified by a full-wave diode bridge rectifier. A unidirectional current I_{dc} passes through the link between the rectifier and the naturally commutated inverter and is sometimes referred to as the d.c. link current. This current is opposed by the average voltage E_{av} of the controlled bridge inverter. Filter inductor L_f acts to smooth the link current and also to absorb the ripple voltage from both the diode rectifier and the

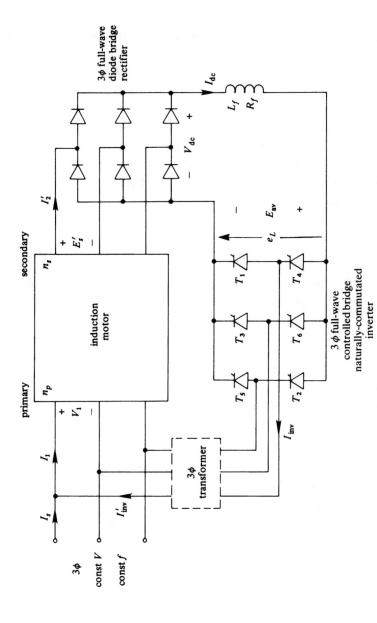

Fig. 10.3 Schematic diagram of an induction motor slip-energy recovery scheme. (Polarity of E_{av} shown for inverter operation.)

inverter. The inverter output currents act to reduce the necessary value of the supply current. SCRs are invariably used as inverter switches because gate turn-off devices are unnecessary and also because SER systems are usually high-power applications.

The power that was lost in the external resistors in the scheme of Fig. 9.38 is no longer used because the speed control does not depend on resistance R_2. In fact, secondary circuit resistance R_2 is made as low as possible so that the secondary copper loss $|I_s|^2 R_2$, dissipated inside the motor frame, is also minimal. This represents a saving of some of the energy that was originally dissipated as heat in the rotor resistance or primary voltage control schemes and leads to the name 'slip-energy recovery'. The inverter voltage on the d.c. side is controlled by the SCR firing-angle and acts, in effect, as a controllable voltage injected into the motor secondary windings. This form of drive operates from a standard constant voltage, constant frequency, three-phase supply that is defined by equations (7.1)–(7.3).

10.2.1 Torque–speed relationship

The secondary e.m.f. E_2' that will occur at the motor brushes when no secondary current is drawn can be deduced from Fig. 9.1(a) or Fig. 10.1(a),

$$E_2' = SE_1 \frac{n_s}{n_p} \tag{10.2}$$

The motor secondary current in the presence of a full-wave diode bridge across the brushes looks rather like the waveform $i_a(\omega t)$ of Fig. 7.2(f) but is modified because the motor leakage reactances act as source reactances. The load on the bridge is highly inductive due to the filter inductor. A set of measured current waveforms is given in Fig. 10.4.

Current waveform I_2' in the circuit of Fig. 10.3 is dominated by its fundamental (i.e. slip frequency) component. Hence, to a first approximation, one can use sinusoidal circuit analysis. The per-phase e.m.f. E_s' at the secondary terminals in Fig. 10.3, while secondary current is flowing, can be deduced from Fig. 10.1(a), and is given by

$$\begin{aligned} E_s' &= E_2' - I_2'(R_2' + jSX_2') \\ &= SE_1 \frac{n_s}{n_p} - I_2'(R_2' + jSX_2') \end{aligned} \tag{10.3}$$

Fig. 10.1 shows valid equivalent circuits for the motor of an SER drive. The primary current I_1 in the system of Fig. 10.3 is less distorted than the secondary current I_2', Fig. 10.4, and is also dominated by its fundamental (i.e.

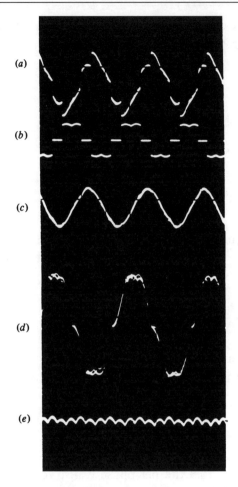

Fig. 10.4 Measured current waveforms for an experimental 200 V, 50 Hz, 3 h.p., SER drive. $N = 400$ r.p.m. ($S = 0.734$), $\alpha = 110°$: (a) supply current, (b) inverter current, (c) motor primary current, (d) motor secondary current, (e) d.c. link current ripple (ref. TP3).

supply frequency) component. Primary side e.m.f. E_1 of Fig. 10.1(a) can therefore be expressed as

$$E_1 = V_1 - I_1(R_1 + jX_1) \tag{10.4}$$

The average output voltage from a full-wave diode bridge with inductive loading and containing supply reactance ωL_s ohms per phase is given by putting $\alpha = 0$ in (7.8) and introducing a negative term $-3\omega L_s I_{av}/\pi$, which represents the subtractive voltage drop across ωL_s (reference 18). If the terminology I_{dc} is now used for the output current and V_{dc} for the average output voltage then in terms of the per-phase r.m.s. input voltage E'_s,

$$V_{dc} = \frac{3}{\pi}(\sqrt{3}\sqrt{2}E'_s - \omega L_s I_{dc}) \tag{10.5}$$

Incorporating (10.3) and (10.4) into (10.5) gives

$$V_{dc} = \frac{3}{\pi}\left[\sqrt{6}S\frac{n_s}{n_p}\{V_1 - I_1(R_1 + jX_1)\} - \sqrt{6}I'_2(R'_2 + jSX'_2) - jX_s I_{dc}\right] \tag{10.6}$$

If the series voltage drops in the equivalent circuits of Fig. 10.1 are small compared with the terminal voltages, which is especially true at light loads, then (10.6) reduces to

$$V_{dc} = \frac{3\sqrt{6}}{\pi} S \frac{n_s}{n_p} V_1 = 1.35 S \frac{n_s}{n_p} E \tag{10.7}$$

where E = r.m.s. line voltage at the supply point.

The average output voltage of the inverter E_{av} is given by (7.72), quoted here,

$$E_{av} = \frac{3\sqrt{3}}{\pi} E_m \cos\alpha \tag{10.8}$$

On the a.c. side of the inverter in Fig. 10.3 the per-phase r.m.s. voltage is denoted by V_1 so that $E_m = \sqrt{2}V_1$. If the average voltage $I_{dc}R_f$ across the filter inductor is small then $|V_{dc}| = |E_{av}|$. Equating (10.8) and (10.7) gives

$$\frac{3}{\pi}\sqrt{6}V_1 \cos\alpha = \frac{3\sqrt{6}}{\pi} S \frac{n_s}{n_p} V_1$$

$$S = S_0 = \frac{n_p}{n_s}|\cos\alpha| \tag{10.9}$$

where S_0 is the no-load value of the slip.

The very important approximate result (10.9) shows that the drive no-load speed depends only on SCR firing-angle α.

Neglecting supply (i.e. motor) series impedance effects, the rectifier bridge r.m.s. terminal voltage is related to its output average voltage V_{dc} by

$$E'_s = \frac{\pi}{3\sqrt{6}} V_{dc} \tag{10.10}$$

Voltage E'_s can be expressed in terms of firing-angle α by combining (10.10) with (10.8), noting that $E_m = \sqrt{2}V_1$,

$$E'_s = V_1 \cos \alpha \tag{10.11}$$

In the equivalent circuit referred to primary turns, Fig. 10.1(b), the 'injected' voltage E_s is related to the actual voltage E'_s by (9.6), giving

$$E_s = \frac{1}{S}\frac{n_p}{n_s} E'_s \tag{10.12}$$

or

$$E_s = \frac{n_p}{n_s}\frac{|\cos \alpha|}{S} V_1 = \frac{S_0}{S} V_1 \tag{10.13}$$

If the series impedances are neglected then (10.9) applies and $E_s = V_1$, which is true by inspection of the referred equivalent circuit Fig. 10.1(b). The injected voltage E_s is in time-phase with V_1 for $0 \leq \alpha \leq 90°$ and in antiphase for inverter operation, where $90° \leq \alpha \leq 180°$.

With primary voltage control or secondary resistance control, it was shown in Chapter 9 that the air-gap power P_g of the motor divides into two components SP_g and $(1-S)P_g$ as shown in Fig. 9.2. It is still true that a fraction $(1-S)P_g$ of the air-gap power is converted to mechanical output power and transferred to the load or dissipated as friction and windage power. With an SER system, however, the mechanical power is no longer given by (9.15) but is a function of the power circulating in the d.c. link. A component of power is still dissipated as heat in the secondary winding resistances. A power flow diagram is shown in Fig. 10.5. The power P_{link} passing from the motor into the d.c. link reduces the input power needed to sustain a given mechanical output power. An SER drive is therefore not limited by the restriction of (9.18) and some improvement of the inherently low induction motor efficiency can be realised at low speeds.

If the secondary winding copper loss is small, due to low R_2, and the friction and windage effect is ignored, all of the air-gap power is divided between P_{mech} and P_{link} in Fig. 10.5. The total power crossing the air-gap is therefore, approximately,

$$\begin{aligned} P_g &\simeq P_{link} + P_{mech} \\ &\simeq V_{dc}I_{dc} + TN \\ &\simeq V_{dc}I_{dc} + TN_1(1-S) \end{aligned} \tag{10.14}$$

But the air-gap power still represents the developed power TN_1 in synchronous watts. Combining (10.14) with (9.20) gives

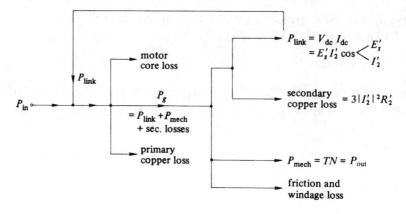

Fig. 10.5 Motor power flow for an induction motor SER system.

$$T = \frac{V_{dc}I_{dc}}{SN_1} \tag{10.15}$$

Combining (10.15) with (10.7) gives

$$T = \frac{3\sqrt{6}}{\pi} \frac{n_s}{n_p} \frac{V_1}{N_1} I_{dc} = 1.35 \frac{E}{N_1} \frac{n_s}{n_p} I_{dc} \tag{10.16}$$

where

$$E = \sqrt{3}V_1$$

Equation (10.16) shows that torque T and d.c. link current I_{dc} are directly proportional, independently of speed. An equation for torque versus slip is obtained by combining (10.16) with (7.73) and (10.7)

$$T = \frac{(1.35E)^2}{R_f N_1} \frac{n_s}{n_p} \left(\frac{n_s}{n_p} S + \cos\alpha \right) \tag{10.17}$$

It can be seen that with constant firing-angle the torque increases linearly with slip. The effect of increased α is to displace the torque characteristic to the left in the torque–speed plane, as seen in the measured characteristics of Fig. 10.6. These are seen to be completely different from those of a conventional induction motor, shown in Fig. 9.8 and Fig. 9.39. In fact, the characteristics of Fig. 10.6 are similar in form to those of a separately excited d.c. motor with armature voltage control, as described in (5.13) and Fig. 5.5. Note that the approximation of (10.9) presumes that $R_f = 0$ and cannot be combined with (10.17).

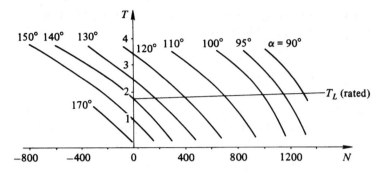

Fig. 10.6 Measured torque (N m) versus speed (r.p.m.) for a 200 V, 50 Hz, 3 h.p., SER drive (ref. TP3).

10.2.2 Current relationships

The equivalent circuits of Figs. 9.1, 9.4 and 10.1(b) refer to sinusoidal operation at supply frequency. If the currents are nonsinusoidal, as for the present case of an SER drive, such circuits represent the fundamental frequency component of the current.

The fundamental component of the actual secondary current $|I'_{2_1}|$ in Fig. 10.1(a) is obtained from the ampere-turn balance equation (9.7), repeated here

$$|I'_{2_1}| = \frac{n_p}{n_s} |I_{2_1}| \tag{10.18}$$

In order to obtain the r.m.s. value of the secondary current, from a knowledge of its fundamental component, it is necessary to know the current waveform. If overlap effects are neglected the waveform has the rectangular shape of Fig. 7.13(a). The assumption of rectangular waveform gives a pessimistic result because the actual waveshape, Fig. 10.4(d), has a higher fundamental value than a rectangular wave of the same height. The height of the secondary current waveform is the d.c. link current I_{dc}.

Using (7.76), with an appropriate change of terminology in terms of (10.18), gives

$$|I_{2_1}| = \frac{n_s}{n_p} |I'_{2_1}| = \frac{\sqrt{2} \times \sqrt{3}}{\pi} \frac{n_s}{n_p} I_{dc} = 0.78 \frac{n_s}{n_p} I_{dc} \tag{10.19}$$

The r.m.s. value of a rectangular wave of pulse height I_{dc} is given by (7.24). In the terminology of (10.19) this becomes

$$|I_2| = \frac{n_s}{n_p}|I_2'| = \frac{n_s}{n_p}\sqrt{\frac{2}{3}}I_{dc} \tag{10.20}$$

Since the r.m.s. and the fundamental values of the referred secondary current I_2 are known the sum of the current higher harmonics is given by

$$\sqrt{\sum_{h=3}^{\infty} I_{2_h}'} = \sqrt{|I_2|^2 - |I_{2_1}|^2}$$

$$= \frac{n_s}{n_p}I_{dc}\sqrt{\frac{2}{3} - \frac{6}{\pi^2}} \tag{10.21}$$

$$= 0.242\frac{n_s}{n_p}I_{dc}$$

The value of the fundamental secondary current referred to primary turns I_{2_1} can also be obtained directly by reference to Fig. 10.1(b) and (10.13)

$$I_{2_1} = \frac{V_1 - E_s}{Z_2}$$

$$= \frac{V_1\left(1 - \frac{n_p}{n_s}\frac{|\cos\alpha|}{S}\right)}{\left(R_1 + \frac{R_2}{S}\right) + j(X_1 + X_2)} \tag{10.22}$$

The magnitude $|I_{2_1}|$ is obtained from (10.22) in the customary way for sinusoidal functions:

$$|I_{2_1}| = \frac{V_1\left(1 - \frac{n_p}{n_s}\frac{|\cos\alpha|}{S}\right)}{\sqrt{\left(R_1 + \frac{R_2}{S}\right)^2 + (X_1 + X_2)^2}} \tag{10.23}$$

$$\psi_2 = \tan^{-1}\left(\frac{X_1 + X_2}{R_1 + \frac{R_2}{S}}\right) \tag{10.24}$$

Note that since the numerator is an approximation, expression (10.23) gives reasonable results over only a small range of speeds.

The inverter output current is also a rectangular wave of height I_{dc}, Fig. 7.13. If the filter inductor is relatively small the waveform is more nearly that

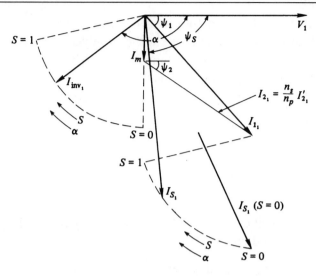

Fig. 10.7 Phasor diagram of the fundamental current components on the supply side of an SER drive. Constant torque (i.e. constant motor current). Core losses are neglected.

of a resistively loaded bridge. The oscillogram of Fig. 10.4(b) is seen to be similar to the theoretical waveform of Fig. 7.2(f). This waveform is closely satisfied by (7.76) so that the fundamental inverter current is given by

$$I_{inv_1} = 0.78\, I_{dc} \angle -\alpha \qquad (10.25)$$

The fundamental inverter current may also be expressed in terms of slip and firing-angle by combining (10.25) with (7.73) and (10.7)

$$|I_{inv_1}| = \frac{18}{\pi^2} \frac{V_1}{R_f} \left(\frac{n_s}{n_p} S + \cos\alpha \right) \qquad (10.26)$$

Under the approximate condition of (10.9), $|I_{inv_1}| = 0$ because the d.c. link current is zero.

A phasor diagram of the fundamental current components is given in Fig. 10.7 for the condition of constant load torque. From (10.16) it is seen that constant torque requires constant link current and hence constant inverter current magnitude. From (7.76) it is seen that constant link current implies constant secondary current magnitude. As the speed changes, the magnitude and phase-angle of the secondary circuit impedance change. Even with constant secondary current there is then some variation of the primary current. The inverter firing-angle varies over the range $\alpha = 90°$ to about $\alpha = 165°$ (rather than the theoretical limit $\alpha = 180°$) to allow for overlap and for SCR turn-off time. At standstill the mechanical power is zero, all of the slip power (neglecting losses) is returned to the supply and becomes equal to the primary input power. Therefore, at standstill, the drive operates at zero power factor and the input phase-angle ψ_s is 90° lagging.

Currents I_m and I_w in the circuit of Fig. 10.1(b) are given by

$$I_m = \frac{V_1}{jX_m} = \frac{V_1}{X_m} \angle -90° \tag{10.27}$$

$$I_w = \frac{V_1}{R_w} \angle 0° \tag{10.28}$$

The fundamental component of the primary current I_{1_1} is given by the phasor sum of I_{2_1}, I_m and I_w, as shown in Fig. 10.7, where

$$I_{1_1} \cos \psi_1 = I_{2_1} \cos \psi_2 = I_w \text{ (if present)} \tag{10.29}$$

$$I_{1_1} \sin \psi_1 = I_{2_1} \sin_2 + I_m \tag{10.30}$$

Harmonic components of the motor primary current are usually negligible. The input current from the supply is found by the phasor addition of the primary motor current and the inverter current

$$\bar{I}_{s_1} = \bar{I}_{1_1} + \bar{I}_{inv_1} \tag{10.31}$$

where

$$I_{s_1} \cos \psi_{s_1} = I_{1_1} \cos \psi_1 - I_{inv_1} |\cos \alpha| \tag{10.32}$$

$$I_{s_1} \sin \psi_{s_1} = I_{1_1} \sin \psi_1 + I_{inv_1} \sin \alpha \tag{10.33}$$

10.2.3 Power, power factor and efficiency

It may be seen in the phasor diagram of Fig. 10.7 that decrease of the motor speed (i.e. increase of slip and of firing-angle) tends to reduce the magnitude of the fundamental input current but also to further retard the displacement angle ψ_{s_1}.

The per-phase input power to the circuit of Fig. 10.3 is given by

$$P_{in} = V_1 I_{s_1} \cos \psi_{s_1} \tag{10.34}$$

The per-phase power entering the motor is given by

$$P_{motor} = P_{in} + P_{link} = V_1 I_{1_1} \cos \psi_{1_1} \tag{10.35}$$

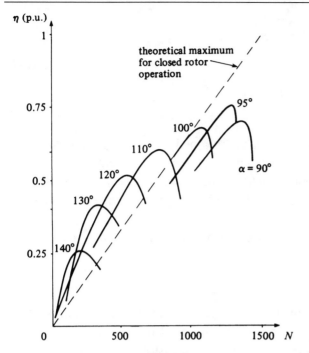

Fig. 10.8 Measured efficiency (p.u.) versus speed (r.p.m.) for a 200 V, 50 Hz, 3 h.p., SER drive (ref. TP3).

It is seen in (10.35) that $P_{motor} > P_{in}$ due to the link power, which may be written

$$P_{link} = V_1 I_{inv_1} |\cos \alpha| \tag{10.36}$$

Now, from (10.14), a useful approximate result is

$$P_{link} = \frac{S}{1-S} P_{mech} = V_{dc} I_{dc} \tag{10.37}$$

The link power is a fraction SP_g of the air-gap power, whereas the mechanical power is a fraction $(1 - S)$ of the air-gap power. Equation (10.37) usually provides a ready means of calculating the mechanical power transferred. For the calculation of drive efficiency it is necessary to take account of the various losses associated with winding resistances and also the core loss. From Fig. 10.5 one can write

$$P_{motor} = P_{link} + P_{mech} + 3(I_w^2 R_w + |I_2|^2 R_2 + |I_1|^2 R_1) + I_{dc}^2 R_f \tag{10.38}$$

The mechanical power and the losses must be injected through the input terminals so that the drive efficiency η may be written

$$\eta = \frac{P_{\text{mech}}}{P_{\text{in}}}$$

$$= \frac{P_{\text{mech}}}{P_{\text{mech}} + P_{\text{losses}}} \qquad (10.39)$$

$$= \frac{P_{\text{mech}}}{P_{\text{motor}} - P_{\text{link}}} \left(> \frac{P_{\text{mech}}}{P_{\text{motor}}} \right)$$

Some efficiency–speed characteristics obtained on an experimental 3 h.p. drive are shown in Fig. 10.8, corresponding to the torque–slip characteristics of Fig. 10.6. The improvement of low-speed efficiency, compared with voltage control, is very evident and this substantiates the last statements of (10.39).

The power factor of the drive, seen from the supply point, is defined as

$$PF = \frac{P_{\text{in}}}{3V_1 I_s} \qquad (10.40)$$

where I_s is the r.m.s. value of the supply current. Now the harmonics of the supply current are due mainly to the harmonics of the inverter current since the motor primary current contains only slight distortion

$$I_s = \sqrt{I_{s_1}^2 + \sum_{h=3}^{\infty} I_{\text{inv}_h}^2}$$

$$= \sqrt{(\bar{I}_{1_1} + \bar{I}_{\text{inv}_1})^2 + \sum_{h=3}^{\infty} I_{\text{inv}_h}^2} \qquad (10.41)$$

Since the inverter current waveform closely resembles the rectangular wave of Fig. 7.13, the summation of its harmonics is given by the form of (10.21). Combining this with (10.16) gives

$$\sqrt{\sum_{h=3}^{\infty} I_{\text{inv}_h}^2} = 0.242\, I_{\text{dc}} = 0.103\, \frac{n_p}{n_s} \frac{TN_1}{V_1} \qquad (10.42)$$

The drive displacement factor is given by

$$\text{displacement factor} = \cos \psi_{s_1} \qquad (10.43)$$

It is not normally possible to obtain an accurate value of the displacement angle ψ_{s_1} by calculation.

As the SCR firing-angle is retarded, the drive power factor becomes progressively lower (i.e. worse), as shown in Fig. 10.9. In effect the drive consists of two main components, the inverter and the motor, both of which draw

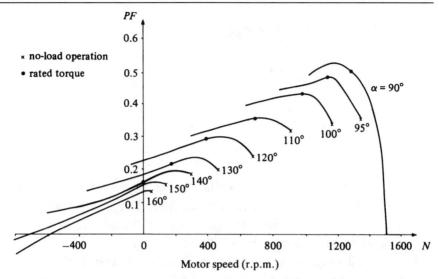

Fig. 10.9 Measured power factor versus speed (r.p.m.) for a 50 Hz, 3 h.p., SER drive (ref. TP3).

lagging voltamperes. The result is a poor power factor at all speeds and this constitutes the main disadvantage of the system.

The reactive voltampere requirement of the diode bridge has to be provided from its a.c. side and hence passes through the motor. The lagging voltampere requirement of the inverter is high at high speeds but reduces as the speed decreases. At low speeds, however, the real power absorption is low, which acts to further depress the power factor. The power factor reaches its maximum possible value if the inverter firing-angle is a maximum at the lowest point of the speed range.

10.2.4 Speed range, drive rating and motor transformation ratio

The equations in the preceding part of Section 10.2 demonstrate that the motor primary/secondary transformation ratio n_p/n_s is an important 'figure of merit' or design constant. For example, the minimum operating speed is directly proportional to n_p/n_s whereas the torque per ampere and the inverter current and power are inversely proportional to n_p/n_s. The speed regulation is seen from (10.17) to vary with the inverse square of n_p/n_s.

A design specification invariably defines the desired speed range and therefore the minimum working speed. Good design dictates that in order to achieve the highest realisable power factor at the specified minimum speed, the SCR firing-angle should then have its maximum usable value of about 170°. Combining (9.2) and (10.9) gives

$$N_{\min} = N_1\left(1 - \frac{n_p}{n_s}|\cos\alpha|\right)$$
$$= N_1\left(1 - 0.985\frac{n_p}{n_s}\right) \tag{10.44}$$

The condition of minimum speed is also the condition of maximum slip

$$S_{\max} = \frac{N_1 - N_{\min}}{N_1} = |\cos\alpha_m|\frac{n_p}{n_s} = 0.985\frac{n_p}{n_s} \tag{10.45}$$

A transformation ratio of unity would enable $S_{\max} = 0.985$ and give a speed range of 0.015–1.0 per unit. If the desired speed range is restricted to 2:1 (i.e. full speed down to one-half speed), $S_{\max} = 0.5$ and the necessary motor transformation ratio is $n_p/n_s = 0.5/0.985 = 0.51$.

For the full range of speed control, down to standstill, $S_{\max} = 1$ and the range of SCR firing-angles is, from (10.44),

$$90° \leq \alpha \leq \left|\cos^{-1}\left(\frac{n_s}{n_p}\right)\right| \tag{10.46}$$

At the condition of (10.45) the referred slip-ring voltage E'_s in Fig. 10.1 has its maximum value (10.11)–(10.13) almost equal to the supply voltage, the d.c. link voltage is a maximum (10.7), the link current is a maximum (7.73), the developed torque at that speed is a maximum (10.16), the inverter current and link power are maxima (10.26) and the efficiency is therefore maximised (10.39).

Very often in system design the induction motor already exists and its transformation ratio is fixed. Moreover, as pointed out in Chapter 9, the ratio $n_p/n_s > 1.0$, which results in a low value for the slip-ring voltage E'_s and the bridge rectifier output voltage V_{dc}. Equation (10.45) cannot be applied. To balance the voltages $|V_{dc}|$ and $|E_{av}|$ in the link the inverter a.c. side voltage must be reduced by introducing a step-down transformer between the supply and the inverter, as shown in Fig. 10.3. If this transformer has n_{T_p} supply-side turns/phase and n_{T_i} inverter-side turns/phase the inverter average voltage is given by

$$E_{av} = \frac{3\sqrt{3}}{\pi}\frac{n_{T_i}}{n_{T_p}}E_m\cos\alpha$$
$$= \frac{3\sqrt{6}V_1}{\pi}\frac{n_{T_i}}{n_{T_p}}\cos\alpha \tag{10.47}$$

Equating (10.47) with the voltage V_{dc} from (10.7) gives

$$S = \frac{n_p}{n_s}\frac{n_{T_i}}{n_{T_p}}|\cos\alpha| \tag{10.48}$$

If maximum slip S_{max} utilises the condition of maximum SCR firing-angle then equation (10.48) becomes

$$S_{max} = \frac{n_p}{n_s}\frac{n_{T_i}}{n_{T_p}} 0.985 \tag{10.49}$$

Comparison of (10.49) and (10.45) shows that the introduction of the transformer enables the designer to freely set the value of the maximum slip for any given induction motor.

The range of necessary firing-angles, from full speed down to any desired minimum value represented by S_{max} is, from (10.48)

$$90° \le \alpha \le \left|\cos^{-1}\left(S_{max}\frac{n_s}{n_p}\frac{n_{T_p}}{n_{T_i}}\right)\right| \tag{10.50}$$

The current returned to the supply in the presence of the step-down transformer, I'_{inv} in Fig. 10.3, is related to the inverter current by the relation

$$I'_{inv} = \frac{n_{T_i}}{n_{T_p}} I_{inv} \tag{10.51}$$

Combining (10.51) with (10.25) gives, for the r.m.s. fundamental component,

$$I'_{inv_1} = \frac{n_{T_i}}{n_{T_p}} 0.78 I_{dc} \angle -\alpha \tag{10.52}$$

The power passing through the link was shown in (10.14) to be (nearly) the fraction S of the air-gap power. An SER drive working in the subsynchronous region of quadrant I of the torque–speed plane has an inverter rating given, nominally, by

$$[kVA]_{inv} = S[kVA]_{motor} \tag{10.53}$$

For full speed range operation the inverter is required to have the same rating as the motor. Equation (10.53) indicates the significant cost saving available if a restricted speed range is used. The inverter has to be rated for the a.c. line voltage or for the step-down voltage when a transformer is used. The bridge rectifier has to be rated for the maximum slip-ring voltage which is proportional to S_{max}. Both the rectifier and the inverter must have equal current ratings corresponding to the maximum torque demand. Application of the criteria of (10.45) or (10.49) helps to minimise the rectifier and inverter ratings. Due to the presence of current harmonics in the motor the efficiency is reduced and it is necessary to derate the motor by about 10%. If a drive is designed for restricted speed range operation, which is the most efficient and economical application, it is usually started from rest and run up to its minimum control speed by the conventional secondary resistance method. The slip-energy recovery drive can be used with closed-loop control similar, in principle, to that of Fig. 6.30.

10.2.5 Filter inductor

The output voltages of the bridge rectifier, Fig. 7.4, and the controlled inverter, Fig. 7.12, both contain zero-order harmonics (i.e. d.c. values) plus higher harmonics of order $n = 6, 12, 18$. For the diode bridge rectifier it can be shown that the r.m.s. value of the nth slip harmonic load voltage is given by

$$V_{n_R} = \frac{\sqrt{2}}{n^2 - 1} V_{dc} \tag{10.54}$$

Combining (10.54) with (10.7) gives

$$V_{n_R} = \frac{\sqrt{2}}{n^2 - 1} \frac{3\sqrt{6}}{\pi} S \frac{n_s}{n_p} V_1 \tag{10.55}$$

For the three-phase controlled inverter it can be shown that the r.m.s. value of the nth supply harmonic load voltage is

$$V_{n_I} = \frac{\sqrt{2}}{n^2 - 1} E_{av_0} \sqrt{n^2 \sin^2 \alpha + \cos^2 \alpha} \tag{10.56}$$

At standstill V_{n_R} and V_{n_I} have the same frequency. These two harmonic voltages exist in the d.c. link simultaneously and, for the worst case condition, may add arithmetically. Under this condition the net harmonic driving voltage is

$$V_{n_R} + V_{n_I} = \frac{3\sqrt{6}V_1}{\pi} \frac{\sqrt{2}}{n^2 - 1} \left(\frac{n_s}{n_p} + \sqrt{n^2 \sin^2 \alpha + \cos^2 \alpha} \right) \tag{10.57}$$

If the filter resistance is neglected the harmonic current is given by

$$I_n = \frac{V_{n_R} + V_{n_I}}{nX_f} \tag{10.58}$$

The lowest order a.c. harmonic is for $n = 6$. The worst case sixth harmonic current, at standstill, is obtained by combining (10.57) and (10.58)

$$I_6 = \frac{3\sqrt{6}V_1\sqrt{2}}{\pi 6\omega L_f(n^2-1)} \left(\frac{n_s}{n_p} + \sqrt{n^2 \sin^2\alpha + \cos^2\alpha}\right)$$

$$= \frac{0.016 V_1}{\omega L_f} \left(\frac{n_s}{n_p} + \sqrt{36 \sin^2\alpha + \cos^2\alpha}\right) \quad (10.59)$$

This sixth harmonic current flows not only in the link but also in the motor windings and in the supply. The motor leakage reactances also oppose the flow of this current but their value is small compared with X_f. The designer must now specify the maximum level of harmonic current that can be permitted and this can be conveniently done by quoting I_6 as a percentage of the link direct current.

Let

$$I_6 = k I_{dc} \text{ (rated)} \quad (10.60)$$

Then

$$\omega L_f = \frac{0.016 V_1}{k I_{dc} \text{ (rated)}} \left(\frac{n_s}{n_p} + \sqrt{36 \sin^2\alpha + \cos^2\alpha}\right) \quad (10.61)$$

For the condition $S = 1$ it is likely that firing-angle α will be large, say $\alpha = 130°$. Then if $k = 0.1$ and $n_s/n_p \simeq 1$,

$$\omega L_f \simeq \frac{V_1}{I_{dc} \text{ (rated)}} \quad (10.62)$$

Equation (10.62) represents a very approximate estimation and much greater accuracy is obtained by the use of (10.61).

The coil resistance is defined by the first statement of (7.73). In a typical drive, however, V_{dc} and E_{av} are very similar in magnitude. The arithmetic difference (since they are of opposite polarity) is a very small number and is unreliable for the calculation of R_f. Equation (10.17) gives a good approximation.

The coil design involves specification of the physical size (e.g. area) and the number of turns. This, in effect, defines the length of wire needed. The wire dimension is defined by the maximum current in the link. From a knowledge of the wire gauge and length the filter resistance can be calculated.

10.2.6 Worked examples

Example 10.2
A 440 V, 50 Hz, 50 kW, three-phase induction motor is used as the drive motor in an SER system. It is required to deliver constant (rated) motor torque over the full range from 100 r.p.m. to the rated speed of 1420 r.p.m.
The motor equivalent circuit parameters are

$$R_1 = 0.067\,\Omega, \quad R_\omega = 64.2\,\Omega$$
$$R_2 = 0.04\,\Omega, \quad X_m = 19.6\,\Omega$$
$$X_1 + X_2 = 0.177\,\Omega, \quad n_p/n_s = 1.15$$

Calculate the motor currents, efficiency and power factor at 100 r.p.m.

Solution. The motor torque for 50 kW, 1420 r.p.m. is

$$T = T_{\text{rated}} = \frac{50\,000}{1420 \times \frac{2\pi}{60}} = 336\,\text{N m}$$

Since the rated speed is 1420 r.p.m. it can be presumed that this is a four-pole motor with a synchronous speed of 1500 r.p.m. at 50 Hz

$$N_1 = 1500\,\text{r.p.m.} = 1500 \times \frac{2\pi}{60} = 157\,\text{rad/s}$$

The per-phase r.m.s. primary voltage V_1 is

$$V_1 = \frac{440}{\sqrt{3}} = 254\,\text{V}$$

From (10.16) it is found that the d.c. link current to support the torque is given by

$$I_{\text{dc}} = \frac{\pi N_1 T}{3\sqrt{6} V_1} \frac{n_p}{n_s}$$

$$= \frac{\pi \times 157 \times 336 \times 1.15}{3 \times \sqrt{6} \times 254} = 102.1\,\text{A}$$

At rated torque, full-speed operation the slip speed is 80 r.p.m. and the torque–slip characteristic has a slope of 336/80 N m/r.p.m., Fig. 10.10. Since the inverter average voltage is zero for $\alpha = 90°$, the d.c. link current is then restrained only by the link resistor R_f, as indicated in (7.73). This represents the nearest possible approach to conventional closed-secondary operation and it can be presumed that rated operation at 1420 r.p.m. occurs for $\alpha = 90°$. At the desired operating speed of 100 r.p.m. the no-load slip is $1320/1500 = 0.88$, Fig. 10.10, since from (10.17), the slope of the torque–slip characteristic is constant.

From the approximate relationship (10.9) the firing-angle for $S_0 = 0.88$ is

$$|\cos \alpha| = \frac{0.88}{1.15} = 0.765$$

$$\therefore \alpha = -\cos^{-1} 0.765 = -140°$$

The magnitude of the r.m.s. fundamental referred secondary current is obtained from (10.19)

$$|I_{2_1}| = \frac{0.78}{1.15} 102.1 = 69.25 \text{ A}$$

The magnitude of the total r.m.s. referred secondary current is, from (10.20),

$$|I_2| = \frac{1}{1.15}\sqrt{\frac{2}{3}} 102.1 = 72.5 \text{ A}$$

In the referred equivalent circuit of Fig. 10.1(b) the fundamental referred secondary current is given by (10.22). Using the actual operational value of slip, $S = 1400/1500 = 0.933$, at 100 r.p.m., gives the approximate value

$$I_{2_1} = \frac{254\left(1 - \dfrac{1.15 \times 0.765}{0.933}\right)}{\left(0.067 + \dfrac{0.04}{0.933}\right) + j0.177}$$

$$= \frac{254 \times 0.057}{0.11 + j0.177}$$

$$= 69.47 \angle -58.2° \text{ A}$$

The parallel branch currents in Fig. 10.1(b) are given by

$$I_m = \frac{254}{19.6} = 12.96 \angle -90° \text{ A}$$

$$I_w = \frac{254}{64.2} = 3.96 \angle 0° \text{ A}$$

Adding the branch currents, using (10.29), (10.30), gives

$$I_{1_1} \cos \psi_1 = 69.47 \cos 58.2° + 3.96$$
$$= 36.6 + 3.96 = 40.56 \text{ A}$$
$$I_{1_1} \sin \psi_1 = 69.47 \sin 58.2° + 12.96$$
$$= 59 + 12.96 = 71.96 \text{ A}$$

$$\therefore \psi_1 = -\tan^{-1}\left(\frac{71.96}{40.56}\right)$$
$$= -\tan^{-1}(1.775) = -60.6°$$

$$|I_{1_1}| = \frac{71.96}{\sin 60.6°} = 82.6 \text{ A}$$

At 100 r.p.m. the direct voltage in the link is obtained from (10.7) using the actual slip $S = 0.933$

$$V_{dc} = \frac{3\sqrt{6}}{\pi} \times \frac{0.933}{115} \times 254 = 482 \text{ V}$$

The power in the link is therefore

$$P_{link} = 482 \times 102.1 = 49.214 \text{ kW}$$

From (10.37) the mechanical power transferred is

$$P_{mech} = \frac{1 - 0.933}{0.933} \times 4.9214 = 3534 \text{ W}$$

This may be confirmed by taking the torque–speed product, at 100 r.p.m.,

$$P_{mech} = 336 \times 100 \times \frac{2\pi}{60} = 3519 \text{ W}$$

The system copper losses are specified in (10.38), but R_f is not known and this component is neglected. Any harmonic components of the motor primary current may be neglected, so

$$P_{losses} = 3[(3.96)^2 64.2 + (72.5)^2 0.04 + (82.6)^2 0.067]$$
$$= 3(1006.8 + 210.25 + 457)$$
$$= 3(1674) = 5022 \text{ W}$$

The drive efficiency is therefore

$$\eta = \frac{3534}{3534 + 5022} = 41.3\%$$

This value is optimistic because the $I_{dc}^2 R_f$ loss is neglected.

For the motor with a closed rotor the maximum theoretical efficiency is $100/1500 = 6.7\%$, so that the use of SER has given great improvement.

In a fundamental components phasor diagram of the form of Fig. 10.7 the power input corresponds to an in-phase component of current

$$I_{s_1} \cos \psi_s = \frac{3534 + 5022}{3 \times 254} = 11.23 \text{ A}$$

The inverter current has a fundamental r.m.s. value given by (10.25),

$$I_{inv_1} = 0.78 \times 102.1 \angle -\alpha$$
$$= 79.64 \angle -140° = -61 - j51.2 \text{ A}$$

Equation (10.42) gives the sum of the inverter current harmonics,

$$\sum_{h=3}^{\infty} I_{inv_h}^2 = (0.242 \times 102.1)^2 = 610.5 \text{ A}^2$$

The cartesian co-ordinates of the supply current may be found from (10.31)–(10.33)

$I_{s_1} \cos \psi_{s_1} = 40.56 - 61 = -20.44$ A
$I_{s_1} \sin \psi_{s_1} = 71.96 + 51.2 = 123.16$ A
$I_{s_1} = 124.8 \angle -99.4°$ A

Therefore

$|I_{s_1}| = 124.8$ A

Adding the inverter current harmonics gives, from (10.41),

$|I_s| = \sqrt{(124.8)^2 + (610.5)} = 127.2$ A

This is seen to be significantly greater than the fundamental motor current, as expected at this high slip.

The negative value for $I_{s_1} \cos \psi_s$ above is physically impossible and is seen to be inconsistent with the value $+11.23$ A obtained by power calculations. The discrepancy arises because of the considerable approximations and simplifications made in the analysis. An accurate calculation of power factor is not possible by the phase-angle method. Using (10.40) gives

$$PF = \frac{8556}{3 \times 254 \times 127.2} = 0.088$$

Example 10.3

Calculate the necessary parameters of the filter choke for the drive application of Example 10.2.

Solution. The choke is required to carry a current of 102.1 A. By the approximate method of (10.62) the reactance is

$$X_f = \frac{254}{102.1} \simeq 2.488 \text{ ohms}$$

The more accurate relationship of (10.61) gives, with k chosen as 0.1 and assuming $\alpha = 90°$ at the highest possible speed,

$$X_f = \frac{0.016 \times 254}{0.1 \times 102.1} \left(\frac{1}{1.15} + \sqrt{36 \sin^2 90° + \cos^2 90°} \right)$$
$$= 0.398(0.8696 + 6)$$
$$= 2.73 \text{ ohms}$$

At 50 Hz, the filter inductance is

$$L_f = \frac{2.73}{2\pi 50} = 8.73 \text{ mH}$$

The ideal value of resistor R_f is zero. An approximate value for R_f can be obtained from (10.17) by using the actual slip of 0.933

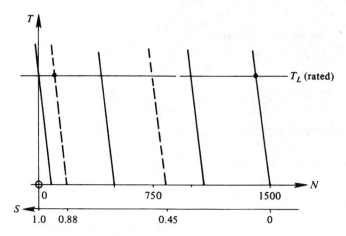

Fig. 10.10 Torque–speed characteristics for the drive of Examples 10.2, 10.4 and 10.5.

$$R_f = \frac{(1.35 \times 440)^2}{336 \times 157} \frac{1}{1.15} \left(\frac{0.933}{1.15} + \cos 140° \right)$$
$$= 5.816(0.8113 - 0.765) = 0.27\,\Omega$$

The value of R_f shows that it may be justifiably neglected in circuit calculations such as that of (7.73). But, in Example 10.2, a link power loss of $(102.1)^2 0.27 = 2815\,\text{W}$ was ignored so that the efficiency was optimistic. A more accurate value would be

$$\eta = \frac{3534}{3534 + 5022 + 2815} = 31.1\%$$

The choice of $k = 0.1$ for the harmonic current levels suggests that the drive motor should be derated by $1/(0.9)^2$ or approximately 20%.

Example 10.4
The SER drive of Example 10.2 is to be used for the speed control of a constant (rated) torque load over the speed range 1000–1420 r.p.m. Conventional starting equipment is used to drive the motor to 1000 r.p.m., when it is switched to SER control. Determine the necessary range of SCR firing-angles, the efficiency and power factor at (i) 1000 r.p.m., (ii) 1420 r.p.m.

Solution. For rated torque operation at 1000 r.p.m., the no-load speed is 1080 r.pm., Fig. 10.10, and $S_0 = 420/1500 = 0.28$. This is realised by a firing-angle, from (10.9), of

$$|\cos\alpha| = \frac{0.28}{1.15} = 0.243$$
$$\alpha = -\cos^{-1} 0.243 = -104°$$

The low value of 104° at the maximum slip condition means that the drive power factor will be poor.

It is assumed that the rated torque of 336 N m at 1420 r.p.m. requires a firing-angle of 90°. As in Example 10.2,

$$V_1 = 254 \text{ V/phase}$$
$$I_{dc} = 102.1 \text{ A (constant)}$$
$$|I_{2_1}| = 69.25 \text{ A}$$
$$|I_2| = 72.5 \text{ A}$$
$$I_m = 12.96 \angle -90° \text{A}$$
$$I_w = 3.96 \angle 0° \text{A}$$
$$|I_{inv_1}| = 79.64 \text{ A}$$
$$\sum_{h=3}^{\infty} I_{inv_n}^2 = 610.5 \text{ A}^2$$

From (10.24), at 1000 r.p.m.,

$$S = 0.333$$

$$\psi_2 = -\tan^{-1}\left(\frac{0.177}{0.067 + \frac{0.04}{0.333}}\right) = -\tan^{-1} 0.946 = -43.4°$$

At 1420 r.p.m.,

$$S = 80/1500 = 0.0533$$

$$\psi_2 = -\tan^{-1}\left(\frac{0.177}{0.067 + \frac{0.04}{0.0533}}\right) = -\tan^{-1} 0.216 = -12.2°$$

The cartesian components of phasor current I_{1_1} can be calculated using (10.29), (10.30)

At 1000 r.p.m.,

$$I_{1_1} \cos\psi_1 = 69.25 \cos 43.4° + 3.96$$
$$= 50.3 + 3.96 = 54.3 \text{ A}$$
$$I_1 \sin\psi_1 = 69.25 \sin 43.4° + 12.96$$
$$= 47.6 + 12.96 = 60.54 \text{ A}$$
$$\therefore I_1 = 81.32 \angle -48.1° \text{A}$$

At 1420 r.p.m.,

$I_1 \cos \psi_1 = 69.25 \cos 12.2° + 3.96 = 71.65 \, \text{A}$

$I_1 \sin \psi_1 = 69.25 \sin 12.2° + 12.96 = 27.6 \, \text{A}$

$\therefore I_1 = 76.78 \angle - 21° \text{A}$

The link voltage is obtained from (10.7).
At 1000 r.p.m.,

$$V_{dc} = \frac{3\sqrt{6}}{\pi} \times \frac{0.333}{0.15} \times 2.54 = 172.2 \, \text{V}$$

$P_{link} = 172.2 \times 102.1 = 17\,581 \, \text{W}$

$$P_{mech} = \frac{1 - 0.333}{0.333} \times 17\,581 = 35\,167 \, \text{W}$$

At 1420 r.pm.,

$$V_{dc} = \frac{3\sqrt{6}}{\pi} \times \frac{0.0533}{0.15} \times 254 = 27.54 \, \text{V}$$

$P_{link} = 27.54 \times 102.1 = 2812 \, \text{W}$

$$P_{mech} = \frac{1 - 0.0533}{0.0533} \times 2812 = 49\,946 \, \text{W}$$

The motor losses are given in (10.38). For all values of slip $I_{dc}^2 R_f = 2815 \, \text{W}$, as in Example 10.3.

At 1000 r.p.m.,

$P_{losses} = 2815 + 3[(3.96)^2 64.2 + (72.5)^2 0.04 + (81.32)^2 0.067]$

$= 2815 + 3(1006.8 + 210.25 + 443.1)$

$= 2815 + 4983 = 7798 \, \text{W}$

$$\eta = \frac{35\,167}{35\,167 + 7798} = 81.85\%$$

At 1420 r.p.m.,

$P_{losses} = 2815 + 3[1006.8 + 210.25 + (76.78)^2 0.067]$

$= 2815 + 4836 = 7651 \, \text{W}$

$$\eta = \frac{49\,946}{49\,946 + 7651} = 86.7\%$$

The inverter current has an r.m.s. fundamental value defined by (10.25).
At 1000 r.p.m.,

$I_{inv_1} = 79.64 \angle - 104° = -19.27 - j77.3 \, \text{A}$

$I_{s_1} \cos \psi_{s_1} = 54.3 - 19.27 = 35 \, \text{A}$

$-I_{s_1} \sin \psi_{s_1} = 60.54 + 77.3 = 137.84 \, \text{A}$

$I_{s_1} = 142.2 \angle 75.8° \text{A}$

10.2 SER system   431

This gives a displacement factor cos 75.8 = 0.24 which is less than the figure for the power factor below – a physical impossibility. The discrepancy arises because of the considerable approximations in calculating ψ_{s_1}.
From (10.41),

$|I_s| = 144.3$ A

From (10.40),

$$PF = \frac{35\,167}{3 \times 254 \times 144.3} = 0.32$$

At 1440 r.p.m.,

$I_{inv_1} = 79.64 \angle -90° = -j79.64$ A

$I_{s_1} \cos \psi_{s_1} = 71.65$ A

$-I_{s_1} \sin \psi_{s_1} = 27.6 + 79.64 = 107.24$ A

$I_{s_1} = 129 \angle -56.25°$ A

Therefore $|I_{s_1}| = 129$ A and displacement factor $\cos \psi_{s_1} = 0.555$.
From (10.41),

$|I_s| = 131.3$ A

From (10.40),

$$PF = \frac{49\,946}{3 \times 254 \times 131.3} = 0.5$$

The supply current distortion factor is therefore $0.5/0.555 = 0.9$, which looks about right for waveform $i_s(\omega t)$ in Fig. 10.4(a).

Example 10.5
The SER drive of Example 10.2 is to be used for a constant (rated) torque load over the speed range 1000–1420 r.p.m. A lossless three-phase transformer is connected between the inverter and the supply so as to maximise the power factor at 1000 r.p.m. Determine the necessary transformer turns ratio and the power factors at 1000 r.p.m. and 1420 r.p.m., compared with the values in Example 10.4.

Solution. For operation with rated torque at 1000 r.p.m. the no-load speed is 1080 r.p.m., Fig. 10.10, $S_0 = 420/1500 = 0.28$. The actual operating slip is $500/1500 = 0.333$.
  If the maximum possible firing-angle of 170° is used then $\cos 170° = 0.985$. The necessary transformer turns ratio is, from (10.49),

$$\frac{n_{T_i}}{n_{T_p}} = \frac{0.333}{1.15} \times \frac{1}{0.985} = 0.294$$

The output voltage of the inverter is now reduced by a factor 0.294 but all of the motor electrical properties are unchanged. For example, the inverter average voltage (10.47) is found to be

$$E_{av} = \frac{3\sqrt{6}}{\pi} \times 254 \times 0.294 \times 0.985 = 172 \text{ V}$$

This is seen to be very nearly equal to the value V_{dc} obtained in Example 10.4 at 1000 r.p.m., when the SCR firing-angle was 104°.

The current returned to the supply is now given by (10.52), for the fundamental component,

$$I'_{inv_1} = 0.294 \times 0.78 \times 102.1 \angle -170°$$
$$= 23.4 \angle -170° = -23.04 - j4.06 \text{ A}$$

since I_{dc} is unchanged at 102.1 A.

Phasor current I'_{inv_1} is now combined with the (unchanged) value of I_1 from Example 10.4, in the manner of (10.32), (10.33)

$$I_{s_1} \cos \psi_{s_1} = 54.3 - 23.04 = 31.26 \text{ A}$$
$$I_{s_1} \sin \psi_{s_1} = 60.54 + 4.06 = 64.6 \text{ A}$$
$$\therefore I_{s_1} = 71.74 \angle -64.2° \text{ A}$$

The step-down transformer is assumed to be lossless and its series reactance is neglected. Current I'_{inv} therefore still contains the proportion of harmonics defined in (10.42)

$$\sqrt{\sum_{h=3}^{\infty} I^2_{inv_h}} = 0.242 \frac{n_{T_i}}{n_{T_p}} I_{dc} = 7.26 \text{ A}$$

The r.m.s. supply current is, from (10.41),

$$|I_s| = \sqrt{71.74^2 + 7.26^2} = 72.1 \text{ A}$$

which compares with a value 144.3 A at the same speed without the transformer (Example 10.4).

The power is unchanged so that the power factor at 1000 r.p.m. is now, from (10.40),

$$PF = \frac{35\,167}{3 \times 254 \times 72.1} = 0.64$$

which compares with the value 0.32 in Example 10.4.

At 1420 r.p.m.,

$$I'_{inv} = 23.4 \angle -90° = -j23.4 \text{ A}$$

The fundamental supply current is now given by

$I_{s_1} \cos \psi_{s_1} = 71.65 \, \text{A}$

$I_{s_1} \sin \psi_{s_1} = 27.6 + 23.4 = 51 \, \text{A}$

$\therefore I_{s_1} = 87.95 \angle -35.5° \, \text{A}$

From (10.41),

$|I_s| = 88.2 \, \text{A}$

Therefore, in (10.40),

$$PF = \frac{49\,946}{3 \times 254 \times 88.2} = 0.743$$

which compares with $PF = 0.5$ in the absence of the transformer.

10.3 PROBLEMS

10.1 A three-phase, wound-rotor induction motor has an e.m.f. injected into its secondary windings for speed control purposes. If this injected e.m.f. has the form $E'_s = K_1/S$ in the equivalent circuit of Fig. 10.1(a), deduce the form of the resulting torque–speed characteristics if $-1 \leq K_1 \leq 1$.

10.2 Slip-energy recovery in a three-phase induction motor drive is a method of attempting to improve the inherent inefficiency of low-speed operation when a constant supply frequency is used.
   (a) Explain and sketch a characteristic of efficiency versus speed for constant frequency operation of a conventional induction motor, indicating the point for full-load operation.
   (b) Draw an induction motor SER scheme incorporating a three-phase line-commutated SCR inverter. Briefly explain the action of the scheme. What is the value of the inverter average voltage when the a.c. line voltage is 400 V and the SCR firing-angle is 120°? Explain, using a phasor diagram, the effect on drive power factor of increasing the firing-angle.

10.3 Sketch the form of the torque–speed characteristics obtained with an SER drive, showing interesections with a load line of the form $T_L = kN^2$, where T_L is the load torque and N is the motor speed. What is, roughly, the speed regulation at 50% speed in your diagram?

10.4 Sketch a phasor diagram representing the r.m.s. values of the fundamental components of the currents in an SER drive when SCR firing-angle $\alpha = -140°$. If the motor primary current is $I_{1_1} \angle -\psi_1$ and the inverter current is I_{inv_1} show that the supply current I_{s_1} is given by

$$|I_{s_1}| = \sqrt{|I_1|^2 + |I_{\text{inv}_1}|^2 - 2|I_{1_1}||I_{\text{inv}_1}|\cos(\psi_1 + \alpha)}$$

10.5 A fan with a load characteristic $T_L = kN^2$ is to be driven by an SER drive incorporating a 440 V, 50 Hz, 100 kW induction motor. It is required to deliver rated torque at the rated speed of 1440 r.p.m. and to provide smooth speed control down to 750 r.p.m. The motor equivalent circuit parameters, referred to primary turns, are

$R_1 = 0.052\,\Omega,$ $X_m = 10\,\Omega$
$R_2 = 0.06\,\Omega$ $X_1 + x_2 = 0.29\,\Omega$
$R_w = 100\,\Omega$ $n_p/n_s = 1.2$

Calculate the motor efficiency and power factor at 750 r.p.m. Friction and windage effects may be neglected. It is assumed that the motor is started from rest and run up to 750 r.p.m. by the secondary resistance method. (Hint: Calculate the no-load slip consistent with operation at 750 r.p.m.)

10.6 Calculate the ratings of the inverter SCRs for operation of the SER drive of Problem 10.5 at 750 r.p.m.

10.7 Determine approximate values for the resistance and inductance of the filter choke to be used in the SER drive of Problem 10.5, if a harmonic derating of 10% is allowed.

10.8 The induction motor of Problem 10.5 is to be used in an SER scheme to deliver the rated motor torque over a speed range 1200–1440 r.p.m. Filter resistance R_f is $0.094\,\Omega$. Use the most advantageous values of SCR firing-angle and calculate the efficiency and power factor at 1200 r.p.m. and at 1420 r.p.m. (Hint: Use the magnitude of the secondary current consistent with I_{dc}).

10.9 The SER drive of Problem 10.8 is now modified by the inclusion of a step-down transformer between the supply and the inverter. Determine the necessary rating and turns ratio of the transformer to give the maximum realisable power factor at 1200 r.p.m. What are the operating power factors compared with the corresponding values in Problem 10.8?

10.10 Calculate the necessary ratings of the inverter SCRs for operation of the SER drive of Problem 10.9.

10.11 If the SER drive of Problem 10.9 has a filter resistance of $R_f = 0.094\,\Omega$, calculate the speed regulation of a typical open-loop torque–speed characteristic.

11

Induction motor speed control by the use of adjustable voltage, adjustable frequency step-wave inverters

11.1 THREE-PHASE INDUCTION MOTOR WITH CONTROLLED SINUSOIDAL SUPPLY VOLTAGES OF ADJUSTABLE FREQUENCY

11.1.1 Theory of operation

It was shown in equation (9.1) of Chapter 9 that the synchronous speed of a three-phase induction motor is proportional to the supply frequency f_1. A motor speed can be smoothly adjusted from zero up to its rated operating speed and higher by increase of the supply frequency. If the supply frequency is reduced from its rated value while the primary voltage is kept constant, (9.30), (9.31) show that the motor flux must increase. But operation above the design flux level would result in excessive core losses and high magnetising current due to an undesirably high level of magnetic saturation. In order to maintain operation at the rated flux density when the speed is varied it is necessary to vary the primary e.m.f. E_1 proportionately with variation of f_1. From (9.30),

$$\frac{E_1}{f_1} = 4.44\hat{\Phi}_m n = \text{constant} \tag{11.1}$$

E.m.f. E_1 may be varied indirectly by appropriate variation of the terminal voltage V_1.

The equivalent circuits of Fig. 9.1 are valid for operation at different frequencies. For the range of frequencies usually used in motor speed control, $0 < f_1 < 100\,\text{Hz}$, skin effects are negligible and the primary resistances R_1 and R_{FW} may be considered constant. The motor reactances X_1, X_2 and X_m are proportional to the excitation frequency f_1. Magnetising inductance L_m is constant if the air-gap flux is unchanged.

Combining (11.1) with (9.21) for the circuit of Fig. 9.1 gives

$$\frac{E_1}{f_1} = \frac{V_1}{f_1} - I_1\left(\frac{R_1}{f_1} + j\frac{X_1}{f_1}\right) \tag{11.2}$$

The value $I_1 X_1/f_1$ remains constant but $I_1 R_1/f_1$ becomes large at low values of speed (and frequency). Over most of the speed range E_1/f_1 can be kept almost constant by varying the terminal voltage so that V_1/f_1 is constant. At low speeds, however, the volts/frequency ratio V_1/f_1 has to be boosted to compensate for the voltage drop across R_1 in Fig. 9.1.

A convenient expression for the motor torque is given from Fig. 9.1(a) by (9.20),

$$T = \frac{|I_1|^2}{N_1} \frac{R_2}{S} \text{ Nm/phase}$$
$$= \frac{1}{N_1} \frac{|E_1|^2 R_2 S}{R_2^2 + S^2 X_2^2} \tag{11.3}$$

If (11.3) is combined with (9.1)(b) the torque can be expressed in terms of supply angular frequency ω_1,

$$T = \frac{p}{\omega_1} \frac{|E_1|^2 R_2 S}{R_2^2 + S^2 X_2^2} \text{ Nm/phase} \tag{11.4}$$

If E_1 is constant, the slip at which peak torque occurs may be obtained by differentiating (11.4) with respect to slip and equating to zero

$$S_m = \frac{R_2}{X_2} = \frac{R_2}{2\pi f_1 L_2} \tag{11.5}$$

This form differs from the corresponding expression (9.27), which was derived from the approximate equivalent circuit of Fig. 9.4. Substituting (11.5) into (11.4) gives an expression for the maximum torque T_m,

$$T_m = \frac{p}{\omega_1} \frac{|E_1|^2}{2X_2} = \frac{p}{8\pi^2 L_2}\left(\frac{E_1}{f_1}\right)^2 \tag{11.6}$$

If E_1 and f_1 are adjusted proportionately to keep E_1/f_1 constant, it can be seen from (11.6) that T_m is also constant because both ω_1 and X_2 vary directly with f_1. For operation near to synchronous speed $S^2 X_2^2 \ll R_2^2$, so that

$$T_{S\to 0} = \frac{p}{\omega_1} \frac{|E_1|^2}{R_2} S \text{ Nm/phase} \tag{11.7}$$

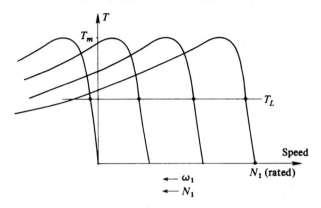

Fig. 11.1 Torque–speed characteristics for adjustable frequency and constant E/f.

The torque–speed characteristics for constant E_1/f_1 are shown in Fig. 11.1 and are seen to constitute a good adjustable speed drive. At low speeds the effect of primary resistance R_1 causes a significant reduction of T_m if the ratio V_1/f_1 is kept constant. This is illustrated by the use of (9.28) combined with (9.1(b))

$$T_m = \frac{\frac{|V_1|^2}{2N_1}}{R_1 + \sqrt{R_1^2 + X^2}}, \text{ where } X = X_1 + X_2$$

$$= \frac{\frac{|V_1|^2}{2N_1 R_1}}{1 + \sqrt{1 + \left(\frac{X}{R_1}\right)^2}}$$

$$= \frac{\frac{p|V_1|^2}{4\pi R_1}}{f_1 \left[1 + \sqrt{1 + \left(\frac{X}{R_1}\right)^2}\right]} \text{ N m/phase} \quad (11.8)$$

The reactance/resistance ratio, at rated frequency, increases with the size of motor. Typical values are

$$\frac{X}{R_1} \simeq 16 \text{ for large motors}$$

$$\frac{X}{R_1} \simeq 5 \text{ for medium size motors}$$

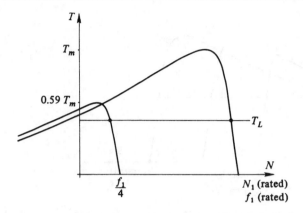

Fig. 11.2 Reduction of peak torque due to primary resistance in Example 11.1.

In (11.8) both V_1 and X are directly proportional to f_1. Typical reduction of T_m as f_1 reduces is calculated in Example 11.1 below and shown in Fig. 11.2. It is obvious that considerable increase of the V_1/f_1 ratio is required in order to maintain high peak torque at low speeds. To increase the torque without changing the speed, the frequency is kept fixed while V_1 is increased.

Although variation of the primary frequency causes proportionate variation of the synchronous speed, the form of the torque–speed characteristics is unchanged if E_1/f_1 is constant, as shown in Fig. 11.1. With constant load torque the slip speed $N_1 - N$ is the same at all values of speed. Per-unit slip S (9.2) is low over the whole speed range for all load conditions and the secondary circuit copper loss SP_g is therefore low.

High efficiency is maintained, even at low speeds, as seen in (9.19), but the efficiency–speed variation is no longer described by Fig. 9.3.

The use of variable frequency produces results that are dramatically different from the use of constant frequency and these differences are contrasted in Table 11.1. In effect, the ability to operate an induction motor with smoothly adjustable primary frequency confers all the advantages of normal high-speed operation (low current, high efficiency, fairly good power factor, good speed regulation, low slip, high flux) upon operation at any chosen speed.

The operating frequency f_2 of the secondary windings can be expressed in terms of the slip speed $N_1 - N$ by combining (9.1(b)) and (9.3):

$$f_2 = Sf_1 - \frac{p}{2\pi}(N_1 - N) \tag{11.9}$$

Therefore, if $N_1 - N$ is maintained constant the secondary winding frequency remains constant with adjustable frequency control. This is true, for example,

Table 11.1 *Comparative operation of three-phase induction motor.*

State of parameter or variable	Fixed frequency control (speed adjustment by voltage reduction)	Adjustable frequency control (with V/f constant)
Constant	f_1, N_1	Φ, f_2 (with constant load torque)
Variable	$f_2, V_1, \Phi, N, S, \eta$	$f_1, V_1, N_1, N, S, \eta$

with the constant torque load of Fig. 11.1. With a fan load, however, there is some variation of slip speed due to load level and also due to speed setting.

When the branch currents in the circuit of Fig. 9.1(b) are added, neglecting core loss, the primary phase current is found to be

$$I_1 = \frac{E_1}{jX_m} + \frac{E_1}{\frac{R_2}{S} + jX_2}$$

$$= \frac{E_1}{X_m} \left[\frac{SX_m R_2 - j\{R_2^2 + S^2 X_2(X_2 + X_m)\}}{R_2^2 + S^2 X_2^2} \right] \quad (11.10)$$

Substituting $S = f_2/f_1$ from (11.9) into (11.10),

$$I_1 = \left(\frac{E_1}{f_1}\right) \frac{1}{2\pi L_m} \left[\frac{2\pi f_2 R_2 L_m - j\{R_2^2 + 4\pi^2 f_2^2 L_2(L_2 + L_m)\}}{R_2^2 + 4\pi^2 f_2^2 L_2^2} \right] \quad (11.11)$$

Equation (11.11) shows that in the presence of constant air-gap flux the primary current is independent of supply frequency. If f_2 is constant in (11.11) then the primary current is constant at all speeds.

A corresponding expression can be realised for the approximate equivalent circuit of Fig. 9.4 by adding (9.22) and (9.24). This expression (11.12) contains a multiplier V_1/f_1 but is a complicated function not only of f_2 but also of f_1. In other words, even with constant f_2 the primary current does not remain constant if V_1/f_1 is fixed:

$$I_1 = \left(\frac{V_1}{f_1}\right) \frac{1}{2\pi L_m} \left[\frac{2\pi f_1 f_2 L_m (f_1 R_2 + f_2 R_1) - jx^2}{(f_1 R_2 + f_2 R_1)^2 + 4\pi^2 f_1^2 f_2^2 (L_1 + L_2)^2} \right] \quad (11.12)$$

where

$$x^2 = (f_1 R_2 + f_2 R_1)^2 + 4\pi^2 f_1^2 f_2^2 (L_1 + L_2)(L_1 + L_2 + L_m)$$

The phase-angle ϕ_1 between the primary phase-voltage V_1 and the phase-current I_1 can be deduced from expression (11.12):

$$\tan\phi_1 = \frac{(f_1 R_2 + f_2 R_1)^2 + 4\pi^2 f_1^2 f_2^2 (L_1 + L_2)(L_1 + L_2 + L_m)}{2\pi f_1 f_2 L_m (f_1 R_2 + f_2 R_1)} \quad (11.13)$$

Since the slip is now small at all speeds, then $f_2 \ll f_1$. Expression (11.13) then reduces to

$$\tan\phi_1 \simeq \frac{R_2}{2\pi f_2 L_m} \quad (11.14)$$

The input phase-angle is approximately constant at all speeds especially for a constant torque load and so, therefore, is the power factor.

It may be deduced from the equivalent circuits of Fig. 9.1 that the phase-angle of the referred secondary current I_2 with respect to e.m.f. E_1 is given by

$$\tan\phi_2 = \frac{X_2}{R_2} = \frac{2\pi f_2 L_2}{R_2} \quad (11.15)$$

The input phase-angle may therefore be written

$$\tan\phi_1 \simeq \frac{L_2}{L_m} \cot\phi_2 \quad (11.16)$$

11.1.2 Worked examples

Example 11.1
A three-phase induction motor drives a constant (rated) torque load over a 4:1 speed range by frequency control with the ratio V_1/f_1 maintained constant. If the motor reactance/primary resistance ratio $X/R_1 = 5$ calculate the effect on the peak torque of operating at the lowest speed. What value of V_1/f_1 is required at one-quarter frequency if the peak torque is to remain constant?

Solution. The peak torque, with constant V_1/f_1, is given by (11.8). At the lowest operating speed of one-quarter of rated speed the variables in (11.8) have the following values:

$$V_{1_{1/4}} = \frac{V_1}{4}, \quad X_{1/4} = \frac{X}{4}, \quad \frac{X_{1/4}}{R_1} = \frac{5}{4}, \quad f_{1/4} = \frac{f_1}{4}$$

Therefore, the peak torque at one-quarter frequency, compared with rated frequency, is

$$\frac{T_{m_{1/4}}}{T_m} = \frac{\dfrac{p|V_1/4|^2/4\pi R_1}{(f_1/4)[1+\sqrt{1+(5/4)^2}]}}{\dfrac{p|V_1|^2/4\pi R_1}{f_1(1+\sqrt{1+5^2})}}$$

$$= \frac{\dfrac{\frac{1}{16}}{\frac{1}{4}\times 1+\sqrt{1+1.56}}}{\dfrac{1}{1+\sqrt{1+25}}} = 0.587$$

The consequent reduction of peak torque is shown in Fig. 11.2. To maintain unchanged peak torque the V_1/f_1 ratio (i.e. the terminal voltage) at one-quarter frequency must be increased. Let the low-speed terminal voltage $= kV_1$ (rather than $V_1/4$), where V_1 is the rated value. If $T_{m_{1/4}} = T_m$ then

$$\frac{\dfrac{p|kV_1|^2}{4\pi R_1}}{\dfrac{f_1}{4}\left[1+\sqrt{1+\left(\dfrac{5}{4}\right)^2}\right]} = \frac{\dfrac{p|V_1|^2}{4\pi R_1}}{f_1(1+\sqrt{1+5^2})}$$

from which

$k = 0.326$

Compared with the nominal ratio $V_{1_{1/4}} = 0.25V_1$ the value has to be boosted to $V_{1_{1/4}} = 0.326V_1$, which represents about 25% increase.

Example 11.2
A 450 V, 50 Hz, 1450 r.p.m., 25 kW, star-connected three-phase induction motor delivers constant (rated) torque at all speeds. The motor equivalent circuit parameters at rated frequency are $R_1 = 0.10\,\Omega$, $R_2 = 0.17\,\Omega$, $X_1 = 0.3\,\Omega$, $X_2 = 0.5\,\Omega$, $X_m = 23.6\,\Omega$. Smooth speed variation is obtained by primary frequency control with simultaneous variation of the terminal voltage to maintain constant air-gap flux. Calculate the motor current, power factor and efficiency at one-fifth of rated speed.

Solution. Since the rated speed is 1450 r.p.m. it may be assumed to be a four-pole motor with a rated synchronous speed

$$N_1 = 1500\,\text{r.p.m.} = 1500 \times \frac{2\pi}{60} = 157.1\,\text{rad/s}$$

The lower operating speed is specified as

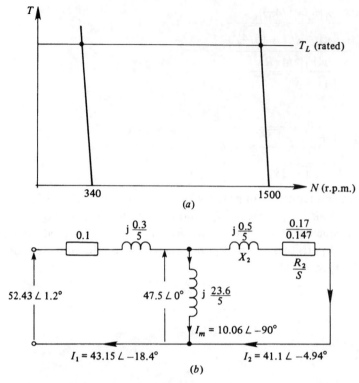

Fig. 11.3 Adjustable frequency drive of Example 11.2: (*a*) torque–speed characteristics, (*b*) per-phase equivalent circuit at 290 r.p.m.

$$N = \frac{1450}{5} = 290 \text{ r.p.m.} = 30.37 \text{ rad/s}$$

If the torque–speed characteristics are parallel, Fig. 11.3, the slip speed at all operating points is

$$N_1 - N = 50 \text{ r.p.m.} = 5.24 \text{ rad/s}$$

The synchronous speed for one-fifth of rated speed, with rated torque, is 340 r.p.m. or

$$N_{1_{1/5}} = 340 \times \frac{2\pi}{60} = 35.6 \text{ rad/s}$$

Secondary frequency f_2 is constant for all operating speeds. From (11.9),

$$f_2 = (N_1 - N)\frac{p}{2\pi} = 5.24 \times \frac{2}{2\pi} = 1.67 \text{ Hz}$$

At $N = 290$ r.p.m. the per-unit slip is

$$S = \frac{N_1 - N}{N_1} = \frac{50}{340} = 0.147$$

The rated torque is

$$T = \frac{P_{out}}{N} = \frac{25\,000}{1450 \times \frac{2\pi}{60}} = 164.6 \text{ N m}$$

From (9.20) the referred secondary current at 290 r.p.m. is

$$|I_2| = \sqrt{\frac{164.6}{3} \times 35.6 \times \frac{0.147}{0.17}} = 41.1 \text{ A}$$

From (11.15),

$$\phi_2 = -\tan^{-1}\left(\frac{0.147}{0.17} \times \frac{0.5}{5}\right) = \tan^{-1} 0.086 = -4.94°$$

Phase-angle ϕ_2 is with reference to e.m.f. E_1.

Internal e.m.f. E_1 in the equivalent circuit of Fig. 11.3(b) is therefore given by

$$E_1 = I_2 Z_2$$
$$= 41.1 \angle -4.94° \left(\frac{0.17}{0.147} + j\frac{0.5}{5}\right)$$
$$= 41.1 \angle -4.94° \times 1.156 \angle 4.94°$$
$$= 47.5 \angle 0°$$

Magnetising current I_m is obtained by the use of Fig. 11.3(b),

$$I_m = \frac{E_1}{jX_m} = \frac{47.5 \angle 0°}{\frac{23.6}{5} \angle +90°} = 10.06 \angle -90° \text{ A}$$

The primary phasor current I_1 is therefore

$$I_1 = I_m + I_2$$
$$= 41.4 \cos 4.94° - j(10.06 + 41.1 \sin 4.94°)$$
$$= 40.95 - j13.6$$
$$= 43.15 \angle -18.4° \text{ with reference to } E_1$$

The voltage drop across the primary winding impedance is

$$I_1 Z_1 = 41.15 \angle -18.4° \left(0.1 + j\frac{0.3}{5}\right)$$
$$= 41.15 \angle -18.4° \times 0.117 \angle 31°$$
$$= 5.05 \angle 12.6°$$
$$= 4.93 + j1.1$$

The supply voltage is given by

$$V_1 = E_1 + I_1 Z_1$$
$$= 47.5 + 4.93 + j1.1 = 52.43 \angle 1.2°$$

This checks with the value of the per-phase terminal voltage that is nominally used for one-fifth speed operation

$$V_1 = \frac{450}{\sqrt{3}} \times \frac{1}{5} = 52 \text{ V}$$

The phase-angle between V_1 and I_1 is $18.4° + 1.2° = 19.6°$, so that the power factor is

$$PF = \cos 19.6° = 0.942$$

This is the order of power factor that would be obtained at full-load rated speed with a voltage controlled motor.

Use of the approximate expression (11.14) gives a value

$$\tan \phi_1 = \left(\frac{0.17 \times 2\pi \times 50}{2\pi \times 1.67 \times 23.6} \right) = 0.216$$
$$\therefore \phi_1 \simeq 12.2°$$

Neglecting core loss and friction and windage effects the motor losses are due to winding dissipation:

$$P_{\text{loss}} = 3(|I_1|^2 R_1 + |I_2|^2 R_2)$$

At 290 r.p.m.,

$$P_{\text{loss}} = 3[(43.15)^2 0.1 + (41.1)^2 0.17]$$
$$= 3(186.2 + 287.2) = 1420 \text{ W}$$

With constant torque the output power is proportional to the speed. At 290 r.p.m. the efficiency is

$$\eta = \frac{P_{\text{out}}}{P_{\text{out}} + P_{\text{loss}}} = \frac{5000}{6420} = 77.9\%$$

11.2 THREE-PHASE, STEP-WAVE VOLTAGE SOURCE INVERTERS WITH PASSIVE LOAD IMPEDANCE

The basic form of voltage source inverter (VSI) consists of a three-phase controlled rectifier providing adjustable direct voltage V_{dc} as input to a three-phase, force-commutated inverter, Fig. 11.4. With a d.c. supply there is no natural commutation available caused by cycling of the supply voltages, and other methods of device switch-off have to be employed.

11.2 Voltage source inverters

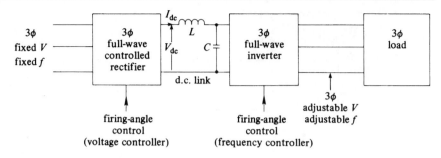

Fig. 11.4 Basic form of voltage source step-wave inverter.

The skeleton inverter consists of six semiconductor rectifier devices shown as generalised switches S in Fig. 11.5(a). In high power applications the switches are most likely to be SCRs, in which case they must be switched-off by forced quenching of the anode voltages. This adds greatly to the complexity and cost of the inverter design and reduces the reliability of its operation.

If the inverter devices are GTOs, Fig. 11.5(b), they can be extinguished using negative gate current. Various forms of transistor switches such as BJTs, Fig. 11.5(c), and IGBTs, Fig. 11.5(d), can be extinguished by control of their base currents, as discussed in Chapters 1–3. In Fig. 11.5 the commutating circuitry is not shown. It is assumed in the following analysis that each switch can be opened or closed freely.

From the power circuit point of view all versions of the skeleton inverter of Fig. 11.5 are identical. In each case the frequency of the generated voltages depends on the frequency of gating of the switches and the waveform of the generated voltages depends on the switching mode and on the load impedance. To distinguish this form of frequency-changer from other forms of inverter or cycloconverter it is often referred to as a 'd.c. link inverter'. The voltage source d.c. link inverter operates either as a stepped-wave inverter or as a pulse-width modulated inverter. Each of these forms is discussed extensively below. The switching sequence restrictions discussed in Section 7.2 of Chapter 7 also apply here to forced-commutation inverters. Note that the numerical notation of the switches in Fig. 11.5 is precisely the same as for the controlled rectifier and the naturally commutated inverter in Chapter 7. Certain authors use a slightly different notation of numbering.

Many different voltage waveforms can be generated by the use of appropriate switching patterns in the circuit of Fig. 11.5. An invariable requirement in three-phase systems is that the three-phase voltages be identical in form but phase-displaced by 120° electrical from each other. This does not

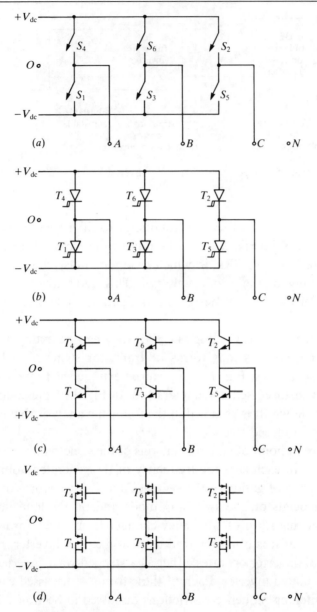

Fig. 11.5 Skeleton switching circuit of voltage source inverter: (*a*) generalised switches, (*b*) GTOs, (*c*) BJTs, (*d*) IGBTs.

necessarily create a balanced set of load voltages, in the sinusoidal sense of summing to zero at every instant of the cycle, but it reduces the possibility of gross voltage unbalance.

11.2 Voltage source inverters

A voltage source inverter is best suited to loads which have a high impedance to harmonic currents, such as a series tuned circuit or an induction motor. The series inductance of such loads often results in operation at a low power factor.

11.2.1 Stepped-wave inverter voltage waveforms

For the purpose of voltage waveform fabrication it is convenient to switch the devices of Fig. 11.5 sequentially at intervals of 60° electrical or one-sixth of a period. The use of a d.c. supply having equal positive and negative voltage values $\pm V_{dc}$ is common.

11.2.1.1 Two simultaneously conducting switches

If two switches conduct at any instant, a suitable switching pattern is defined in Fig. 11.6, for no-load operation. The devices are switched in numerical order and each remains in conduction for 120° electrical. Phase voltages v_{AN}, v_{BC}, v_{CN} consist of rectangular pulses of height $\pm V_{dc}$ which sum to zero at every instant and are sometimes called quasi-square waves. The corresponding line voltages are six-stepped waves of maximum height $\pm 2V_{dc}$. If equal resistors R are now connected in star to the load terminals A, B, C of Fig. 11.5 the conduction pattern of Fig. 11.7 ensues for the first half-period.

In interval $0 < \omega t < \pi/3$,

$$\left.\begin{array}{l} v_{AN} = -I_L R = -\dfrac{2V_{dc}}{2R} R = -V_{dc} \\ v_{BN} = 0 \\ v_{CN} = I_L R = \dfrac{2V_{dc}}{2R} R = +V_{dc} \\ V_{AB} = v_{AN} + v_{NB} = v_{AN} - v_{BN} = -V_{dc} \end{array}\right\} \quad (11.17)$$

In the interval $\pi/3 < \omega t < 2\pi/3$,

$$\left.\begin{array}{l} v_{AN} = 0 \\ v_{BN} = -I_L R = -V_{dc} \\ v_{CN} = +I_L R = +V_{dc} \\ v_{AB} = +V_{dc} \end{array}\right\} \quad (11.18)$$

In the interval $2\pi/3 < \omega t < \pi$,

448 Adjustable frequency control

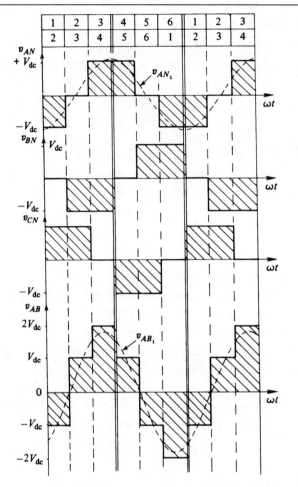

Fig. 11.6 Voltage waveforms with two simultaneously conducting switches. No-load and resistive load.

$$\left.\begin{array}{l} v_{AN} = I_L R = +V_{dc} \\ v_{BN} = -I_L R = -V_{dc} \\ v_{CN} = 0 \\ v_{AB} = 2V_{dc} \end{array}\right\} \quad (11.19)$$

For each interval it is seen that the load current during conduction is

$$I_L = \frac{\pm 2V_{dc}}{2R} = \pm \frac{V_{dc}}{R} \quad (11.20)$$

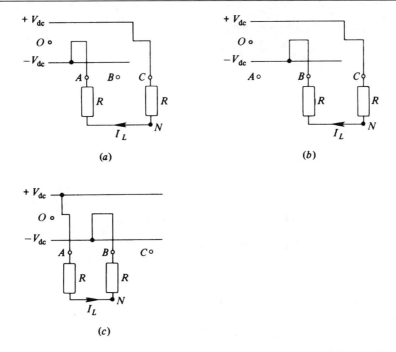

Fig. 11.7 Current conduction pattern for the case of two simultaneously conducting switches: (a) $0 \le \omega t \le 60°$, (b) $60° \le \omega t \le 120°$, (c) $120° \le \omega t \le 180°$.

The results of (11.17)–(11.20) are seen to be represented by the waveforms of Fig. 11.6. For this particular mode of switching the load voltage and current waveforms with star-connected resistive load are therefore identical with the pattern of the open-circuit voltages. The potential of load neutral point N coincides with the potential of the supply mid-point O.

Phase voltage waveform v_{AN} in Fig. 11.6 is given by an expression

$$v_{AN}(\omega t) = V_{dc}\big|_{120°}^{240°} - V_{dc}\big|_{0°,300°}^{60°,360°} \qquad (11.21)$$

This has the r.m.s. value

$$V_{AN} = \sqrt{\frac{1}{2\pi}\int_0^{2\pi} v_{AN}^2(\omega t)\,d\omega t} = \sqrt{\frac{2}{3}}\,V_{dc} = 0.816 V_{dc} \qquad (11.22)$$

The fundamental Fourier coefficients of waveform $v_{AN}(\omega t)$ are found to be

$$a_1 = \frac{1}{\pi} \int_0^{2\pi} v_{AN}(\omega t) \cos \omega t \, d\omega t$$
$$= -\frac{2\sqrt{3}}{\pi} V_{dc}$$
(11.23)

$$b_1 = \frac{1}{\pi} \int_0^{2\pi} v_{AN}(\omega t) \sin \omega t \, d\omega t$$
$$= 0$$
(11.24)

From (11.23), (11.24) it is seen that the fundamental phase voltage has a peak value $(2\sqrt{3}/\pi)V_{dc}$ or $1.1 V_{dc}$ and this is sketched in Fig. 11.6. The distortion factor of the phase voltage is given by

$$\text{distortion factor} = \frac{V_{AN_1}}{V_{AN}} = \frac{\frac{c_1}{\sqrt{2}}}{V_{AN}} = \frac{3}{\pi}$$
(11.25)

Line voltage $v_{AB}(\omega t)$ in Fig. 11.6 is defined by the relation

$$v_{AB}(\omega t) = V_{dc}\big|_{60°,180°}^{120°,240°} - V_{dc}\big|_{0,240°}^{60°,300°} + 2V_{dc}\big|_{120°}^{180°} - 2V_{dc}\big|_{300°}^{360°}$$
(11.26)

This is found to have fundamental Fourier coefficients of value

$$a_1 = -\frac{3\sqrt{3}}{\pi} V_{dc}$$
$$b_1 = +\frac{3}{\pi} V_{dc}$$
$$\therefore c_1 = \frac{6}{\pi} V_{dc}, \quad \psi_1 = -\tan^{-1}\sqrt{3} = -60°$$
(11.27)

The fundamental component of $v_{AB}(\omega t)$ is therefore given by

$$v_{AB_1}(\omega t) = \frac{6}{\pi} V_{dc} \sin(\omega t - 60°)$$
(11.28)

It is seen in Fig. 11.6 that $v_{AB_1}(\omega t)$ leads $v_{AN_1}(\omega t)$ by 30°, as in a balanced three-phase system, and the magnitude $|V_{AB_1}|$ is $\sqrt{3}$ times the magnitude $|V_{AN_1}|$.

With a firing pattern of two simultaneously conducting switches the load voltages of Fig. 11.6 are not retained with inductive load. Instead, the load voltages become irregular with dwell periods that differ with load phase-angle. Because of this, the pattern of two simultaneously conducting switches has only limited application.

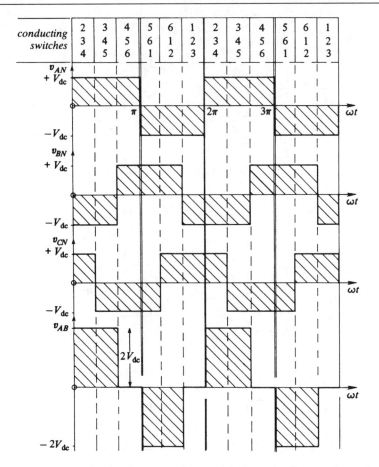

Fig. 11.8 No-load voltage waveforms with three simultaneously conducting switches.

11.2.1.2 Three simultaneously conducting switches

A further load voltage waveform is generated if a mode of switching is used whereby three switches conduct at any instant. Once again the switching devices conduct in numerical sequence but now each with a conduction angle of 180° electrical. At any instant of the cycle three switches with consecutive numbering are in conduction simultaneously. The pattern of waveforms obtained on no-load is shown in Fig.11.8. With equal star-connected resistors the conduction patterns of Fig. 11.9 are true for the first three 60° intervals of the cycle, if the load neutral N is isolated.

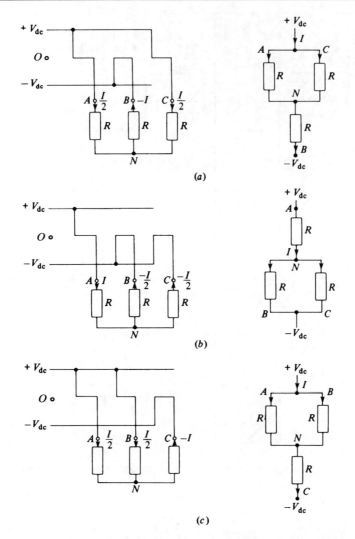

Fig. 11.9 Current conduction pattern for the case of three simultaneously conducting switches. Star-connected R load: (a) $0 \leq \omega t \leq 60°$, (b) $60° \leq \omega t \leq 120°$, (c) $120° \leq \omega t \leq 180°$.

For each interval,

$$I = \frac{2V_{dc}}{R + \dfrac{R}{2}} = \frac{4V_{dc}}{3R} \tag{11.29}$$

In the interval $0 < \omega t < \pi/3$,

$$V_{AN} = V_{CN} = \frac{I}{2}R = \frac{2}{3}V_{dc}$$
$$V_{BN} = -IR = -\frac{4}{3}V_{dc} \tag{11.30}$$
$$V_{AB} = V_{AN} - V_{BN} = 2V_{dc}$$

In the interval $\pi/3 \leqslant \omega t \leqslant 2\pi/3$,

$$V_{AN} = IR \frac{4}{3}V_{dc}$$
$$V_{BN} = V_{CN} = -\frac{I}{2}R = -\frac{2}{3}V_{dc} \tag{11.31}$$
$$V_{AB} = 2V_{dc}$$

In the interval $2\pi/3 \leq \omega t \leq \pi$,

$$V_{AN} = V_{BN} = \frac{I}{2}R = \frac{2}{3}V_{dc}$$
$$V_{CN} = -IR = -\frac{4}{3}V_{dc} \tag{11.32}$$
$$V_{AB} = 0$$

The voltage waveforms obtained with star-connected resistive load are plotted in Fig. 11.10. The phase voltages are seen to be different from the corresponding no-load values but the line voltages remain unchanged. Although the no-load phase voltages do not sum to zero, the load currents, with three-wire star-connection, must sum to zero at every instant of the cycle. In Fig. 11.10 the phase voltage v_{AN} is given by

$$v_{AN}(\omega t) = \frac{2}{3}V_{dc}\Big|_{0,120°}^{60°,180°} - \frac{2}{3}V_{dc}\Big|_{180°,300°}^{240°,360°} + \frac{4}{3}V_{dc}\Big|_{60°}^{120°} - \frac{4}{3}V_{dc}\Big|_{240°}^{300°} \tag{11.33}$$

It can be seen by inspection that the fundamental component of $v_{AN}(\omega t)$ is in time-phase with it, so that

$$a_1 = 0$$
$$\psi_1 = \tan^{-1}\frac{a_1}{b_1} = 0 \tag{11.34}$$

Fundamental Fourier coefficient b_1 for the load peak phase voltage is found to be

$$b_1 = c_1 = \frac{4}{\pi}V_{dc} \tag{11.35}$$

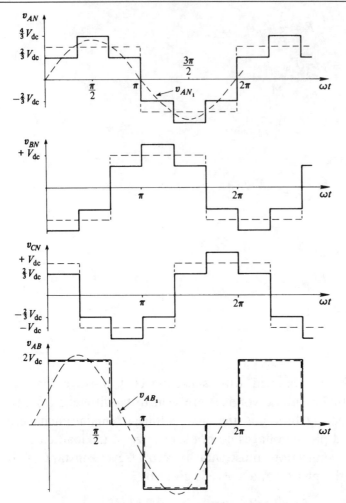

Fig. 11.10 Voltage waveforms with three simultaneously conducting switches. Star-connected R load with isolated neutral.

The corresponding Fourier coefficients for line voltage $v_{AB}(\omega t)$ are given by

$$a_1 = \frac{2\sqrt{3}}{\pi} V_{dc}$$
$$b_1 = -\frac{6}{\pi} V_{dc}$$
$$c_1 = \frac{4}{\pi}\sqrt{3} V_{dc} = \sqrt{3} \times \text{the phase value}$$
$$\psi_1 = -\tan^{-1}\frac{1}{\sqrt{3}} = -30°$$

(11.36)

Table 11.2. *Properties of step waves.*

Phase voltage wave form	Properties of the phase voltage waveform					Corresponding line voltage waveform
	Fundamental comp.		Total r.m.s.	Distortion factor	THD	
	Peak	R.m.s.				
(waveform)	$\frac{4}{\pi}V_{dc} = 1.273 V_{dc}$	$\frac{4}{\sqrt{2}\pi}V_{dc} = \frac{2\sqrt{2}}{\pi}V_{dc}$	V_{dc}	$\frac{2\sqrt{2}}{\pi} = 0.9$	$\sqrt{\frac{\pi^2}{8} - 1} = 0.483$	(waveform)
(waveform)	$\frac{2\sqrt{3}}{\pi}V_{dc} = 1.1 V_{dc}$	$\frac{\sqrt{6}}{\pi}V_{dc} = 0.78 V_{dc}$	$\sqrt{\frac{2}{3}}V_{dc}$	$\frac{3}{\pi} = 0.955$	$\sqrt{\frac{\pi^2}{9} - 1} = 0.311$	(waveform)
(waveform)	$\frac{6}{\pi}V_{dc} = 1.91 V_{dc}$	$\frac{6}{\sqrt{2}\pi}V_{dc} = 1.35 V_{dc}$	$\sqrt{2}V_{dc}$	$\frac{3}{\pi} = 0.955$	$\sqrt{\frac{\pi^2}{9} - 1} = 0.311$	(waveform)
(waveform)	$\frac{2}{\pi}V_{dc}$	$\frac{\sqrt{2}}{\pi}V_{dc}$	$\frac{1}{\sqrt{2}}V_{dc}$	$\frac{2}{\pi} = 0.637$	$\sqrt{\frac{\pi^2}{4} - 1} = 1.212$	(waveform)

The fundamental components of the load voltages, plotted in Fig. 11.10, show that, as with a three-phase sinusoidal system, the line voltage leads its corresponding phase voltage by 30°. The r.m.s. value of phase voltage $v_{AN}(\omega t)$ is found to be

$$V_{AN} = \sqrt{\frac{1}{\pi} \int_0^\pi v_{AN}^2(\omega t)\, d\omega t} = \frac{2\sqrt{2}}{3} V_{dc} = 0.943 V_{dc} \tag{11.37}$$

Combining (11.35) and (11.37) gives the distortion factor of the phase voltage,

$$\text{distortion factor} = \frac{V_{AN_1}}{V_{AN}} = \frac{\frac{c_1}{\sqrt{2}}}{V_{AN}} = \frac{3}{\pi} \tag{11.38}$$

This is seen to be identical to the value obtained for the corresponding waveform of Fig. 11.6. Although the distortion factors are identical, waveform $v_{AN}(\omega t)$ of Fig. 11.10 has a slightly greater fundamental value (11.35) than the corresponding value for $v_{AN}(\omega t)$ of Fig. 11.6, given by (11.23). The switching mode that utilises three simultaneously conducting switches is therefore potentially more useful for motor speed control applications. The properties of relevant step-waves and square-waves are summarised in Table 11.2.

It can be deduced from the waveforms of Fig. 11.9 that load neutral point N is not at the same potential as the supply neutral point O. While these points remain isolated, a difference voltage V_{NO} exists that is square-wave in form, with amplitude $\pm V_{dc}/3$ and of frequency three times the inverter switching frequency. If the two neutral points are joined a neutral current will flow that is square-wave in form, of amplitude $\pm V_{dc}/R$ and of three times the inverter switching frequency.

11.2.2 Measurement of harmonic distortion

The extent of waveform distortion for an alternating waveform can be defined in a number of different ways. The best known of these, the distortion factor defined by (11.38), was used in connection with the naturally commutated rectifier/inverter in Chapter 7 and the a.c. voltage controller in Chapter 8.

An alternative measure of the amount of distortion is by means of a property known as the total harmonic distortion, or *THD*, which is defined as

$$THD = \sqrt{\frac{V_{AN}^2 - V_{AN_1}^2}{V_{AN_1}^2}} = \frac{V_{AN_h}}{V_{AN_1}} \qquad (11.39)$$

For a pure sinusoid $V_{AN_1} = V_{AN}$ and the THD then has the ideal value of zero. The numerator of (11.39) is seen to represent the effective sum of the non-fundamental or higher harmonic components, V_{AN_h}.

A comparison of (11.38) and (11.39) shows that, for any wave,

$$\text{distortion factor} = \frac{V_{AN_1}}{V_{AN}} = \sqrt{\frac{1}{1 + (THD)^2}} \qquad (11.40)$$

11.2.3 Harmonic properties of the six-step voltage wave

The six-step phase voltage waveforms of Fig. 11.10 are defined by the Fourier series

$$v_{AN}(\omega t) = \frac{4}{\pi} V_{dc} \left(\sin \omega t + \frac{1}{5} \sin 5\omega t + \frac{1}{7} \sin 7\omega t \right.$$
$$\left. + \frac{1}{11} \sin 11\omega t + \frac{1}{13} \sin 13\omega t + \ldots \right) \qquad (11.41)$$

Waveform $v_{AN}(\omega t)$ of Fig. 11.10 contains no triplen harmonics and its lowest higher harmonic is of order five with an amplitude equal to 20% of the fundamental. The r.m.s. value of the function (11.41) is given

$$V_{AN} = \frac{4V_{dc}}{\sqrt{2}\pi} \sqrt{1 + \frac{1}{5^2} + \frac{1}{7^2} + \frac{1}{11^2} + \frac{1}{13^2} + \frac{1}{17^2} + \ldots}$$
$$= \frac{1}{\sqrt{2}} \times \frac{4}{\pi} V_{dc} \sqrt{1.079 + \ldots} \qquad (11.42)$$
$$= 0.935 V_{dc} + \ldots$$

which confirms the value obtained by integration in (11.37).

For the step-wave of Fig. 11.10, substituting (11.35) and (11.37) into (11.39) gives

$$THD = \frac{V_{dc}\sqrt{\left(\frac{2\sqrt{2}}{3}\right)^2 - \left(\frac{4}{\pi\sqrt{2}}\right)^2}}{V_{dc}\frac{4}{\pi}\frac{1}{\sqrt{2}}} \quad (11.43)$$

$$= \sqrt{\frac{\pi^2}{9} - 1} = 0.311$$

Harmonic voltage V_{AN_h} is 31.1% of the r.m.s. value of the fundamental component and 29.7% of the total r.m.s. value. Values of THD for other waveforms are given in Table 11.2. In general, if there are N steps/cycle, each occupying $2\pi/N$ radians, the only harmonics present are of the order $h = nN \pm 1$, where $n = 1, 2, 3 \ldots$ For a six-step waveform, Fig. 11.10 for example, $N = 6$ so that $h = 5, 7, 11, 13$ etc. as depicted in (11.41).

11.2.4 Harmonic properties of the optimum twelve-step waveform

A reduction of the harmonic content can be realised by increase of the number of steps in the phase voltage wave. If a twelve-step waveform is used $N = 12$ and $h = 11, 13, 23, 15 \ldots$ Example 11.4 below gives some detail of a certain twelve-step waveform calculation. It is found that the optimum twelve-step waveform, shown in Fig. 11.11, is represented by the Fourier expression

$$v(\omega t) = \frac{\pi}{3} V_m \left(\sin \omega t + \frac{1}{11} \sin 11\omega t + \frac{1}{13} \sin 13\omega t + \frac{1}{23} \sin 23\omega t \ldots \right) \quad (11.44)$$

In each interval of the optimum waveform of Fig. 11.11 the step height corresponds to the average value of the sinusoidal segment. For $0 \leq \omega t \leq \pi/6$, for example, the average value is

$$\text{step height} = \frac{\pi V}{3} \frac{6}{\pi} \int_0^{\pi/6} \sin \omega t \, d\omega t = 0.268 V \quad (11.45)$$

A twelve-step waveform can be fabricated by the use of two six-step inverters with their outputs displaced by 30° or by the series addition of square-wave or pulse-width modulated voltages.

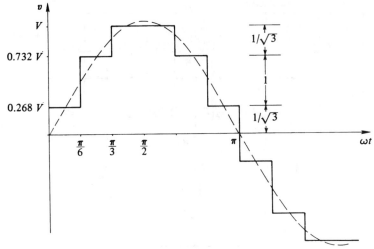

Fig. 11.11 Twelve-step voltage waveform.

11.2.5 Six-step voltage inverter with series R–L load

When a reactive load is connected to a step-wave inverter it becomes necessary to include a set of reverse-connected diodes in the circuit to carry return current, Fig. 11.12. The presence of the diodes immediately identifies the circuit as a VSI rather than a current source inverter (CSI) for which return diodes are unnecessary. In the presence of load inductance with rectifier supply a shunt capacitor must be connected in the d.c. link to absorb the reactive voltamperes because there is no path for reverse current.

11.2.5.1 Star-connected load

In the switching mode where three switches conduct simultaneously the no-load voltages are given by Fig. 11.8. Let these voltages now be applied to the star-connected R–L loads, as in Fig. 11.12. The resulting current undergoes an exponential increase of value. Consider the instant $\omega t = 0$ in the typical steady-state cycle shown in Fig. 11.13. Transistor T_1 has been in conduction for 180° and has just switched off. Transistor T_2 has been in conduction for 120° passing positive current i_C. Transistor T_3 is 60° into its conduction cycle resulting in current i_B that is increasing negatively. Transistor T_4 has just switched on, connecting terminal A to $+V_{dc}$, which will attempt to create positive i_A. The negative current $i_A(0)$ at $\omega t = 0$ is diverted from its previous path through T_1 and passes through diode D_4 to circulate through capacitor C. As soon as $i_A = 0$, D_4 switches off, at point t' in Fig. 11.13, and T_4 takes up the positive current I_A. Even if the load impedance is highly inductive the

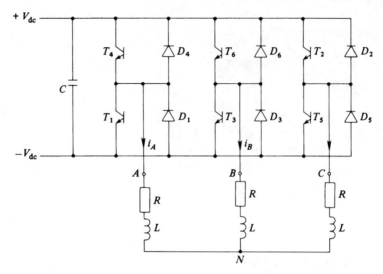

Fig. 11.12 Voltage source transistor inverter, incorporating return current path diodes.

load phase and line voltages largely retain the forms of Fig. 11.8, which are impressed by the supply battery and the inverter switching pattern, independently of load phase-angle.

11.2.5.2 Delta-connected load

Let the voltages of Fig. 11.8, for the case of three simultaneously conducting switches, be applied to a balanced, three-phase, delta-connected load, as in Fig. 11.14. Since the star-connected load of Fig. 11.12 can be replaced by an equivalent delta-connected load the line current waveforms of Fig. 11.13 remain true. The phase current waveforms can be deduced by the application of classical mathematical analysis.

In the interval $0 \leq \omega t \leq 120°$ of Fig. 11.13 a voltage $2V_{dc}$ is impressed across terminals AB so that, with $\cot\phi = R/\omega L$,

$$i_{AB}(\omega t)\Big|_{0<\omega t<120°} = \frac{2V_{dc}}{R}(1 - \varepsilon^{-\cot\phi \cdot \omega t}) + i_{AB}(0)\varepsilon^{-\cot\phi \cdot \omega t} \quad (11.46)$$

In the interval $120° < \omega t < 180°$ of Fig. 11.13 terminals A and B are coincident and load branch AB is short circuited so that

11.2 Voltage source inverters

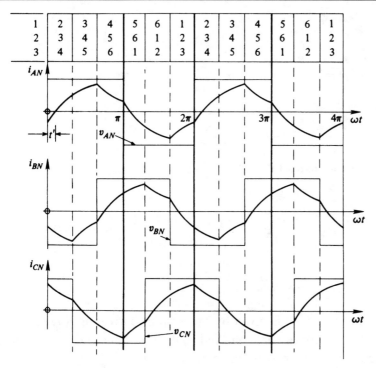

Fig. 11.13 Current waveforms for voltage source, six-step inverter with star-connected series R–L load.

$$i_{AB}(\omega t)\big|_{120°<\omega t<180°} = \left[\frac{2V_{dc}}{R}(1-\varepsilon^{-\cot\phi\cdot 2\pi/3}) + i_{AB}(0)\varepsilon^{-\cot\phi\cdot 2\pi/3}\right]\varepsilon^{-\cot\phi\cdot(\omega t-2\pi/3)} \tag{11.47}$$

Since the current wave possesses half-wave inverse symmetry $i_{AB}(0) = -i_{AB}(\pi) = i_{AB}(2\pi)$. Putting $\omega t = \pi$ in (11.47) and utilising the inverse-symmetry identity gives

$$i_{AB}(0) = -\frac{2V_{dc}}{R}\frac{\varepsilon^{-\cot\phi\cdot\pi/3} - \varepsilon^{-\cot\phi\cdot\pi}}{1+\varepsilon^{-\cot\phi\cdot\pi}} \tag{11.48}$$

Combining (11.48) with (11.46) and (11.47) respectively gives

$$i_{AB}(\omega t)\big|_{0<\omega t<120°} = \frac{2V_{dc}}{R}\left(1 - \frac{1+\varepsilon^{-\cot\phi\cdot\pi/3}}{1+\varepsilon^{-\cot\phi\cdot\pi}}\varepsilon^{-\cot\phi\cdot\omega t}\right) \tag{11.49}$$

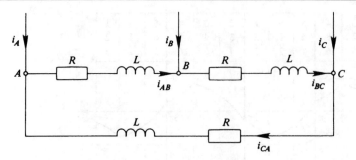

Fig. 11.14 Delta-connected series R–L load.

$$i_{AB}(\omega t)\big|_{120<\omega t<180°} = \frac{2V_{dc}}{R}\left(1 - \frac{1-\varepsilon^{-\cot\phi\cdot 2\pi/3}}{1+\varepsilon^{-\cot\phi\cdot\pi}}\varepsilon^{-\cot\phi(\omega t - 2\pi/3)}\right) \tag{11.50}$$

Current $i_{CA}(\omega t)$ in Fig. 11.14 is given by expressions corresponding to those of (11.49), (11.50) but with the time delayed by $4\pi/3$ radians. The r.m.s. value of the branch current is defined by the expression

$$I_{AB} = \sqrt{\frac{1}{2\pi}\int_0^{2\pi} i_{AB}^2(\omega t)\,d\omega t} \tag{11.51}$$

In elucidating (11.51) it is convenient to use the substitutions

$$K_1 = \frac{1+\varepsilon^{-\cot\phi\cdot\pi/3}}{1+\varepsilon^{-\cot\phi\cdot\pi}}, \quad K_2 = \frac{1-\varepsilon^{-\cot\phi\cdot 2\pi/3}}{1+\varepsilon^{-\cot\phi\cdot\pi}} \tag{11.52}$$

An examination of K_1 and K_2 above shows that

$$K_2 = 1 - K_1\varepsilon^{-\cot\phi\cdot 2\pi/3} \tag{11.53}$$

Substituting (11.49) and (11.50) into (11.51) gives

$$I_{AB}^2 = \frac{2}{2\pi} \times \frac{4V_{dc}^2}{R^2} \left\{ \int_0^{120°} (1 - K_1 \varepsilon^{-\cot\phi \cdot \omega t})^2 \, d\omega t \right.$$

$$\left. + \int_{120°}^{180°} [K_2 \varepsilon^{-\cot\phi \cdot (\omega t - 2\pi/3)}]^2 \, d\omega t \right\}$$

$$= \frac{4V_{dc}^2}{\pi R^2} \left\{ \left[\omega t + \frac{2K_1}{\cot\phi} \varepsilon^{-\cot\phi \cdot \omega t} - \frac{K_1^2 \varepsilon^{-\cot\phi \cdot 2\omega t}}{2\cot\phi} \right]_0^{120°} \right.$$

$$\left. + \left[-\frac{K_2^2 \varepsilon^{-\cot\phi \cdot 2(\omega t - 2\pi/3)}}{2\cot\phi} \right]_{120°}^{180°} \right\}$$

$$= \frac{4V_{dc}^2}{\pi R^2} \left\{ \frac{2\pi}{3} + \frac{2K_1}{\cot\phi} (\varepsilon^{-\cot\phi \cdot 2\pi/3} - 1) - \frac{K_1^2}{2\cot\phi} (\varepsilon^{-\cot\phi \cdot 4\pi/3} - 1) \right.$$

$$\left. - \frac{K_2^2}{2\cot\phi} (\varepsilon^{-\cot\phi \cdot 2\pi/3} - 1) \right\} \quad (11.54)$$

Eliminating the explicit exponential terms between (11.52) and (11.54) gives

$$I_{AB}^2 = \frac{4V_{dc}^2}{\pi R^2} \left\{ \frac{2\pi}{3} + \frac{1}{\cot\phi} \left[\frac{3}{2} - K_2 - 2K_1 + \frac{K_1^2}{2} - \frac{K_2^2(1 - K_2)}{2K_1} \right] \right\} \quad (11.55)$$

Line current $i_A(\omega t)$ in Fig. 11.14 changes in each 60° interval of conduction. In general, $i_A(\omega t) = i_{AB}(\omega t) - i_{CA}(\omega t)$, so that

$$i_A(\omega t)|_{0 < \omega t < 60°} = \frac{2V_{dc}}{R} \left[1 - \frac{(1 + \varepsilon^{-\cot\phi \cdot \pi/3})(2 - \varepsilon^{-\cot\phi \cdot \pi/3})}{1 + \varepsilon^{-\cot\phi \cdot \pi}} \varepsilon^{-\cot\phi \cdot \omega t} \right] \quad (11.56)$$

$$i_A(\omega t)|_{60° < \omega t < 120°} = \frac{2V_{dc}}{R} \left[2 - \frac{(1 + \varepsilon^{-\cot\phi \cdot \pi/3})^2}{1 + \varepsilon^{-\cot\phi \cdot \pi}} \varepsilon^{-\cot\phi(\omega t - \pi/3)} \right] \quad (11.57)$$

$$i_A(\omega t)|_{120° < \omega t < 180°} = \frac{2V_{dc}}{R}$$

$$\times \left[1 - \frac{(1 + 2\varepsilon^{-\cot\phi \cdot \pi/3})(1 + \varepsilon^{-\cot\phi \cdot \pi/3})}{1 + \varepsilon^{-\cot\phi \cdot \pi}} \varepsilon^{-\cot\phi(\omega t - 2\pi/3)} \right] \quad (11.58)$$

A typical pattern of waves, consistent with (11.56)–(11.58) is shown in Fig. 11.15. At any instant the current $i_A(\omega t)$ must be flowing through one of the

464   Adjustable frequency control

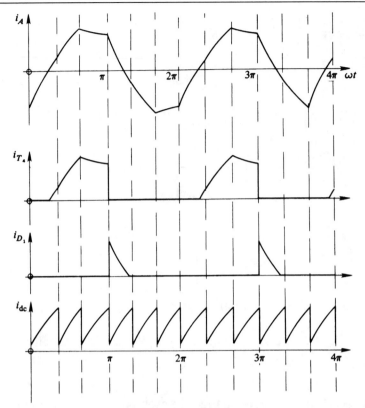

Fig. 11.15 Current waveforms for six-step voltage source inverter with delta-connected series R–L load.

devices T_1, T_4, D_1 or D_4 in the inverter of Fig. 11.12. In the interval $0 \le \omega t \le 60°$, the negative part of $i_A(\omega t)$, up to $\omega t = \pi$, is conducted via transistor T_4. For $\omega t > 180°$, the positive current $i_A(\omega t)$ reduces to zero through diode D_1 and then goes negative via T_1. The properties of both the transistor and the diode currents can be calculated by use of the appropriate parts of equations (11.49)–(11.58). The oscillating unidirectional current in the d.c. link, Fig. 11.15, consists of a repetition of the current $i_A(\omega t)$ in the interval $60° \le \omega t \le 120°$. For the interval, $0 \le \omega t \le 60°$, $i_{dc}(\omega t)$ is defined by

$$i_{dc}(\omega t) = \frac{2V_{dc}}{R}\left(2 - K_3 \varepsilon^{-\cot \phi \cdot \omega t}\right) \qquad (11.59)$$

where

$$K_3 = \frac{(1+\varepsilon^{-\cot\phi\cdot\pi/3})^2}{1+\varepsilon^{-\cot\phi\cdot\pi}} \tag{11.60}$$

This link current will become negative for part of the cycle if the load is sufficiently inductive. The boundary condition for the start of negative link current is if $i_{dc}(\omega t) = 0$ at $\omega t = 0$, which occurs when $K_3 = 2$. This happens for load with a power factor smaller than 0.525 lagging. The average value of $i_{dc}(\omega t)$ in the interval $0 \leq \omega t \leq 60°$ and therefore in all the intervals is given by

$$\begin{aligned}I_{dc} &= \frac{3}{\pi}\int_0^{60°} \frac{2V_{dc}}{R}(2 - K_3\varepsilon^{-\cot\phi\cdot\omega t})\,d\omega t \\ &= \frac{3}{\pi}\frac{2V_{dc}}{R}\left[2\omega t + \frac{K_3}{\cot\phi}\varepsilon^{-\cot\phi\cdot\omega t}\right]_0^{60°} \\ &= \frac{3}{\pi}\frac{2V_{dc}}{R}\left[\frac{2\pi}{3} + \frac{K_3}{\cot\phi}(\varepsilon^{-\cot\phi\cdot\pi/3} - 1)\right]\end{aligned} \tag{11.61}$$

11.2.6 Worked examples

Example 11.3
An ideal d.c. supply of constant voltage V supplies power to a three-phase force-commutated inverter consisting of six ideal transistor switches. Power is thence transferred to a delta-connected resistive load of R ohms per branch. The mode of inverter switching is such that two transistors are in conduction at any instant of the cycle. Deduce and sketch waveforms of the phase and line currents.

Solution. The load is connected so that the system currents have the notation shown in Fig. 11.14. The triggering sequence is given at the top of Fig. 11.6. At any instant of the cycle two of the three terminals, A, B, C, will be connected to the supply, which has a positive rail $+V$ while the other rail is zero potential. The load effectively consists of two resistors R in series shunted by another resistor R.
In the interval $0 \leq \omega t \leq \pi/3$, for example, transistors T_1 and T_2 are conducting so that

$$i_C = -i_A = \frac{V}{2R/3} = \frac{3}{2}\frac{V}{R}$$

$$i_B = 0$$

$$i_{CA} = \frac{2}{3}i_C = \frac{V}{R}$$

$$i_{BC} = i_{AB} = -\frac{1}{3}i_C = -\frac{1}{2}\frac{V}{R}$$

In the interval $\pi/3 < \omega t < 2\pi/3$, transistors T_2 and T_3 are conducting, resulting in the isolation of terminal A so that

$$i_C = -i_B = \frac{3}{2}\frac{V}{R}$$

$$i_A = 0$$

$$i_{BC} = -\frac{V}{R}$$

$$i_{CA} = i_{AB} = +\frac{1}{2}\frac{V}{R}$$

In the interval $2\pi/3 \le \omega t \le \pi$, transistors T_3 and T_4 are in conduction so that terminal B has the negative rail potential of zero while terminal A is connected to the $+V$ rail, so that

$$i_C = 0$$

$$i_A = -i_B = \frac{3}{2}\frac{V}{R}$$

$$i_{AB} = \frac{V}{R}$$

$$i_{CA} = i_{BC} = -\frac{1}{2}\frac{V}{R}$$

The pattern of waveforms produced, Fig. 11.16, is that of a six-step phase (i.e. branch) current but a square-wave line current. In fact the pattern of waveforms is identical in form, but with different amplitude scaling, to that obtained with a star-connected load of R ohms/phase in Fig. 11.10, when three transistors conduct simultaneously.

Example 11.4

The voltage waveform of a certain type of twelve-step inverter is given in Fig. 11.17. For this waveform calculate the fundamental value, the total r.m.s. value and the distortion factor.

Solution. The waveform of Fig. 11.17 is defined by the relationship

$$e(\omega t) = \frac{E_m}{3}\bigg|_{0,4\pi/5}^{\pi,5\pi} + \frac{2E_m}{3}\bigg|_{\pi/5,3\pi/5}^{2\pi/5,4\pi/5} + E_m\bigg|_{2\pi/5}^{3\pi/5}$$

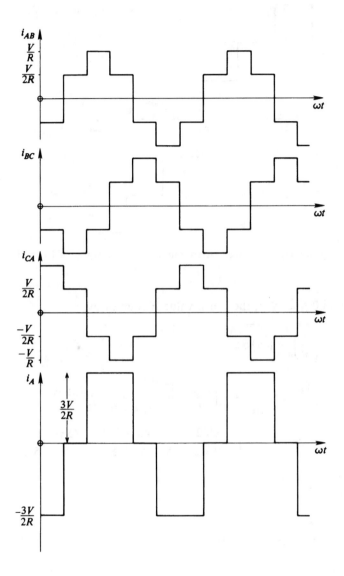

Fig. 11.16 Waveforms of voltage source inverter with delta-connected R load (Example 11.3).

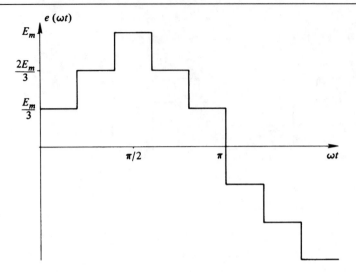

Fig. 11.17 Voltage waveform of twelve-step voltage source inverter of Example 11.4.

For the interval $0 \leq \omega t \leq \pi$, the r.m.s. value E is given by

$$E = \sqrt{\frac{1}{\pi}\int_0^\pi e^2(\omega t)\,d\omega t}$$

$$E^2 = \frac{1}{\pi}\left[\frac{E_m^2}{9}\omega t\right]_{0,4\pi/5}^{\pi,5\pi} + \frac{1}{\pi}\left[\frac{4E_m^2}{9}\omega t\right]_{\pi/5,3\pi/5}^{2\pi/5,4\pi/5} + \frac{1}{\pi}[E_m^2\omega t]_{2\pi/5}^{3\pi/5}$$

$$= \frac{E_m^2}{\pi}\left[\frac{1}{9}\left(\frac{\pi}{5}-0+\pi-\frac{4\pi}{5}\right) + \frac{4}{9}\left(\frac{2\pi}{5}-\frac{\pi}{5}+\frac{4\pi}{5}-\frac{3\pi}{5}\right) + \left(\frac{3\pi}{5}-\frac{2\pi}{5}\right)\right]$$

$$= \frac{E_m^2}{\pi}\left(\frac{1}{9}\frac{2\pi}{5}+\frac{4}{9}\frac{2\pi}{5}+\frac{\pi}{5}\right)$$

$$= E_m^2\left(\frac{2}{45}+\frac{8}{45}+\frac{9}{45}\right) = E_m^2\,\frac{19}{45}$$

$\therefore E = 0.65 E_m$

It is obvious that the fundamental component of waveform $e(\omega t)$ in Fig. 11.17 is symmetryical with respect to the waveform itself. Therefore

$$\psi_1 = \tan^{-1}\left(\frac{a_1}{b_1}\right) = 0$$

and

$a_1 = 0$

The fundamental component b_1 is given by

$$b_1 = \frac{1}{\pi} \int_0^{2\pi} e(\omega t) \sin \omega t \, d\omega t$$

$$= \frac{2}{\pi} \int_0^{\pi} e(\omega t) \sin \omega t \, d\omega t$$

In this case

$$b_1 = \frac{2}{\pi} \left(-\frac{E_m}{3} \cos \omega t \Big|_{0,4\pi/5}^{\pi,5\pi} - \frac{2E_m}{3} \cos \omega t \Big|_{\pi/5,3\pi/5}^{2\pi/5,4\pi/5} \right.$$

$$\left. - E_m \cos \omega t \Big|_{2\pi/5}^{3\pi/5} \right)$$

$$= \frac{2E_m}{\pi} \left[-\frac{1}{3} \left(\cos \frac{\pi}{5} - \cos 0 + \cos \pi - \cos \frac{4\pi}{5} \right) \right.$$

$$\left. - \frac{2}{3} \left(\cos \frac{2\pi}{5} - \cos \frac{\pi}{5} + \cos \frac{4\pi}{5} - \cos \frac{3\pi}{5} \right) - \cos \frac{3\pi}{5} + \cos \frac{2\pi}{5} \right]$$

$$= \frac{2E_m}{\pi} \left(+\frac{2}{3} + \frac{1}{3} \cos \frac{\pi}{5} - \frac{1}{3} \cos \frac{4\pi}{5} + \frac{1}{3} \cos \frac{2\pi}{5} - \frac{1}{3} \cos \frac{3\pi}{5} \right)$$

$$= \frac{2E_m}{3\pi} (+2 + 0.809 + 0.809 + 0.309 + 0.309)$$

$$= \frac{2E_m}{3\pi} (4.24) = \frac{2.82 E_m}{\pi} = 0.9 E_m$$

$$\text{distortion factor} = \frac{E_1}{E} = \frac{0.9}{\sqrt{2} \times 0.65} = 0.98$$

Example 11.5

A six-step voltage source inverter is supplied with power from an ideal battery of constant voltage $V = 150$ V. The inverter has a delta-connected series R–L load where $R = 15\,\Omega$, $X_L = 25$ ohms at 50 Hz. Calculate the r.m.s. current in the load, the power transferred and the average value of the supply current, at 50 Hz.

Solution. In this example an inverter of the form of Fig. 11.12 supplies power to a load with the connection of Fig. 11.14. The pattern of phase or branch currents $i_{AB}(\omega t)$, $i_{AB}(\omega t)$, $i_{CA}(\omega t)$ is similar in form to the load currents with star-connected load shown in Fig. 11.13. The line currents have the typical form $i_A(\omega t)$ given in Fig. 11.15. The branch current $i_{AB}(\omega t)$ is defined by (11.49) and (11.50), where the voltage is now V (rather than $2V_{dc}$)

$$\phi = \tan^{-1} \frac{\omega L}{R} = \tan^{-1} 1.67 = 59.1°$$

Adjustable frequency control

$\cot \phi = \cot 59.1 = 0.6$

$\varepsilon^{-\cot \phi \cdot \pi/3} = \varepsilon^{-0.63} = 0.533$

$\varepsilon^{-\cot \phi \cdot 2\pi/3} = \varepsilon^{-1.26} = 0.285$

$\varepsilon^{-\cot \phi \cdot \pi} = \varepsilon^{-1.88} = 0.152$

$\varepsilon^{-\cot \phi \cdot 4\pi/3} = \varepsilon^{-2.51} = 0.081$

Now in (11.52),

$$K_1 = \frac{1+\varepsilon^{-\cot \phi \cdot \pi/3}}{1+\varepsilon^{-\cot \phi \cdot \pi}} = \frac{1.533}{1.152} = 1.33$$

$$K_2 = \frac{1-\varepsilon^{-\cot \phi \cdot 2\pi/3}}{1+\varepsilon^{-\cot \phi \cdot \pi}} = \frac{0.715}{1.152} = 0.621$$

Substituting into (11.55) gives

$$I_{AB}^2 = \frac{150^2}{\pi \times 15^2}\left[2.094 + 1.67\left(1.5 - 0.621 - 2.66 + \frac{1.77}{2} - \frac{0.386 \times 0.379}{2.66}\right)\right]$$

$$I_{AB} = 10\sqrt{\frac{1}{\pi}(2.094 - 1.67 \times 0.951)} = 4.014\,\text{A}$$

The total power dissipation is

$$P = 3I^2 R$$
$$= 3(4.014)^2 15 = 725\,\text{W}$$

The average value of the link current may be obtained by integrating (11.59) between the limits zero and $\pi/3$:

$$I_{\text{dc}} = \frac{3V}{\pi R}\left[\frac{2\pi}{3} + \frac{K_3}{\cot \phi}(\varepsilon^{-\cot \phi \cdot \pi/3} - 1)\right]$$

In this case, from (11.52),

$$K_3 = \frac{(1+\varepsilon^{-\cot \phi \cdot \pi/3})^2}{1+\varepsilon^{-\cot \phi \cdot \pi}} = \frac{(1.533)^2}{1.152} = 2.04$$

Therefore,

$$I_{\text{dc}} = \frac{3}{\pi}\frac{150}{15}\left[2.094 + \frac{2.04}{0.6}(0.533 - 1)\right]$$
$$= \frac{30}{\pi}(2.094 - 1.588) = 4.83\,\text{A}$$

The power entering the inverter through the link is

$P_{in} = VI_{dc} = 150 \times 4.83 = 725\,\text{W}$

which agrees with the value of the load power.

11.3 THREE-PHASE, STEP-WAVE VOLTAGE SOURCE INVERTERS WITH INDUCTION MOTOR LOAD

For the calculation of steady-state motor performance when supplied by a variable frequency, step-wave, voltage source inverter, the technique outlined in Section 9.2.3 may be used. The applied voltage is resolved into a time harmonic series and each individual harmonic is separately applied to its respective equivalent circuit, such as those of Fig. 9.10 and Fig. 9.11. If the system nonlinearities, such as magnetic saturation, are ignored the overall steady-state response is obtained by algebraic summation of the separate harmonic responses.

11.3.1 Motor currents

With a fixed terminal voltage the magnitude of its fundamental component is constant. The magnitude of the fundamental component of the associated current is determined largely by the load torque and the motor speed using, for example, the circuit of Fig. 9.10 with $n = 1$. Since the harmonic slip $S_n \simeq 1$ for all motor speeds the harmonic currents are virtually constant at all speeds as implied by the equivalent circuits of Fig. 9.1(b)(c). The primary current I_n for the nth higher harmonic of positive or negative sequence is

$$I_{n \neq 1} = \frac{V_{1_n}}{n(X_1 + X_2)} \quad (11.62)$$

If the load is delta-connected a path also exists in the delta loop for currents of zero sequence nature, in addition to the positive sequence and negative sequence components defined by (11.62). If the per-phase impedance to zero sequence currents is X_0 the zero sequence current is

$$I_n = \frac{V_{1_n}}{nX_0} \quad (11.63)$$

It is seen from (11.41) that the magnitude of the nth harmonic component of a six-step voltage wave is $1/n$ of the fundamental component:

$$V_{1_n} = \frac{V_1}{n} \tag{11.64}$$

Combining (11.62) and (11.64), denoting the fundamental primary voltage by the terminology V_1, gives

$$I_n = \frac{V_1}{n^2(X_1 + X_2)}, \quad \text{for } n > 1 \tag{11.65}$$

Each current harmonic in (11.65) will lag its corresponding voltage by 90°. There are no triplen harmonics (of order $n = 3, 9, 12, \ldots$) in the lines of a balanced three-wire load and the r.m.s. line current I is therefore

$$I = \sqrt{I_1^2 + I_5^2 + I_7^2 + I_{11}^2 + I_{13}^2 + \cdots}$$

or

$$I^2 = I_1^2 + \sum_{n=5}^{\infty} I_n^2$$

$$= I_1^2 + \frac{V_1^2}{(X_1 + X_2)^2} \sum_{n=5}^{\infty} \frac{1}{n^4}$$

$$= I_1^2 = \frac{V_1^2}{(X_1 + X_2)^2} \left(\frac{1}{5^4} + \frac{1}{7^4} + \frac{1}{11^4} + \frac{1}{13^4} + \frac{1}{17^4} + \cdots \right)$$

$$\therefore I_{6\text{-step}}^2 = I_1^2 + \frac{0.0022 V_1^2}{(X_1 + X_2)^2} \tag{11.66}$$

Since the fundamental current I_1 at full load is several times larger than the no-load value, the constant value of harmonic current $\sqrt{\Sigma I_n^2}$ is only a small proportion of the rated fundamental current. Typically, at full load with a six-step voltage waveform, $I = |1.02 \text{ to } 1.1|I_1$.

For a twelve-step waveform the harmonic distortion of the voltage wave is less than for a six-step waveform and the r.m.s. current flowing in the primary side of the equivalent circuit is, using (11.44),

$$I_{12\text{-step}}^2 = I_1^2 = \frac{0.000\,11 V_1^2}{(X_1 + X_2)^2} \tag{11.67}$$

A typical current waveform obtained when a six-step inverter is applied to a star-connected induction motor is shown in Fig. 11.18. In this case the angle of lag of the fundamental current component is about 73°, representing a power factor of approximately 0.3. The current wave contains a spike with a maximum value 1.75 times the peak fundamental component, which is the

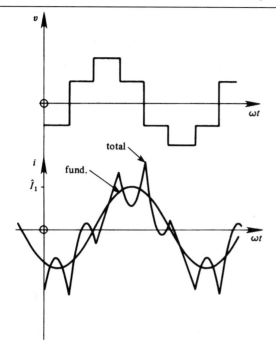

Fig. 11.18 Waveforms with six-step voltage source inverter applied to induction motor load.

order of result obtained with a typical motor reactance $(X_1 + X_2)$ of about 0.1 per unit.

11.3.2 Motor losses and efficiency

With a step-wave inverter drive both the applied voltage and the motor current are nonsinusoidal. The harmonic components of the voltage and current give rise to losses additional to the losses that occur with sinusoidal operation at the same torque and speed. In the primary and secondary windings the harmonic current components cause copper loss and heating. The motor secondary resistance R_2 is increased with harmonic frequency due to skin effect. For example, from (9.35), the fifth and seventh primary time harmonics cause a secondary harmonic current of order $n = 6$ at rated speed (low slip).

In a 50 Hz machine this secondary harmonic has a frequency of 300 Hz and at such frequency the resistance R_{2n} is two to three times its d.c. value R_2. Secondary circuit copper loss is often the most significant contribution to the total losses. For a small induction motor driven by a six-step inverter the total r.m.s. current is about 5% greater than the fundamental at full load.

The copper loss is therefore $(1.05)^2 I^2(R_1 + R_2) = 1.1$ times the sinusoidal value, if R_2 is presumed constant. If R_1 and R_2 are equal and R_2 is increased threefold, the total motor resistance is doubled and the copper loss is then probably greater than the sinusoidal (fundamental) value. For the equivalent circuit of Fig. 9.11(c), the total copper loss, for a three-phase motor, is

$$P_{loss} = 3(I_1^2 + \Sigma I_n^2)(R_1 + R_{2n}) \tag{11.68}$$

Core losses due to harmonic main fluxes are negligible but end leakage effects and skew leakage effects (cage motors only) may become significant at harmonic frequencies. With a six-step voltage the harmonic core loss is usually less than 10% of the fundamental value.

The losses in a sinusoidally driven motor are roughly in the proportions 40% due to copper losses, 40% due to core losses and 20% due to friction and windage. If the copper losses are increased by 25%, the core losses by 10% while the friction and windage is unchanged, the total losses are increased by 14%. For a motor with a 'sinusoidal' efficiency of (say) 85% when the output power is P_{out} the losses constitute, from (9.19), a proportion $(0.15/0.85)P_{out}$ or $0.176P_{out}$. Increase of the losses by 14% results in a loss of $1.14(0.176P_{out}) = 0.2P_{out}$. The resulting efficiency is now 83.33%. A reasonable estimate of the harmonic losses therefore suggests that the efficiency is reduced by about 2% for all load conditions. Opinions differ on the extent, if any, by which an induction motor should be derated when driven by a step-wave inverter. The general, rather conservative, practice is to derate by 10%. If a motor has a particularly low leakage reactance the incidence of harmonic currents is greater, the harmonic losses are also greater and the efficiency is correspondingly less.

In the presence of nonsinusoidal voltage and current the input power to the motor is

$$P_{in} = 3(V_1 I_1 \cos \psi_1 + V_5 I_5 \cos \psi_5 + V_7 I_7 \cos \psi_7 + \ldots) \tag{11.69}$$

Substituting (11.64) and (11.65) into (11.69) gives

$$P_{in} = 3V_1 \left(I_1 \cos \psi_1 + \frac{I_5}{5} \cos \psi_5 + \frac{I_7}{7} \cos \psi_7 + \ldots \right)$$

$$= 3V_1 \left\{ I_1 \cos \psi_1 + \frac{V_1}{(X_1 + X_2)} \left[\frac{\cos \psi_5}{5^3} + \frac{\cos \psi_7}{7^3} + \ldots \right] \right\}$$

$$= 3V_1 \left[I_1 \cos \psi_1 + \frac{0.014 V_1}{(X_1 + X_2)} \right]$$

$$\tag{11.70}$$

if $\cos\psi_5$, $\cos\psi_7$ etc. have their maximum possible values of unity.

11.3.3 Motor torque

A unidirectional harmonic torque is generated by the interaction between an air-gap flux and a secondary current component of the same harmonic frequency. It is found that such torque harmonics are usually negligibly small and can be ignored. For the nth time harmonic current the synchronous speed is n times the fundamental value:

$$N_n = nN_1 = \frac{n\omega_1}{p} \tag{11.71}$$

where N_1, ω_1 and p are defined in equations (9.1). The torque for the fundamental component of the motor current is given by (9.20). From the equivalent circuit of Fig. 9.11(c) the nth harmonic torque is

$$T_n = \frac{3}{nN_1} I_n^2 \frac{R_{2n}}{S_n} \tag{11.72}$$

Eliminating S_n between (11.72) and (9.33) gives

$$T_n = \frac{3I_n^2 R_{2n}}{nN_1 \mp N} \tag{11.73}$$

For operation near to rated synchronous speed the fundamental slip approaches zero and $N = N_1$ (very nearly). The unidirectional harmonic torque is then

$$T_{n|S \to 0} = \frac{3I_n^2 R_{2n}}{N_1(n \mp 1)} \tag{11.74}$$

For forward rotating fields of order 1, 7, 13, etc. the negative sign applies in (11.74) and the torque is positive. With reverse rotating fields of order 5, 11, 17, etc. the positive sign applies but the torque is negative.

If the harmonic reactance $n(X_1 + X_2)$ in Fig. 9.11(c) is much greater than the resistance $(R_1 + R_{2n}/S_n)$, which is usually true, the relationship of (11.65) is valid. Incorporating this into the harmonic torque expressions gives

$$T_n = \frac{3V_1^2 R_{2n}}{n^4(X_1 + X_2)^2(nN_1 \mp N)} \tag{11.75}$$

and

$$T_{n|S \to 0} = \frac{3V_1^2 R_{2n}}{n^4(X_1 + X_2)^2 N_1(n \mp 1)} \tag{11.76}$$

In addition to the unidirectional harmonic torques described above a pulsating torque is sometimes developed for motors with a six-step inverter drive due to interaction between the fundamental flux and the secondary harmonic currents. As noted in Section 11.3.2 above, the fifth and seventh time harmonic primary currents result in a sixth harmonic secondary current. This, in turn, results in a torque ripple of six times the fundamental frequency superimposed on the steady-state unidirectional torque. With low inertia motors the torque ripple, which may be as much as 10% of rated torque, may be reproduced as a speed oscillation about the mean value.

11.3.4 Worked examples

Example 11.6
A three-phase, four-pole, 18 kW, 300 V, star-connected induction motor is driven at 50 Hz by a six-step voltage source inverter supplied from a d.c. supply of 200 V. The motor equivalent circuit parameters for 50 Hz operation are

$R_1 = 0.1\,\Omega$, $R_2 = 0.17\,\Omega$, $X_1 = 0.3\,\Omega$, $X_2 = 0.5\,\Omega$, $X_m =$ large.

Calculate the r.m.s. current and the harmonic copper losses when this motor operates at 1450 r.p.m., 50 Hz. Estimate the motor efficiency compared with sinusoidal operation.

Solution. At 1450 r.p.m. the per-unit slip is

$$S = \frac{1500 - 1450}{1500} = \frac{50}{1500} = 0.033$$

The input impedance to the circuit of Fig. 9.10 with $n = 1$, $R_0 =$ large and $X_m =$ large is

$$Z = R_1 + \frac{R_2}{S} + j(X_1 + X_2)$$
$$= 0.1 + \frac{0.17}{0.033} + j0.8$$
$$= 5.2 + j0.8 = 5.26 \angle 8.75°\,/\text{phase}$$

If the inverter is operated in the mode of three simultaneously conducting switches the phase voltage has the six-step waveform of Fig. 11.10 in which the maximum voltage is, for an inverter of input $0 \rightarrow 200\,\text{V}$,

$$V_{\max} = \frac{4}{3} V_{dc} = \frac{800}{3} = 266.6\,\text{V}$$

This has a fundamental component of peak value, given by (11.35)

11.3 Step-wave inverter – motor load

$$V_{1_{max}} = \frac{4}{\pi} V_{dc} = \frac{4}{\pi} 200 = 254.6 \text{ V}$$

The r.m.s. fundamental phase voltage is therefore

$$V_1 = \frac{V_{1_{max}}}{\sqrt{2}} = \frac{254.6}{\sqrt{2}} = 180 \text{ V/phase}$$

The corresponding fundamental current is therefore

$$I_1 = \frac{V_1}{Z_1} = \frac{180}{5.26} = 34.22 \text{ A}$$

From (11.66) the total r.m.s. current I is

$$I^2 = I_1^2 + \frac{0.0022 V_1^2}{(X_1 + X_2)^2}$$
$$= 1171 + 111.38$$
$$\therefore I = \sqrt{1282.4} = 35.81 \text{ A}$$

The total higher harmonic current is therefore

$$\sqrt{\sum I_n^2} = \sqrt{I^2 - I_1^2} = \sqrt{1282.4 - 1171} = 10.55 \text{ A}$$

The total harmonic copper loss, assuming constant R_2, is

$$P_{loss} = 3(\Sigma I_n^2)(R_1 + R_2)$$
$$= 3(10.55)^2 0.27 = 90.16 \text{ W}$$

The total motor copper loss, neglecting magnetising current, is

$$P_{copper\ loss} = 3(35.81)^2 0.27 = 1038.7 \text{ W}$$

If this copper loss represents (say) 40% of the total motor losses these then amount to

$$P_{total\ loss} = \frac{1039}{0.4} = 2598 \text{ W}$$

Now the input power to the motor is given by (11.69).

The motor input power due to the fundamental component of the current is

$$P_{in_1} = 3 \times 180 \times 34.22 \cos 8.75° = 18\,264 \text{ W}$$

Using the approximate relationship of (11.70) gives, for the total input power,

$$P_{in} = P_{in_1} + 3 \frac{0.014 V_1^2}{(X_1 + X_2)}$$
$$= 18\,264 + 1701 = 19\,965 \text{ W}$$

The estimated motor efficiency is therefore, from (9.19),

$$\eta = \frac{P_{in} - P_{total\ losses}}{P_{in}}$$
$$= \frac{19\,965 - 2598}{19\,965} = 86.9\%$$

For sinusoidal operation, with only the fundamental components of voltage and current present, the total estimated loss is

$$P_{total\ loss} = \frac{3(34.22)^2 0.27}{0.4} = 2371\ W$$

Taking only the input power due to the fundamental current the efficiency is then

$$\eta = \frac{18\,264 - 2371}{18\,264} = 87\%$$

By this estimation method it is seen that there is no significant change of efficiency due to the use of a six-step voltage waveform.

Example 11.7

The motor of Example 11.6 is now operated at 725 r.p.m., 25 Hz. Calculate the r.m.s. current and efficiency at the new condition.

Solution. Operation at 725 r.p.m. from a 25 Hz supply (which implies a synchronous speed of 750 r.p.m.) means that the per-unit slip is unchanged:

$$S = \frac{750 - 725}{750} = 0.033$$

But a slip-speed of 25 r.p.m., compared with 50 r.p.m. at 50 Hz operation, implies that the torque at 725 r.p.m. is one-half the value for 50 Hz operation, as shown in Fig. 11.19. At 725 r.p.m., 25 Hz the d.c. applied voltage can be presumed to have been reduced from 200 V to 100 V d.c. and the r.m.s. motor phase voltage is then

$$V_1 = \frac{180}{2} = 90\ V/phase$$

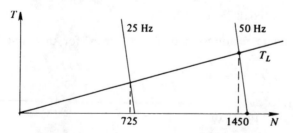

Fig. 11.19 Induction motor torque–speed characteristics in Example 11.7.

11.3 Step-wave inverter – motor load

At 25 Hz the motor input impedance is

$$Z = 5.2 + j0.4 = 5.22 \angle 4.4° \, \Omega/\text{phase}$$

The fundamental current is now

$$I_1 = \frac{V_1}{Z_1} = \frac{90}{5.22} = 17.24 \, \text{A}$$

From (11.66) the total r.m.s. current becomes

$$I^2 = 17.24^2 + 0.0022 \left(\frac{90}{0.4}\right)^2$$

$$= 297.2 + 111.4$$

$$\therefore I = 20.2 \, \text{A}$$

Once again assuming secondary resistance R_2 to be unchanged the copper loss is

$$P_{\text{copper loss}} = 3(20.2)^2 0.27 = 330.5 \, \text{W}$$

If this copper loss is still 40% of the total loss, then

$$P_{\text{total loss}} = \frac{330.5}{0.4} = 826.3 \, \text{W}$$

The input power associated with the fundamental (now 25 Hz) component is

$$P_{\text{in}_1} = 3 \times 90 \times 17.24 \cos 4.4° = 4641 \, \text{W}$$

Using (11.70) gives an approximate value for the total input power,

$$P_{\text{in}} = P_{\text{in}_1} + \frac{3 \times 0.014 V_1^2}{(X_1 + X_2)}$$

$$= 4641 + \frac{3 \times 0.014 \times 90^2}{0.4}$$

$$= 4641 + 850.5 = 5491.6 \, \text{W}$$

The estimated motor efficiency with inverter drive is

$$\eta = \frac{5492 - 826.3}{5492} = 84.9\%$$

Considering only the fundamental component results in a motor loss

$$P_{\text{total loss}} = \frac{3(17.24)^2 0.27}{0.4} = 602 \, \text{W}$$

The efficiency is then

$$\eta = \frac{5438 - 602}{5438} = 88.9\%$$

For one-half speed operation with one-half of rated torque the use of the inverter causes a reduction of efficiency of about 4%.

Table 11.3 *Current higher harmonics in Example 11.8.*

n	f(Hz)	$n(X_1+X_2)$	$\dfrac{V_1}{n}$	$\dfrac{V_1}{n^2(X_1+X_2)}$
1	50	0.8	180	not relevant
5	250	4	36	9
7	350	5.6	25.71	4.59
11	550	8.8	16.36	1.86
13	650	10.4	13.85	1.33

Example 11.8

For the induction motor drive of Example 11.6 deduce the waveform of the primary current if the motor is operating at 1450 r.p.m. and 50 Hz.

Solution. For all relevant harmonic frequencies the harmonic impedance $n(X_1+X_2)$ is presumed to be much greater than the resistance $R_1 + R_{2n}/S_n$ so that the equivalent circuit of Fig. 9.11(c) may be used. This will introduce a slight error in the phase displacement of the calculated harmonic currents. The equivalent circuit of Fig. 9.1 is used, for the condition $R_w = X_m =$ very large, to give the fundamental current component. This is seen from Example 11.6 to be

$$i_1(\omega t) = 34.22\sqrt{2}\sin(\omega t - 9°)$$

The value $V_1/(X_1+X_2)$, which is the blocked rotor current of the motor, has a value $180/0.8 = 225$ A at 50 Hz. The current higher harmonics, obtained from (11.65), are shown in Table 11.3.

The resultant current is then defined by a series:

$$\begin{aligned}i(\omega t) &= \sqrt{2}[34.22\sin(\omega t - 9°) + 9\sin(3\omega t - 90°) \\ &\quad + 4.59\sin(5\omega t - 90°) + 1.86\sin(11\omega t - 90°) + \ldots] \\ &= \sqrt{2}[34.22\sin(\omega t - 9°) + 9\sin 3(\omega t - 30°) \\ &\quad + 4.59\sin 5(\omega t - 18°) + 1.86\sin 11(\omega t - 8.2°) + \ldots]\end{aligned}$$

The waveform $i(\omega t)$, considering terms up to the 13th harmonic, is given in Fig. 11.20. The sharp inflections and spikes obtained in practice, as for example in Fig. 11.18, are not reproduced in the linear analysis. Although the harmonic analysis of Section 11.3 gives quite accurate results for the levels of current, power, torque and efficiency the only effective method of accurately determining the current waveform is to measure it experimentally.

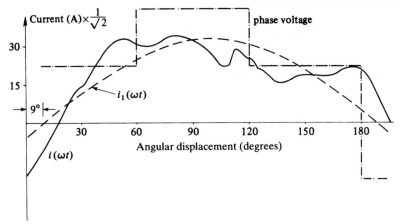

Fig. 11.20 Induction motor current waveform in Example 11.8.

Example 11.9
For the inverter-induction motor drive of Example 11.6 calculate the harmonic torques due to the 5th and 7th harmonic currents. Show that, for operation at 1450 r.p.m., 50 Hz, the harmonic torques are negligible.

Solution. Assuming that the total estimated motor loss of 2598 W, in Example 11.6, is accurate the output power is then

$P_{out} = P_{in} - P_{total\ loss} = 19\,965 - 2598 = 17\,367\,W$

In SI units the motor speed is

$$N = 1450 \text{ r.p.m.} = \frac{1450}{60} \times 2\pi = 151.8 \text{ rad/s}$$

The output torque, including core loss and friction and windage effects, is therefore

$$T = \frac{P_{out}}{N} = \frac{17\,367}{151.8} = 114.4\,\text{N m}$$

The alternative method of calculating the fundamental torque, using (9.20), neglects the effects of core loss and friction and windage and tends to give a very optimistic result. If the value of R_2 is unchanged the 5th harmonic torque, at 1450 r.p.m., from (11.75), is

$$T_5 = \frac{3(180)^2 0.17}{5^4(0.8)^2 \left(8950 \frac{2\pi}{60}\right)} = 0.044\,\text{N m}$$

Even if $R_{2n} = 2R_2$ or $3R_2$ the 5th harmonic torque is negligible.
Similarly, the 7th harmonic torque is

$$T_7 = \frac{3(180)^2 0.17}{7^4 (0.8)^2 \left(9050 \frac{2\pi}{60}\right)} = 0.011 \,\text{N m}$$

The resultant torque at 1450 r.p.m. due to the fundamental current plus the 5th and 7th harmonic is

$$T = T_1 - T_5 + T_7$$
$$= 114.4 - 0.044 + 0.011 = 114.37 \,\text{N m}$$

It is seen that the net harmonic torque of 0.033 N m is 0.03% of the fundamental value and can be ignored.

11.4 PROBLEMS

Induction motor with adjustable frequency, sinusoidal voltages

11.1 An adjustable voltage, adjustable frequency inverter is to be used for induction motor speed control. Use the motor equations to deduce the type of torque–speed characteristics commonly used with this mode of control. How is high efficiency maintained at low speeds?

What relationship between voltage and frequency is pertinent to low-speed operation?

11.2 A three-phase squirrel-cage induction motor is to be driven from an adjustable voltage, adjustable frequency sine-wave inverter. Use the motor equations to discuss variable frequency control of motor speed and explain how good speed regulation and high efficiency are maintained at low speeds.

Sketch a set of motor torque–speed characteristics for (a) supply frequency f_1, (b) $f_2 = \frac{1}{2} f_1$, and (c) $f_3 = \frac{1}{3} f_1$, showing intersections with a load of constant torque. If the motor reactance/primary resistance ratio $X/R_1 = 5$, calculate the effect on the peak torque of operating at frequency f_3 compared with operation at frequency f_1.

11.3 Discuss and explain the disadvantages of low-speed operation of a three-phase induction motor driven by sinusoidal voltages of supply frequency. Compare the torque–speed characteristics of a rotor resistance controlled motor with those obtained by the use of an adjustable voltage, adjustable frequency inverter with sinusoidal output voltage waveform. What are the advantages and disadvantages of wide-range control by means of an adjustable frequency, square-wave inverter?

11.4 A three-phase induction motor is driven from an adjustable voltage, adjustable frequency supply. Show that if the air-gap flux is kept constant then the value of the primary current is independent of the supply frequency.

11.5 A three-phase induction motor operates from a sinusoidal voltage source of frequency f_1 and phase voltage V_1. The per-phase equivalent circuit parameters R_1, R_2, X_1, X_2 and X_m are defined in accordance with Fig. 9.4. Show that the primary phase current is given by

$$I_1 = \left(\frac{V_1}{f_1}\right) \frac{f_1}{X_m}$$

$$\times \left[\frac{X_m f_2(f_1 R_2 + f_2 R_1) - j\{(f_1 R_2 + f_2 R_1)^2 + f_2^2(X_1 + X_2)(X_1 + X_2 + X_m)\}}{(f_1 R_2 + f_2 R_1)^2 + f_2^2(X_1 + X_2)^2}\right]$$

11.6 A three-phase induction motor has the equivalent circuit parameters shown in Example 11.2 and is required to deliver rated torque at one-fifth rated speed. Calculate the internal e.m.f. E_1 at this speed and use (11.11) to obtain the primary current. Calculate the power factor in terms of voltage, current and power and check this with the value obtained by the circuit analysis method. Sketch a phasor diagram of the voltages and currents.

11.7 For the induction motor drive described in Example 11.2, calculate the power factor and efficiency for operation at one-half of rated speed.

11.8 The three-phase induction motor of Example 11.2 is to be used to drive a pump load with a characteristic $T_L = KN^2$, where N is the shaft speed. The motor is required to deliver its rated torque at the rated speed of 1450 r.p.m. An adjustable frequency, adjustable voltage sine-wave inverter is used to drive the motor, keeping the air-gap flux constant. Calculate the motor currents, power, power factor and efficiency at one-half speed compared with rated speed operation.

Stepped-voltage waveforms

11.9 Sketch the circuit diagram of a three-phase, force-commutated inverter incorporating six SCRs and six diodes. The commutation system should not be shown. Two SCRs only conduct at any instant and power is transferred from the d.c. source, voltage $\pm V$, into a balanced three-phase resistive load.

Explain the sequence of SCR firing over a complete cycle and sketch a resulting per-phase load voltage waveform consistent with your firing pattern.

11.10 Sketch the skeleton circuit of the basic six-switch, force-commutated inverter with direct supply voltage $\pm V$. The switching mode to be used is that where three switches conduct simultaneously at every instant of the cycle. Deduce and sketch consistent waveforms of the output phase voltages v_{AN}, v_{BN}, v_{CN} (assuming phase sequence ABC) and the line voltage v_{AB} on open circuit over a complete time cycle, indicating which switches are conducting through each 60° interval. What is the phase difference between the fundamental component v_{AB_1} of the line voltage v_{AB} and the fundamental component v_{AN_1} of the phase voltage v_{AN}? In what ways would a phasor diagram

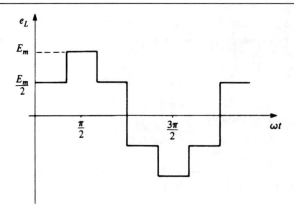

Fig. 11.21 Motor phase voltage waveform in Problem 11.13.

of the fundamental, open-circuit phase voltages give a misleading impression of the actual operation?

11.11 The basic circuit of a six-switch, force-commutated inverter with supply voltage $\pm V$ is shown in Fig. 11.5. The triggering mode to be used is where three switches conduct simultaneously. Deduce and sketch waveforms of the instantaneous phase voltages v_{AN}, v_{BN}, v_{CN}, and the instantaneous line voltage v_{AB} for open-circuit operation with phase sequence ABC. Indicate which of the six switches are conducting during each 60° interval of the cyclic period. If equal resistors R are connected to terminals A, B, C as a star-connected load, deduce and sketch the waveform of phase current i_{AN}.

11.12 In the inverter circuit of Fig. 11.5 the triggering mode to be used is where three switches conduct simultaneously. The load consists of three identical resistors R connected in wye (star).
  (a) If the load neutral point N is electrically isolated from the supply neutral point O, deduce the magnitude, frequency and waveform of the neutral–neutral voltage V_{NO}.
  (b) If the two neutral points N and O are joined, deduce the magnitude, frequency and waveform of the neutral current.

11.13 The stepped waveform of Fig. 11.21 is typical of the phase voltage waveform of a certain type of inverter. Use Fourier analysis to calculate the magnitude and phase-angle of the fundamental component of this waveform. Sketch in correct proportion, the waveform and its fundamental component. What is the half-wave average value of the stepped-wave compared with the half-wave average value of its fundamental component?

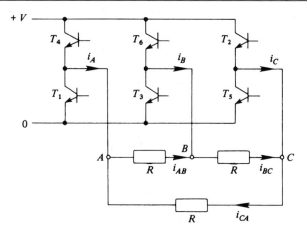

Fig. 11.22 Inverter connection in Problem 11.15.

11.14 A set of no-load, phase voltage waveforms v_{AN}, v_{BN}, v_{CN} produced by a certain type of inverter is given in Fig. 11.8. Sketch, on squared paper, the corresponding no-load line voltages v_{AB}, v_{BC}, v_{CA}. Calculate the magnitude and phase-angle of the fundamental component v_{AB_1} of the line voltage v_{AB} and sketch v_{AB_1} in correct proportion to v_{AB}. What is the half-wave average value of v_{AB} compared with the corresponding average value of v_{AB_1}?

The set of voltages in Fig. 11.8 is applied to a set of equal star-connected resistors of resistance R. Deduce and sketch the waveform of the current in phase A with respect to the open-circuit voltage v_{AN}.

11.15 An ideal d.c. supply of constant voltage V supplies power to a three-phase, force-commutated inverter consisting of six ideal transistors. Power is then transferred to a delta-connected resistive load of R ohms per branch, Fig. 11.22. The mode of inverter switching is such that three transistors are conducting simultaneously at every instant of the cycle. Show that the line current waveforms are of six-step form with a peak height of $2V/R$. Further show that the phase (branch) currents are square-waves of height V/R.

11.16 For the periodic voltage waveform of Fig. 11.23 calculate the fundamental component, the total r.m.s. value, the distortion factor and the displacement factor.

11.17 For the 12-step waveform of Fig. 11.11 show that the step height for the interval $\pi/6 < \omega t < \pi/3$ is given by $0.732V$. Also show that the fundamental component of this waveform has a peak height of $(\pi/3)V$ and a displacement angle $\psi_1 = 0$.

11.18 For the 12-step voltage waveform of Fig. 11.11 calculate the r.m.s. value and hence the distortion factor.

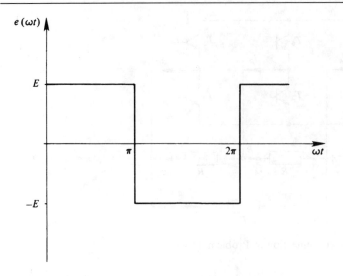

Fig. 11.23 Voltage waveform of Problem 11.16.

11.19 A six-step voltage source inverter is supplied from an ideal battery with terminal voltage $V = 200$ V. The inverter supplies a delta-connected load with a series R–L impedance in each leg consisting of $R = 20\,\Omega$, $X_L = 30\,\Omega$ at the generated frequency. Calculate the r.m.s. load current and the average value of the supply current. Check that, within calculation error, the input power is equal to the load power.

11.20 Repeat Problem 11.19 if the load inductance is removed.

11.21 For the inverter operation of Problem 11.19 calculate the maximum and minimum values of the time-varying link current.

12

Induction motor speed control by the use of adjustable frequency PWM inverters

A requirement for a.c. motor speed control by the use of a variable frequency supply is that the applied voltage or current waveforms contain the minimum possible distortion. The best solution would be an inverter that generated three sinusoidal waveforms of symmetrical phasor form. Such a device would be elaborate and expensive since it would require a large number of switching elements.

The ideal requirement of sinusoidal motor voltages can be closely approximated by the synthesis of voltage waveforms using a technique known as pulse-width modulation (PWM). In this the fundamental component can be controlled in both magnitude and frequency. The harmonic content can be made low and the harmonic order higher than those obtained with six-step or quasi-square-wave inverters. The PWM inverter is now gradually taking over the inverter market in motor control applications.

PWM techniques are characterised by the generation of constant amplitude pulses in which the pulse duration is modulated to obtain some specific waveform. Several different modulation methods may be used. Some of these are discussed in Section 12.1 below.

12.1 PROPERTIES OF PULSE-WIDTH MODULATED WAVEFORMS

12.1.1 Single-pulse modulation

In Fig. 11.6 the phase voltages consist of a fixed duration single pulse in each half-wave. More flexible forms of control would permit variation of this single pulse by (i) fixing the leading edge but varying the trailing edge, (ii) fixing the trailing edge but varying the leading edge or (iii) varying the pulse-width while keeping the pulses symmetrical about $\pi/2$, $3\pi/2$, ... Fig. 12.1(*a*)

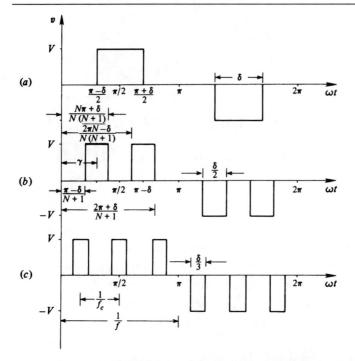

Fig. 12.1 PWM voltage waveforms: (*a*) single-pulse modulation, $N = 1$, (*b*) two-pulse modulation, $N = 2$, (*c*) three-pulse modulation, $N = 3$.

shows a single-pulse waveform of pulse-width δ symmetrical about $\pi/2$ and $3\pi/2$. This waveform has the Fourier series

$$v(\omega t) = \frac{4V}{\pi}\left(\sin\frac{\delta}{2}\sin\omega t - \frac{1}{3}\sin\frac{3\delta}{2}\sin 3\omega t + \frac{1}{5}\sin\frac{5\delta}{2}\sin 5\omega t \ldots\right) \quad (12.1)$$

Pulse width δ has a maximum value of π radians at which the fundamental term in (12.1) is a maximum. An individual harmonic of order n may be eliminated by making $\delta = 2\pi/n$ but this is likely also to reduce the value of the fundamental component. The r.m.s. value of the single-pulse waveform of Fig. 12.1(*a*) is found to be

$$V_{\text{rms}} = V\sqrt{\frac{\delta}{\pi}} \quad (12.2)$$

The *n*th-order harmonic in (12.1) is seen to have a peak value

$$V_n = \frac{4V}{n\pi}\sin\frac{n\delta}{2} \quad (12.3)$$

The distortion factor of the single-pulse waveform is therefore

$$\text{distortion factor} = \frac{\frac{V_1}{\sqrt{2}}}{V_{\text{rms}}} = \frac{2\sqrt{2}}{\sqrt{\pi\delta}}\sin\frac{\delta}{2} \tag{12.4}$$

which has a maximum value of 0.9 when $\delta = \pi$. This is consistent with the data of Table 11.2.

12.1.2 Multiple-pulse modulation

Alternative waveforms containing either two or three symmetrically spaced pulses per half-cycle, of the same periodic frequency, are also shown in Fig. 12.1. For example, Fig. 12.1(b) shows the case of two pulses per half-cycle. When this contains the same area under the curve as that of Fig. 12.1(a) then, since the pulse heights are also equal, the two waveforms have the same r.m.s. value (i.e. the same total harmonic content). But the nth harmonic component of the two-pulse waveform has the amplitude

$$V_n = \frac{8V}{n\pi}\sin n\gamma \sin\frac{n\delta}{4} \tag{12.5}$$

Magnitude V_n in (12.5) obviously depends on both γ and δ. An individual harmonic can be eliminated by making $\gamma = \pi/n$ or $\delta = 4\pi/n$.

The use of two pulses, in Fig. 12.1(b), causes a reduction of the fundamental component compared with equivalent single-pulse operation. Increase in the number of pulses reduces the proportion of higher harmonics to fundamental component, and hence reduces the value of the total harmonic distortion (*THD*).

Let the pulse or carrier frequency be f_c and the overall cycle or modulated frequency be f, where $f_c > f$, as illustrated in Fig. 12.1(c). If the number of equal symmetrical pulses per half-cycle is N then

$$N = \frac{f_c}{2f} = \frac{\omega_c}{2\omega} = \text{integer} \tag{12.6}$$

In terms of integer N the displacement γ in Fig. 12.1(b) is given by

$$\gamma = \frac{2N\pi - \delta}{2N(N+1)} \tag{12.7}$$

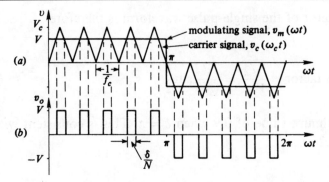

Fig. 12.2 Multiple-pulse voltage waveform obtained from a triangular carrier wave with square-wave modulation.

The magnitude of the fundamental component of the pulse (modulated) voltage wave varies with the duration of the conduction period δ. Control of this fundamental component amplitude may be realised by maintaining constant pulse width but varying the pulse number N. Alternatively, the number of pulses N may be kept constant while the pulse width δ/N is varied.

Waveforms such as those of Fig. 12.1 can be interpreted as modulated voltages created by modulating a triangular carrier wave $v_c(\omega_c t)$, of pulse frequency f_c, by means of an adjustable direct voltage square-wave that constitutes a modulating signal $v_m(\omega t)$, Fig. 12.2. The pulse height V of the resulting modulated signal $v_o(\omega t)$ can be adjusted within the range $0 < V < V_c$ and the pulse width δ/N varied in the range $0 < \delta/N < \pi/N$. It is seen in Fig. 12.2 that the width of the equal pulses is related to the signal voltages by a relation

$$\frac{\delta}{N} = \left(1 - \frac{V}{V_c}\right)\frac{\pi}{N} \tag{12.8}$$

The total r.m.s. value of $v_o(\omega t)$ is still given by (12.2). It is found that the harmonic content of $v_o(\omega t)$ is lower than that of a six-step waveform if $N \geq 0$.

It is a characteristic of all modulated waves that the fundamental frequency component of the output (modulated) wave is equal to the frequency of the modulating wave. Frequency variation of the output signal is therefore obtained by frequency adjustment of the modulating signal. In the early days of solid-state inverter design the technique of square-wave modulating signals, as in Fig. 12.2, was often used because it was easy to implement in terms of analogue electronics. Modern design techniques involving digital electronics and microprocessors now permit the use of more suitable modulating waveforms.

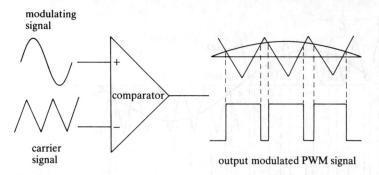

Fig. 12.3 Principle of sinusoidal modulation of a triangular carrier wave.

12.1.3 Sinusoidal modulation

A periodic (carrier) waveform of any waveshape can be modulated by another periodic (modulating) waveform of any other waveshape, of lower frequency. For most waveshape combinations of a carrier waveform modulated by a modulating waveform, however, the resultant modulated waveform would not be suitable for either power applications or for information transmission.

For induction motor speed control the motor voltage waveforms should be as nearly sinusoidal as possible. If nonsinusoidal voltages are used, as with most inverter drives, it is preferable to use waveforms that do not contain- low-order harmonic such as the fifth and the seventh because these can cause ripple disturbances, especially at low speeds. The lower order harmonics of a modulated voltage wave can be greatly reduced if a sinusoidal modulating signal modulates a triangular carrier wave. The pulse widths then cease to be uniform as in Fig. 12.1, but become sinusoidal functions of the angular pulse position, as in Fig. 12.3.

12.1.3.1 Sinusoidal modulation with natural sampling

The principle of PWM is further illustrated in Fig. 12.4. A sinusoidal modulating signal $v_m(\omega t) = V_m \sin \omega t$ is applied to a single-sided triangular carrier signal $v_c(\omega t)$ of maximum height V_c. The natural intersections of $v_m(\omega t)$ and $V_c(\omega t)$ determine both the onset and duration of the modulated pulses so that the pulse pattern is described as being due to natural sampling. The circuitry actuating the turn-on and turn-off of the inverter switches also is controlled by sensing these intersections. In Fig. 12.4 the pulse height V of the pulse-width modulated output signal $v_o(\omega t)$ is determined by the direct voltage on the supply side of the inverter (not by the switching pattern).

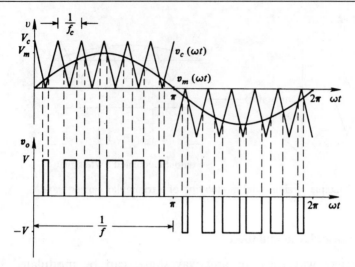

Fig. 12.4 PWM voltage waveform obtained by sinusoidal modulation using natural sampling, $p = 12$, $M = 0.75$.

In PWM waveforms the pulse pattern is dependent on the ratio of the peak modulating voltage V_m to the peak carrier voltage V_c, often called the modulation index or modulation ratio, M.

$$M = \frac{V_m}{V_c} = \text{modulation ratio} \tag{12.9}$$

In Fig. 12.4, for example, the value of M is seen to be about 0.75. Modulation ratio M is usually varied by varying V_m in the presence of fixed carrier wave amplitude V_c, when M is in its usual range $0 \leq M \leq 1$.

A further basic property of PWM waveforms is the ratio between the frequencies of the carrier and modulating waveforms.

$$p = \frac{\text{frequency of carrier wave}}{\text{frequency of modulating wave}} \tag{12.10}$$
$$= \text{carrier ratio}$$
$$= \text{frequency ratio}$$

When p is an integer, as in Fig. 12.4, this is defined as a case of *synchronous modulation*. If p is an odd integer then the modulated waveform contains half-wave symmetry (i.e. the positive and negative half-cycles are symmetrical) and there are no even-order harmonics.

With a large value of p the dominant harmonics of the PWM waveform are high and clustered around the carrier frequency and its sidebands. The filter

Table 12.1 *Major harmonics in a PWM waveform with naturally sampled sinusoidal modulation.*

Frequency ratio p	$n = 1$ $m = 2$ $p \pm 2$	$n = 2$ $m = 1$ $2p \pm 1$	$n = 3$ $m = 2$ $3p \pm 2$
3	5, 1	7, 5	11, 7
9	11, 7	19, 17	29, 25
15	17, 13	31, 29	47, 43
21	23, 19	43, 41	65, 61

action of a motor series inductance is then high and torque pulsations are eliminated, resulting in smooth rotation.

For three-phase operation the triangular carrier wave is usually symmetrical, without d.c. offset. Each half-wave of the carrier is then an identical isosceles triangle. Waveforms for a three-phase inverter are shown in Fig. 12.5, in which frequency ratio $p = 9$ and modulation ratio M is almost unity. For balanced three-phase operation p should be an odd multiple of 3. The carrier frequency is then a triplen of the modulating frequency so that the output modulated waveform does not contain the carrier frequency or its harmonics.

The Fourier analysis of a sinusoidally pulse-width modulated waveform such as those of Fig. 12.5 is very complex and involves Bessel Functions. In general, the harmonics k of the modulated waveform are given by

$$k = np \pm m \tag{12.11}$$

where n is the carrier harmonic order and m is the carrier side-band. The major harmonic orders are shown in Table 12.1 for several values of p. At $p = 15$, for example, the lowest significant harmonic is $k = p - 2 = 13$ and this is of much higher order than the harmonics $k = 5, 7$, obtained with a 6-step waveform. It is found that the $2p \pm 1$ harmonics are dominant in magnitude for values of modulation ratio up to about $M = 0.9$. When $p > 9$ the harmonic magnitudes at a given value of M are independent of p.

The basic skeleton inverter circuit is shown in Fig. 12.6, using the general terminology S for the inverter switches. A pole, or point O, is identified as the centre point of the constant supply voltage V_{dc}. In Fig. 12.5 the PWM pole voltages are measured from the phase terminals A, B, C to point O. By Kirchhoff's Loop Law the line voltages are given by

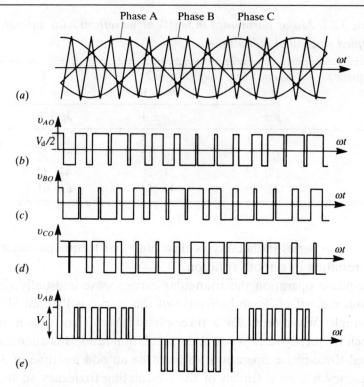

Fig. 12.5 Voltage waveforms for a three-phase sinusoidal PWM inverter, $p = 9$, $M = 0.9$ (ref. 31): (a) comparator voltages, (b), (c), (d) pole voltages, (e) a.c. line voltage, $v_{AB} = v_{AO} - v_{BO}$.

$$v_{AB} = v_{AO} - v_{BO}$$
$$v_{BC} = v_{BO} - v_{CO} \quad (12.12)$$
$$v_{CA} = v_{CO} - v_{AO}$$

For a three-wire, star-connected load the phase currents and voltages, to load neutral, must sum to zero

$$i_{AN} = i_{BN} + i_{CN} = 0 \quad (12.13)$$

$$v_{AN} + v_{BN} + v_{CN} = 0 \quad (12.14)$$

A potential difference v_{NN_s} exists between the load neutral point N and the supply neutral rail N_s given by

$$V_{NN_s} = V_{AN_s} - V_{AN} = V_{BN_s} - V_{BN} = V_{CN_s} - V_{CN} \quad (12.15)$$

Combining (12.14), (12.15) gives

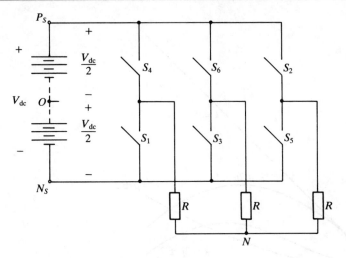

Fig. 12.6 Basic skeleton inverter circuit.

$$V_{NN_s} = \frac{1}{3}(V_{AN_s} + V_{BN_s} + V_{CN_s}) \quad (12.16)$$

Similarly, the instantaneous voltage between the load neutral point N and the supply centre tap O is given by

$$v_{NO} = \frac{1}{3}(v_{AO} + v_{BO} + v_{CO}) \quad (12.17)$$

For a balanced, three-phase, star-connected load the peak value of the fundamental component of the modulated line to neutral voltage V_1 (peak) is proportional to M in the range $0 \leq M \leq 1$, for all values of $p > 9$.

$$V_{1_{(peak)}} = \frac{MV_{dc}}{2} \quad (12.18)$$

The corresponding r.m.s. value of the fundamental component of the modulated line-to-line voltage V_{L_1} (r.m.s.) is given by

$$V_{L_{1(r.m.s.)}} = \frac{\sqrt{3}V_{1_{(peak)}}}{\sqrt{2}}$$

$(0 \leq M \leq 1)$ \quad (12.19)

$$= \frac{\sqrt{3}}{2\sqrt{2}} MV_{dc} = 0.612 MV_{dc}$$

There is no straightforward analytical expression for the related values of the higher harmonic voltage components. Calculated values of these, in the range $0 \leq M \leq 1$ with $p > 9$, are given in Fig. 12.7, quoted from reference 31. If the

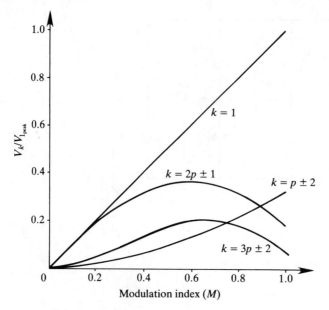

Fig. 12.7 Harmonic component voltage (relative to peak fundamental value) for sinusoidal PWM with natural sampling ($p > 9$)(ref. 31).

modulating wave amplitude in PWM is varied linearly with frequency then the ratio M/f is constant. This represents the desirable condition of a waveform having a constant ratio of fundamental voltage to frequency at the load terminals.

12.1.3.2 Overmodulation in sinusoidal PWM inverters

Increase of the fundamental component of the modulated output voltage V_1, beyond the $M = 1$ value, is possible by making $M > 1$ but V_1 is then no longer proportional to M, Fig. 12.8. In this condition of overmodulation the process of natural sampling no longer occurs. Some intersections between the carrier wave and the modulating wave are lost, as illustrated in Fig. 12.9. The result is that some of the pulses of the original PWM wave are dropped in the manner shown in Fig. 12.10.

In the extreme, when M reaches the value $M = 3.24$, the original forms of PWM waveform in Fig. 12.5 are lost. The phase voltages then revert to the quasi-square waveshape of Table 11.2 or Fig. 11.6 in which harmonics of order 5 and 7 reappear. Variation of the fundamental output voltage versus modulation ratio M is shown in Fig. 12.8. For a pulse voltage V_{dc} (i.e. twice the value given in Fig. 12.6) the r.m.s. fundamental line value of the quasi-square-wave is

12.1 Properties of PWM waveforms

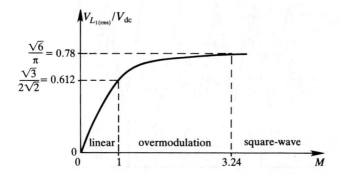

Fig. 12.8 R.m.s. value of fundamental line voltage (relative to V_{dc}) versus modulation ratio for sinusoidal modulation.

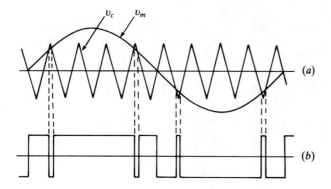

Fig. 12.9 Overmodulation of a triangular carrier wave by a sinusoidal modulating wave, $M = 1.55$.

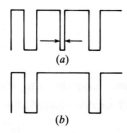

Fig. 12.10 Example of pulse dropping due to overmodulation: (a) containing a minimum pulse, (b) minimum pulse dropped.

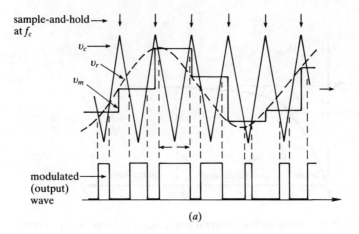

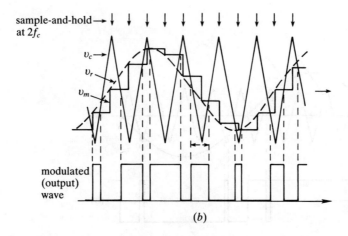

Fig. 12.11 Sinusoidal modulation of a triangular carrier wave using regular sampling, $M = 0.75$, $p = 4.5$, (ref. 13): (a) symmetrical sampling, (b) asymmetrical sampling.

$$V_{1(\text{r.m.s.})} = \frac{4}{\pi} \frac{\sqrt{3}}{2\sqrt{2}} V_{dc}$$
$$= \frac{\sqrt{6}}{\pi} V_{dc} = 0.78 V_{dc} \quad (12.20)$$

Overmodulation increases the waveform harmonic content and can also result in undesirable large jumps of V_1, especially in inverter switches with large dwell times.

Other options for increase of the fundamental output voltage beyond the $M = 1$ value, without increase of other harmonics, are to use a nonsinusoidal reference (modulating) wave such as a trapezoid or a sine-wave plus some third harmonic component.

12.1.3.3 Sinusoidal modulation with regular sampling

As an alternative to natural sampling the sinusoidal reference wave can be sampled at regular intervals of time. If the sampling occurs at instants corresponding to the positive peaks or the positive and negative peaks of the triangular carrier wave, Fig. 12.11, the process is known as uniform or regular sampling. A sample value of the reference sine-wave is held constant until the next sampling instant when a step transition occurs. The stepped version of the reference wave becomes, in effect, the modulating wave. The resulting output modulated wave is defined by the intersections between the carrier wave and the stepped modulating wave.

When sampling occurs at carrier frequency, coincident with the positive peaks of the carrier wave, Fig. 12.11(a), the intersections of adjacent sides of the carrier with the step wave are equidistant about the non-sampled (negative) peaks. For all values of M the modulated wave pulse widths are then symmetrical about the lower (non-sampled) carrier peaks and the process is called *symmetrical regular sampling*. The pulse widths are proportional to the appropriate step height and the pulse centres occur at uniformly spaced sampling times.

When sampling coincides with both the positive and negative peaks of the carrier wave, Fig. 12.11(b), the process is known as *asymmetrical regular sampling*. Adjacent sides of the triangular carrier wave then intersect the stepped modulation wave at different step levels and the resultant modulated wave has pulses that are asymmetrical about the sampling point.

For both symmetrical and asymmetrical regular sampling the output modulated waveforms can be described by analytic expressions. The number of sine-wave values needed to define a sampling step wave is equal to the frequency ratio p (symmetrical sampling) or twice the frequency ratio, $2p$ (asymmetrical sampling). In both cases the number of sample values is much smaller than in natural sampling which requires scanning at sampling instants every degree or half-degree of the modulating sine-wave.

It is common that PWM systems are now implemented by modern digital techniques using PROM (programmable read only memory) and LSI (large-scale integrated) circuits. This is partly to avoid the need for analogue electronic systems with their associated problems such as parameter drift, d.c. offset, reliability of low-speed (i.e. low-frequency) reference oscillators, etc.

The use of regular sampled PWM in preference to naturally sampled PWM requires much less ROM-based computer memory. Also, the analytic nature of regular sampled PWM waveforms makes this approach feasible for implementation using microprocessor based techniques because the pulse widths are easy to calculate.

Some detail of various forms of pulse-width modulation, using a symmetrical triangular carrier wave, are given in Table 12.2.

12.1.4 Optimal pulse-width modulation (harmonic elimination)

An arbitrary pulse-width modulated waveform is shown in Fig. 12.12. The inverter switchings occur at angles defined as $\alpha_1, \alpha_2, \ldots, \alpha_n$ over the repetitive period of 2π radians. Because the waveform contains both half-wave and quarter-wave symmetry a complete cycle can be fully defined by the switching angles for only a quarter-cycle of the waveform.

The switching angles in Fig. 12.12 can be calculated in order that the PWM waveform possesses a fundamental component of a desired magnitude while, simultaneously, optimising a certain performance criterion. For example, the criterion might be to eliminate certain selected harmonics, such as the fifth and/or the seventh, from the waveform. Alternatively, the criterion might be to minimise the total harmonic content and thereby maximise the distortion factor.

From (A7), (A8) of the Appendix it can be inferred, for the quarter-cycle range, that Fourier coefficient $a_n = 0$ and that b_n is given by

$$b_n = \frac{4}{\pi} \int_0^{\pi/2} v(\omega t) \sin n\omega t \, d\omega t \qquad (12.21)$$

For the example in Fig. 12.12, containing two notches (four switchings) per quarter-cycle, the waveform is defined by

$$v(\omega t) = V_{dc} \Big|_{0,\alpha_2,\alpha_4}^{\alpha_1,\alpha_3,\pi/2} - V_{dc} \Big|_{\alpha_1,\alpha_3}^{\alpha_2,\alpha_4} \qquad (12.22)$$

Combining (12.21) and (12.22) gives

$$b_n = \frac{4V_{dc}}{n\pi}[1 - 2\cos n\alpha_1 + 2\cos n\alpha_2 - 2\cos n\alpha_3 + 2\cos n\alpha_4] \qquad (12.23)$$

The pattern of equation (12.23) can be extended to accommodate any desired number of notches or switchings per quarter-wave. Each switching-angle in the quarter-wave represents an unknown to be determined.

Table 12.2 *Techniques of PULSE-WIDTH MODULATION (PWM) using symmetrical triangular carrier wave.*

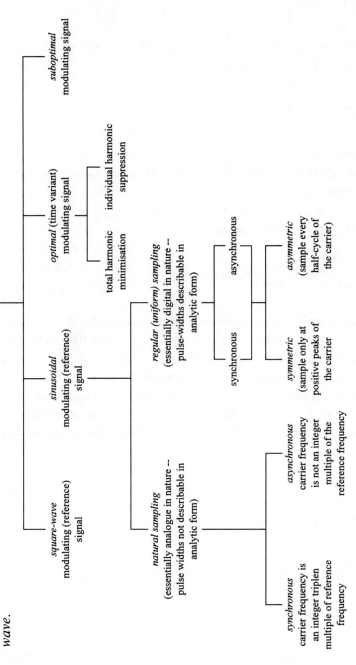

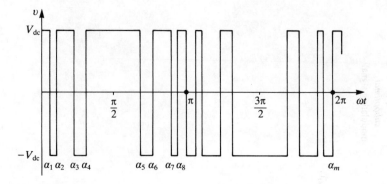

Fig. 12.12 PWM voltage waveform with eight arbitrary switchings per half-cycle.

A generalised form of (12.23) is given by

$$V_n = \frac{4V_{dc}}{n\pi}\left[1 + 2\sum_{i=1}^{m}(-1)^i \cos n\alpha_i\right] \quad (12.24)$$

where m is the number of switchings per quarter-cycle. The solution of (12.24) requires m independent, simultaneous equations; the particular case of Fig. 12.12 and equation (12.23), for example, has $m = 4$. This means that with two notches per quarter-wave it is possible to limit or eliminate four harmonics, one of which may be the fundamental component. In balanced three-phase systems the triplen harmonics are suppressed naturally. It may therefore be logical to suppress the 5, 7, 11 and 13 order harmonics which results in the following equations:

$$b_5 = \frac{4V_{dc}}{5\pi}[1 - 2\cos 5\alpha_1 + 2\cos 5\alpha_2 - 2\cos 5\alpha_3 + 2\cos 5\alpha_4] = 0 \quad (12.25)$$

$$b_7 = \frac{4V_{dc}}{7\pi}[1 - 2\cos 7\alpha_1 + 2\cos 7\alpha_2 - 2\cos 7\alpha_3 + 2\cos 7\alpha_4] = 0 \quad (12.26)$$

$$b_{11} = \frac{4V_{dc}}{11\pi}[1 - 2\cos 11\alpha_1 + 2\cos 11\alpha_2 - 2\cos 11\alpha_3 + 2\cos 11\alpha_4] = 0 \quad (12.27)$$

$$b_{13} = \frac{4V_{dc}}{13\pi}[1 - 2\cos 13\alpha_1 + 2\cos 13\alpha_2 - 2\cos 13\alpha_3 + 2\cos 13\alpha_4] = 0$$

(12.28)

Solution of the four simultaneous equations (12.25)–(12.28) gives the results $\alpha_1 = 10.55°$, $\alpha_2 = 16.09°$, $\alpha_3 = 30.91°$ and $\alpha_4 = 32.87°$. Increase of the number of notches per quarter-cycle increases the number of harmonics that may be suppressed, but has the concurrent effects of reducing the fundamental component and increasing the switching losses.

In general, the set of simultaneous, nonlinear equations describing particular performance criteria need to be solved or optimised using numerical methods. This requires a main-frame computer. The precomputed values of switching angle are then stored in a ROM-based look-up table from which they are accessed by a microprocessor in order to generate the necessary switching pulses. It would not be possible to solve numerically the set of equations in real-time, as would be needed in a motor control application. The larger the number of notchings per quarter-cycle, the more refined becomes the inverter output waveform. This may entail solving a large set of nonlinear equations for which a solution is not always practicable. Furthermore, these equations need to be solved repetitively, once for each desired level of output. Because of this, optimal PWM is yet to be practical at output frequencies below about 10 Hz. Optimum PWM switching strategies are attractive when there are only a few switchings per cycle, as in high power, high voltage inverters involving large switching losses.

12.1.5 PWM voltage waveforms applied to three-phase inductive load

A double-sided triangular carrier wave modulated by a sinusoid results in the pulse waveforms v_A, v_B of Fig. 12.13. If modulating signal V_{m_B} is delayed 120° with respect to v_{m_A} the resulting modulated wave v_B is identical in form to v_A but is also delayed by 120°. The corresponding line voltage $v_{AB}(= v_A - v_B)$ has a fundamental component that leads the fundamental component of v_A by 30°, as in a sinusoidal balanced set of voltages. Note that the positive pulse pattern of $v_{AB}(\omega t)$ is not quite the same as the negative pulse pattern, although the two areas are the same to give zero time average value. This issue is the subject of a Problem in Section 12.5 below.

The application of a PWM voltage waveform to an inductive load results in a current that responds (very nearly) only to the fundamental component. The harmonics of a PWM waveform, including the fundamental, are a complicated function of the carrier frequency ω_c, the modulating (output)

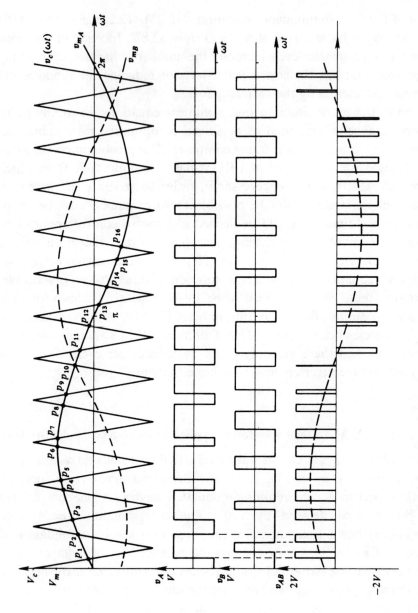

Fig. 12.13 PWM voltage waveforms with sinusoidal natural sampling instants $M = 0.75$, $p = 12$.

frequency ω_m, the carrier amplitude V_c and the modulating wave amplitude V_m. Harmonic components of the carrier frequency are in phase in all three load phases and therefore have a zero sequence nature. With a star-connected load there are no carrier frequency components in the line voltages.

An approximate method of calculating the harmonic content of a PWM waveform is to use graphical estimation of the switching angles, as demonstrated in Example 12.2 below. A precise value of the intersection angles between the triangular carrier wave and the sinusoidal modulating wave can be obtained by equating the appropriate mathematical expressions. In Fig. 12.13, for example, the modulating wave is synchronised to the peak value of the carrier wave. The first intersection p_1 between the carrier $v_c(\omega_m t)$ and modulating wave $v_{m_A}(\omega_m t)$ occurs when

$$\frac{V_m}{V_c} \sin \omega_m t = 1 - \frac{24}{\pi} \omega_m t \qquad (12.29)$$

Intersection p_2 in Fig. 12.13 occurs when

$$\frac{V_m}{V_c} \sin \omega_m t = -3 + \frac{24}{\pi} \omega_m t \qquad (12.30)$$

This oscillating series has the general solution, for the Nth intersection,

$$p_N = (2N - 1)(-1)^{N+1} + (-1)^N \frac{2p}{\pi} \omega_m t \qquad (12.31)$$

where $N = 1, 2, 3, \ldots, 24$.

Expressions similar to (12.29), (12.30) can be obtained for all of the intersections, as shown in Example 12.2. Equations of the form (12.29)–(12.31) are transcendental and require to be solved by iteration.

12.1.6 Worked examples

Example 12.1
An inverter produces a double-pulse notched voltage waveform of the type shown in Fig. 12.1(b). If the peak amplitude of this waveform is V calculate its fundamental component and compare with this the value obtained by the use of a single-pulse waveform of the same total area.

Solution. In the waveform $V(\omega t)$ in Fig. 12.1(b), $N = 2$ and the instantaneous value is therefore

$$v(\omega t) = V \Big|_{n(\pi-\delta)/3,(4\pi-\delta)/6}^{(2\pi+\delta)/6,(2\pi+\delta)/3}$$

Since the waveform is antisymmetrical about $\omega t = 0$ the fundamental component $V_1(\omega t)$ passes through the origin, $\psi_1 = 0$ and $a_1 = 0$. Fourier coefficient b_1 is given by

$$b_1 = \frac{2}{\pi} \int_0^\pi v(\omega t) \sin \omega t \, d\omega t$$

$$= -\frac{2V}{\pi} [\cos \omega t]_{(\pi-\delta)/3,(4\pi-\delta)/6}^{(2\pi+\delta)/6,(2\pi+\delta)/3}$$

$$= \frac{2V}{\pi} \left[\cos\left(\frac{\pi-\delta}{3}\right) - \cos\left(\frac{2\pi+\delta}{6}\right) + \cos\left(\frac{4\pi-\delta}{6}\right) - \cos\left(\frac{2\pi+\delta}{3}\right) \right]$$

$$= \frac{2V}{\pi} \left(\cos\frac{\delta}{3} - \cos\frac{\delta}{6} + \sqrt{3}\sin\frac{\delta}{6} + \sqrt{3}\sin\frac{\delta}{3} \right)$$

In this case, $\delta = 0.4\pi = 72°$. Therefore

$$V_1 = b_1 = \frac{2V}{\pi} [0.914 - 0.978 + \sqrt{3}(0.407 + 0.208)]$$

$$= \frac{2V}{\pi} (1.001) = 0.637 V_{\text{peak}}$$

This value may be verified by the use of (12.5) and (12.7) as shown below. In (12.7), given that $\delta = 72° = 0.4\pi$ rad,

$$\gamma = \frac{4\pi - 0.4\pi}{12} = 0.942 \text{ rad} = 54°$$

Substituting into (12.5) gives, for $n = 1$,

$$V_1 = \frac{8V}{\pi} \sin 54° \sin \frac{72°}{4}$$

$$= \frac{8V}{\pi} \times 0.809 \times 0.309 = 0.64 V_{\text{peak}}$$

This compares with the value from (12.3) for the single-pulse of the same area,

$$V_1 = \frac{4V}{\pi} \sin \frac{72°}{2}$$

$$= \frac{4V}{\pi} \times 0.59 = 0.75 V$$

The use of two pulses per half-cycle has therefore resulted in reduction of the fundamental component.

Example 12.2
A double-sided triangular carrier wave of height V_c is natural sampling modulated by a sinusoidal modulating signal $v_m(\omega t) = V_m \sin \omega t$, where $V_m = 0.6 V_c$. The carrier frequency ω_c is 12 times the modulating frequency

ω_m. Sketch a waveform of the resultant modulated voltage and calculate its principal harmonic components and its r.m.s. value.

Solution. The waveforms are shown in Fig. 12.13. The phase voltage $v_A(\omega t)$ is symmetrical about $\pi/2$ radians and contains only odd harmonics. Since $v_A(\omega t)$ is antisymmetrical about $\omega t = 0$, the Fourier harmonics $a_n = 0$ so that the fundamental output component is in phase with the modulating voltage $v_m(\omega t)$.

It is necessary to determine the intersection points p_1 to p_6.

Point p_1

From (12.31) $p = 12$, $V_m/V_c = 0.6$ so that

$$0.6 \sin \omega t = 1 - \frac{24}{\pi} \omega t$$

which gives

$$\omega t = 7°$$

Point p_2

$$\frac{V_m}{V_c} \sin \omega t = \frac{4N}{\pi} \omega t - 3$$

or

$$\omega t = 24°$$

Point p_3

$$\frac{V_m}{V_c} \sin \omega t = 5 - \frac{4N}{\pi} \omega t$$

$$\omega t = 34.5°$$

Point p_4

$$\frac{V_m}{V_c} \sin \omega t = \frac{4N}{\pi} \omega t - 7$$

$$\omega t = 56°$$

Point p_5

$$\frac{V_m}{V_c} \sin \omega t = 9 - \frac{N}{\pi} \omega t$$

$$\omega t = 63°$$

Point p_6

$$\frac{V_m}{V_c}\sin\omega t = \frac{4N}{\pi}\omega t - 11$$

$$\omega t = 87°$$

For the first quarter-cycle in Fig. 12.13 waveform $v_A(\omega t)$ is given by

$$v_A(\omega t) = V\big|_{7°,34.5°,63°}^{24°,56°,87°} - V\big|_{0°,24°,56°,87°}^{7°,34.5°,63°,90°}$$

Fourier coefficients b_n are given by

$$b_n = \frac{4}{\pi}\int_0^{\pi/2} v_A(\omega t)\sin n\omega t\, d\omega t$$

In this case,

$$\begin{aligned}b_n &= \frac{4V}{n\pi}\left\{[-\cos n\omega t]_{7°,34.5°,63°}^{24°,56°,87°} + [\cos n\omega t]_{0°,24°,56°,87°}^{7°,34.5°,63°,90°}\right\}\\ &= \frac{8V}{n\pi}(\cos n7° + \cos n34.5° + \cos n63°\\ &\quad -\cos n24° - \cos n56° - \cos n87° - 0.5)\end{aligned}$$

It is found that the peak values of Fourier coefficient b_n are

$$b_1 = 0.626V$$
$$b_3 = \frac{8V}{3\pi}(0.038) = 0.0323V$$
$$b_5 = \frac{8V}{5\pi}(0.102) = 0.052V$$
$$b_7 = \frac{8V}{7\pi}(0.323) = 0.118V$$
$$b_9 = \frac{8V}{9\pi}(0.876) = 0.248V$$
$$b_{11} = \frac{8V}{11\pi}(2.45) = 0.567V$$
$$b_{13} = \frac{8V}{13\pi}(2.99) = -0.585V$$
$$b_{15} = \frac{8V}{15\pi}(3.44) = 0.548V$$
$$b_{17} = \frac{8V}{17\pi}(1.52) = -0.23V$$
$$b_{19} = -\frac{8V}{19\pi}(1.22) = -0.164V$$

The r.m.s. value of the waveform is

$$V_A = \frac{V}{\sqrt{2}}\sqrt{b_1^2 + b_3^2 + b_5^2 + \ldots + b_{19}^2}$$

$$= \frac{V}{\sqrt{2}}\sqrt{0.598} = 0.547 V$$

The distortion factor is

$$\text{distortion factor} = \frac{\frac{b_1}{\sqrt{2}}}{V_A} = \frac{0.626}{\sqrt{2} \times 0.547} = 0.809$$

Note that the highest value harmonics satisfy an $n = p \pm 1$ relationship and that the low-order harmonics have small values. This enhances the suitability of the waveform for a.c. motor speed control.

Example 12.3
The PWM voltage waveform $v_A(\omega t)$ of Fig. 12.13 is generated by an inverter that uses a modulating frequency of 50 Hz. If the d.c. supply is 200 V calculate the r.m.s. current that would flow if $v_A(\omega t)$ was applied to a single-phase series R–L load in which $R = 10\,\Omega$ and $L = 0.01$ H.

Solution. At the various harmonic frequencies the load impedance is

$$|Z_h| = \sqrt{R^2 + (n\omega L)^2}$$

which gives

$$Z_1 = \sqrt{10^2 + (2\pi \times 50 \times 0.01)^2} = 10.48\,\Omega$$

$Z_3 = 13.74\,\Omega, \quad Z_{13} = 42\,\Omega$
$Z_5 = 18.62\,\Omega, \quad Z_{15} = 48.17\,\Omega$
$Z_7 = 24.16\,\Omega, \quad Z_{17} = 54.33\,\Omega$
$Z_9 = 30\,\Omega, \quad Z_{19} = 60.52\,\Omega$
$Z_{11} = 35.97\,\Omega$

If the harmonic voltages of Example 12.2 are divided by the respective harmonic impedances above, one obtains the following peak current harmonics:

$$I_1 = \frac{V_1}{Z_1} = \frac{0.626 \times 200}{10.48} = 11.95\,\text{A}$$

$I_3 = 0.47\,\text{A}, \quad I_{13} = 2.79\,\text{A}$
$I_5 = 0.56\,\text{A}, \quad I_{15} = 2.42\,\text{A}$
$I_7 = 0.977\,\text{A}, \quad I_{17} = 0.85\,\text{A}$
$I_9 = 1.65\,\text{A}, \quad I_{19} = 0.54\,\text{A}$
$I_{11} = 3.15\,\text{A}$

510 PWM inverter control

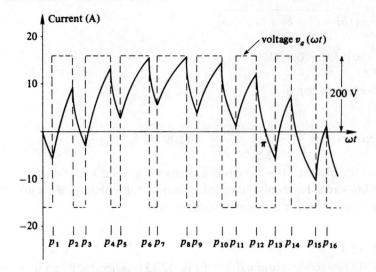

Fig. 12.14 PWM waveforms with series R–L load, from Example 12.4.

The harmonic sum is

$$\sum I_n^2 = I_1^2 + I_3^2 + I_5^2 + \ldots I_{19}^2 = 172\,\text{A}^2$$

This has an r.m.s. value

$$I = \sqrt{\frac{\sum I_n^2}{2}} = 9.27\,\text{A}$$

which compares with a fundamental r.m.s. current of $11.95/\sqrt{2} = 8.45\,\text{A}$.

The current distortion factor is therefore $8.45/9.27 = 0.912$, which is greater (i.e. better) than the corresponding voltage distortion factor of 0.809.

Example 12.4
The PWM waveform $v_A(\omega t)$ of Fig. 12.13 is applied to a single-phase series R–L circuit with $R = 10\,\Omega$ and $L = 0.01\,\text{H}$. Voltage $v_A(\omega t)$ has a frequency of 50 Hz and an amplitude $V = 200\,\text{V}$. Deduce the waveform of the resulting current.

Solution. The waveform $v_A(\omega t)$ is reproduced in Fig. 12.14. If a direct voltage V is applied across a series R–L circuit carrying a current I_o the subsequent rise of current satisfies the relation

$$i(\omega t) = \frac{V}{R}\left(1 - \varepsilon^{-t/\tau}\right) + I_o \varepsilon^{-t/\tau} = \frac{V}{R} - \left(\frac{V}{R} - I_o\right)\varepsilon^{-t/\tau}$$

where I_o is the value of the current at the switching instant.

12.1 Properties of PWM waveforms

In this case $\tau = L/R = \omega L/\omega R = \pi/10\omega$ so that

$$i(\omega t) = 20 - (20 - I_o)\varepsilon^{-10/\pi \cdot \omega t}$$

At switch-on $I_o = 0$ and the current starts from the origin. Consider the current values at the voltage switching points in Fig. 12.14. These time values are given in Example 12.2.

At point p_1,

$\omega t = 7° = 0.122 \text{ rad}$
$i(\omega t) = -20(1 - 0.678) = -6.44 \text{ A}$

At point p_2,

$\omega t = 24° - 7° = 0.297 \text{ rad}$
$i(\omega t) = 20 - (20 + 6.44)0.39 = 9.8 \text{ A}$

At point p_3,

$\omega t = 34.5° - 24° = 0.183 \text{ rad}$
$i(\omega t) = -20 - (-20 - 9.8)0.56 = -3.3 \text{ A}$

At point p_4,

$\omega t = 56° - 34.5° = 0.375 \text{ rad}$
$i(\omega t) = 20 - (20 + 3.3)0.303 = 12.94 \text{ A}$

At point p_5,

$\omega t = 63° - 56° = 0.122 \text{ rad}$
$i(\omega t) = -20 - (-20 - 12.92)0.678 = 2.33 \text{ A}$

At point p_6,

$\omega t = 87° - 63° = 0.419 \text{ rad}$
$i(\omega t) = 20 - (20 - 2.33)0.264 = 15.34 \text{ A}$

Similarly,

At point $p_7, \omega t = 93° - 87° = 0.105 \text{ rad}, i(\omega t) = 5.3 \text{ A}$
At point $p_8, \omega t = 116° - 93° = 0.401 \text{ rad}, i(\omega t) = 15.9 \text{ A}$
At point $p_9, \omega t = 124.5° - 116° = 0.148 \text{ rad}, i(\omega t) = 2.4 \text{ A}$
At point $p_{10}, \omega t = 145.5° - 124.5° = 0.367 \text{ rad}, i(\omega t) = 14.53 \text{ A}$
At point $p_{11}, \omega t = 155° - 145.5° = 0.166 \text{ rad}, i(\omega t) = 0.3 \text{ A}$
At point $p_{12}, \omega t = 172.5° - 155° = 0.305 \text{ rad}, i(\omega t) = 12.5 \text{ A}$
At point $p_{13}, \omega t = 187.5° - 172.5° = 0.262 \text{ rad}, i(\omega t) = -5.9 \text{ A}$
At point $p_{14}, \omega t = 201° - 187.5° = 0.236 \text{ rad}, i(\omega t) = 7.78 \text{ A}$
At point $p_{15}, \omega t = 220.5° - 201° = 0.34 \text{ rad}, i(\omega t) = -10.6 \text{ A}$
At point $p_{16}, \omega t = 229.5° - 220.5° = 0.157 \text{ rad}, i(\omega t) = 1.43 \text{ A}$

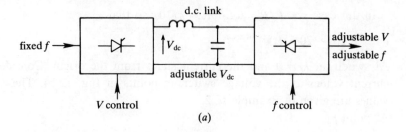

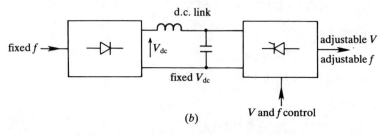

Fig. 12.15 Basic forms of voltage source inverter (VSI): (*a*) step-wave (or quasi-square-wave) VSI, (*b*) PWM VSI.

The time variation of the current, shown in Fig. 12.14, is typical of the current waveforms obtained with PWM voltages applied to inductive and a.c. motor loads.

12.2 THREE-PHASE INDUCTION MOTOR CONTROLLED BY PWM VOLTAGE SOURCE INVERTER (VSI)

12.2.1 Theory of operation

The basic differences of structure between the voltage source, step-wave inverter, such as that of Fig. 10.4, and the voltage source, PWM inverter are given in Fig. 12.15. A step-wave inverter uses a controlled rectifier to give a direct-voltage source of adjustable level at the input to the d.c. link. The voltage level of the inverter output is controlled by the adjustable V_{dc} link voltage, whereas the frequency is controlled independently by the gating of the inverter switches.

A PWM inverter uses a diode bridge rectifier to give a fixed level of V_{dc} at the d.c. link. Both the voltage and frequency of the inverter are controlled by gating of the inverter switches, Fig. 12.15(*b*). The complete assembly of rectifier stage, d.c. link and inverter stage is shown in Fig. 12.16. Since the

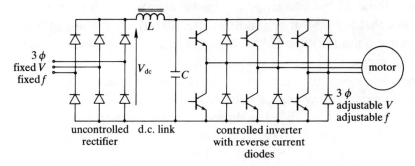

Fig. 12.16 Main features of a PWM VSI with motor load.

output voltage of the diode bridge rectifier is not a pure direct voltage, a filter inductor is included to absorb the ripple component.

The use of a fixed d.c. rail voltage means that several independent inverters can operate simultaneously from the same d.c. supply. At low power levels the use of transistor (rather than thyristor) switches permits fast switching action and fast current and torque transient response, compared with step-wave inverters.

Because the harmonic currents are small and can be made of relatively high order, compared with single-pulse or multiple-pulse modulation, and because the fundamental component is easily controlled, PWM methods are becoming increasingly popular for a.c. motor control. Although the harmonic currents may be small, however, the harmonic heating losses may be considerable through increase of the motor resistances due to the skin effect, as discussed in Section 10.3.1. PWM switching techniques are better suited to power transistor inverters than to thyristor inverters because the commutation losses due to the many switchings are then less significant. Above about 100 Hz the commutation losses with PWM switching become unacceptably large and stepped-wave techniques are used in a.c. motor drives.

When PWM voltage waveforms are applied to an induction motor the motor torque responds largely to the fundamental frequency component. Motor current harmonics are usually small and of high harmonic order, depending on the frequency ratio p and can be calculated using the method of analysis of Section 10.3.1. The harmonics of the PWM applied voltage are often more significant than those of the consequent motor current. This has the result that the eddy-current and hysteresis iron losses, which vary directly with flux and with frequency, are often greater than the copper losses in the windings. The total losses due to harmonics in a PWM driven motor may exceed those of the comparable step-wave driven motor. It is a common practice that a PWM driven motor is derated by an amount of 5–10%.

Torque pulsations in a PWM drive are small in magnitude and are related to high harmonic frequencies so that they can usually be ignored. The input current waveform to a d.c. link-inverter drive is determined mostly by the rectifier action rather than by the motor operation. This has a waveshape similar to that of a full-wave three-phase bridge with passive series resistance–inductance load, so that the drive operates, at all speeds, at a displacement factor near to the ideal value of unity.

12.2.2 Worked example

Example 12.5

The PWM waveform of Fig. 12.5(b), (c), (d) have a height $V = 240$ V and is applied as the phase voltage waveform of a three-phase, four-pole, 50 Hz, star-connected induction motor. The motor equivalent circuit parameters, referred to primary turns, are $R_1 = 0.32\,\Omega$, $R = 0.18\,\Omega$, $X_1 = X_2 = 1.65\,\Omega$, $X_m =$ large. Calculate the motor r.m.s. current at 1440 r.p.m. What are the values of the main harmonic currents?

Solution. A 4-pole, 50 Hz motor has a synchronous speed

$$N_s = \frac{f120}{p} = \frac{50 \times 120}{4} = 1500 \text{ r.p.m.}$$

At a speed of 1440 r.p.m. the per-unit slip is given by (9.3)

$$S = \frac{1500 - 1440}{1500} = 0.04$$

It is seen from Fig. 12.5(a) that the PWM waveform has a modulation ratio $M = 0.9$.

The height $V = 240$ V of the PWM voltage waveform of Fig. 12.5 is the value of the battery (or mean rectified) supply voltage V_{dc} in the circuit of Fig. 12.6. The r.m.s. value of the per-phase applied voltage is therefore, from (12.18),

$$V_{1\text{rms}} = 0.9 \times \frac{240}{2\sqrt{2}} = 76.4 \text{ V/phase}$$

The motor per-phase equivalent circuit is a series R–L circuit. At 1440 r.p.m. this has the input impedance

$$Z_{in_{1440}} = \left(0.32 + \frac{0.18}{0.04}\right) + j1.65$$

$$= 4.82 + j1.65 = 5.095 \angle 18.9° \; \Omega/\text{phase}$$

$$\therefore I_1 = \frac{76.4 \angle 0}{5.095 \angle 18.9} = 15 \angle -18.9° \; \text{A}/\text{phase}$$

With $M = 0.9$ the dominant harmonics are likely to be those of order $2p \pm 1$ and $p \pm 2$. In this case, therefore, with $p = 9$, the harmonics to be considered are $n = 7, 11, 17$ and 19. From (9.34) the harmonic slip values are

$$S_7 = \frac{7-(1-0.04)}{7} = \frac{6.04}{7} = 0.863$$

$$S_{11} = \frac{11+(1-0.04)}{11} = \frac{11.96}{11} = 1.087$$

$$S_{17} = \frac{17+(1-0.04)}{17} = \frac{17.96}{17} = 1.056$$

$$S_{19} = \frac{19-(1-0.04)}{19} = \frac{18.04}{19} = 0.95$$

For the nth harmonic currents the relevant input impedances to the respective equivalent circuits are

$$Z_{in_7} = \left(0.32 + \frac{0.18}{0.863}\right) + j7 \times 1.65 = 0.53 + j11.55 = 11.56 \, \Omega$$

$$Z_{in_{11}} = \left(0.32 + \frac{0.18}{1.087}\right) + j11 \times 1.65 = 0.486 + j18.15 = 18.16 \, \Omega$$

$$Z_{in_{17}} = \left(0.32 + \frac{0.18}{1.056}\right) + j17 \times 1.65 = 0.291 + j28.1 = 28.2 \, \Omega$$

$$Z_{in_{19}} = \left(0.32 + \frac{0.18}{0.949}\right) + j19 \times 1.65 = 0.51 + j31.35 = 31.35 \, \Omega$$

The 7, 11, 17 and 19 order harmonic voltage levels have to be deduced from Fig. 12.7. For $M = 0.9$ the $2p \pm 1$ and $p \pm 2$ levels have the value 0.26 of the peak fundamental value. The $M = 1$ value of the r.m.s. component of the fundamental phase voltage, from (12.18), is

$$V_{1_{\text{rms}}} = \frac{240}{2\sqrt{2}} = 84.86 \, \text{V}/\text{phase}$$

Therefore, the r.m.s. harmonic voltage values

$$V_7 = V_{11} = V_{17} = V_{19} = 0.26 \times 84.86 = 22 \, \text{V}/\text{phase}$$

The appropriate r.m.s. harmonic phase currents are

$$I_7 = \frac{22.1}{11.56} = 1.91 \, \text{A} \qquad I_{17} = \frac{22.1}{28.2} = 0.78 \, \text{A}$$

$$I_{11} = \frac{22.1}{18.16} = 1.22 \, \text{A} \qquad I_{19} = \frac{22.1}{3.135} = 0.7 \, \text{A}$$

The total r.m.s. current is obtained by the customary square-law summation

$$I_{rms} = \sqrt{I_1^2 + I_7^2 + I_{11}^2 + I_{17}^2 + I_{19}^2 + \ldots}$$

$$I_{rms}^2 = 15^2 + (1.91)^2 + (1.22)^2 + (0.78)^2 + (0.7)^2$$
$$= 225 + 3.65 + 1.49 + 0.6 + 0.49$$
$$= 231.23 \, A^2$$

$$I_{rms} = \sqrt{231.23} = 15.25 \, A$$

which is about 2% greater than the fundamental value.

12.3 THREE-PHASE INDUCTION MOTOR CONTROLLED BY PWM CURRENT SOURCE INVERTER (CSI)

A current source is one in which the load current remains constant in the presence of load impedance variations and the terminal voltage changes to satisfy the $V = IZ$ requirement. Such a source can be approximated by a controlled rectifier or chopper with a large d.c. side inductor. The inductor causes the input to the inverter to appear as a d.c. current source, having very large (ideally infinite) impedance. The CSI could be represented by Fig. 12.15(a) with the capacitor removed. Additional smoothing of the current can be realised by the introduction of a current control loop.

12.3.1 Current source inverter with passive load

The current source inverter (CSI) delivers a periodic current of constant r.m.s. value and is best suited to loads that possess a low harmonic impedance. The classic example is a parallel tuned circuit, which is characteristic of induction heating applications where an inductive load is usually shunted by a power factor correction capacitor. This causes operation at high power factor and usually results in a substantially sinusoidal voltage drop even if the applied current is nonsinusoidal.

12.3.2 Current source inverter with induction motor load

With an electrical supply of constant frequency there is no advantage to be gained by the use of constant (rated) current rather than constant (rated) voltage for reduced speed operation. Such operation would result in high magnetic saturation and restricted torque operation. Characteristics for a typical motor are shown in Fig. 12.17 comparing operation at 1 p.u. current

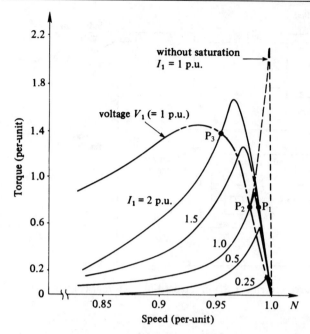

Fig. 12.17 Calculated torque–speed characteristics of a typical induction motor with current source supply (ref. TP 8).

with operation at 1 p.u. voltage. The point of intersection P_2 representing both 1 p.u. current and 1 p.u. voltage, occurs on the positive slope of the current characteristic. Continuous operation is not feasible at point P_1 on the negative part of the slope due to excessive magnetising current and iron losses. At point P_2, however, the motor flux is near to its rated value, the iron losses are normal and the motor operates near to both its rated voltage and current. Since point P_2 is on the positive slope of the torque–speed characteristic, stable operation is not possible on open-loop and some kind of feedback control system is necessary. A widely used system employs a motor voltage control loop which regulates the phase-angle of the rectifier switches, Fig. 12.18, by using the voltage error as a reference signal for the current controller CC.

A current source inverter requires to be force-commutated. Commutation circuits in silicon controlled rectifier CSIs usually consist of diodes, inductors, capacitors and the motor leakage inductance. These major circuit components are gradually increasing in cost making the CSI less competitive compared with the transistor implemented VSI. A CSI is therefore specific to the motor for which it is designed.

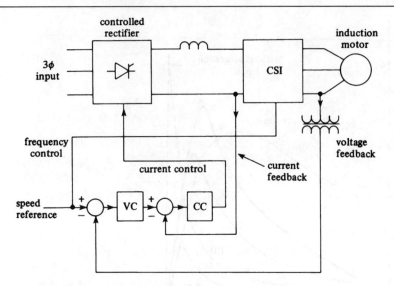

Fig. 12.18 Current source inverter (CSI) in a closed-loop current control system.

An advantage of the CSI over the VSI is its ability to ride through a commutation failure and to return naturally to normal operation. In addition, the use of a controlled current source eliminates the possibility of transient current surges and therefore has the benefit of restricting the overcurrent design capacity of the motor and supply. A current source inverter is able to regenerate back into the supply without reversing the d.c. link current (as needed in the VSI) by simply reversing the polarity of the rectifier output voltage. To ensure a unidirectional d.c. link current the feedback diodes, which are characteristic of VSI circuits such as Fig. 12.6, should be removed.

The equivalent circuits of Fig. 9.1 may also be used for calculating the motor performance with a current source. Terminal voltage V_1 is not an impressed value but is then a voltage drop dependent on the impressed current I_1 and the motor impedance parameters.

12.4 SECONDARY FREQUENCY CONTROL

With a voltage source or a current source the power crossing the air-gap in the circuit of Fig. 9.1 is given by

$$P_g = TN_1 = E_1 I_2 \cos \phi_2 \tag{12.32}$$

Combining (12.32) with (9.1(b)) and (9.30) gives

$$T = \frac{4.44}{2\pi} \hat{\Phi}_m n p I_2 \cos \phi_2 \qquad (12.33)$$

In (12.33) $\hat{\Phi}_m$ is the peak value of the air-gap (mutual) flux. Since the motor torque is determined by the product of the flux and the secondary current it is more convenient to some applications to control the motor current rather than its terminal voltage.

From (9.9) and the circuit of Fig. 9.1 it is seen that

$$I_2 \cos \phi_2 = \frac{S E_1 R_2}{R_2^2 + S^2 X_2^2} \qquad (12.34)$$

In adjustable frequency drives, whether voltage controlled or current controlled, the torque–speed characteristics have only small speed regulation, even on open-loop, and the operating slip is small. In (12.34), when $SX_2 \ll R_2$ and combining with (9.3),

$$I_2 \cos \phi_2 = \frac{S E_1}{R_2} = \frac{f_2}{R_2} \frac{E_1}{f_1} \qquad (12.35)$$

Since the r.m.s. flux Φ_m is proportional to E_1/f_1, for either voltage or current control with adjustable frequency, substituting (12.35) into (12.33) gives

$$T = n p \Phi_m \frac{f_2}{R_2} \frac{E_1}{f_1} = K \Phi_m^2 f_2 \qquad (12.36)$$

The important result of (12.36) is that, with small slip and constant flux, the torque is proportional to the secondary frequency independently of motor speed. Up to rated speed the frequency f_2 may be kept constant to produce a 'constant torque' range of operation, Fig. 12.19. Higher speeds can be obtained by increasing the primary frequency f_1 but the applied voltage V_1 (and hence E_1) must be limited to its rated value. The effect of constant voltage and increased frequency is to reduce the flux and hence the torque. If frequency f_2 is appropriately controlled at high speeds, the torque can be made to vary inversely with speed resulting in a 'constant horsepower' region of operation, as illustrated in Fig. 4.8. Alternatively, constant output power can be realised at speeds below the rated value if f_2 is kept constant while V_1 and f_1 are varied to control the flux and the torque. Control of the motor speed by flux reduction is sometimes referred to as field weakening since it corresponds to field flux control of the speed of a d.c. shunt motor. A secondary frequency or slip frequency controlled drive can only be implemented by a closed-loop control system incorporating tachometric feedback.

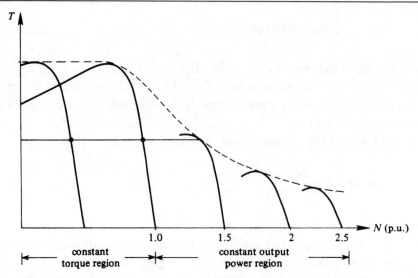

Fig. 12.19 Induction motor torque–speed characteristics with secondary frequency control.

12.5 PROBLEMS

Properties of pulse-width modulated waveforms

12.1 Show that the r.m.s. value of the single-pulse waveform of Fig. 12.1(a) is given by expression (12.2).

12.2 Calculate the values of the fundamental components of the pulse waveforms of Fig. 12.1(a), (b) if $\delta = 108°$.

12.3 A voltage waveform, Fig. 12.20, contains three single-sided pulses in each half-cycle, spaced symmetrically with respect to $\pi/2$. Obtain an expression for the amplitude of the nth harmonic if $\alpha_1 = \pi/6$, $\alpha_2 = \pi/3$ and compare this with the corresponding expression for a single-pulse waveform of the same area. What are the respective fundamental values?

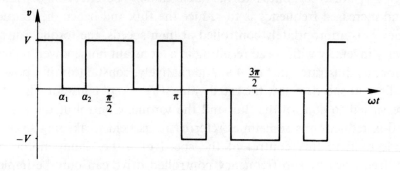

Fig. 12.20 Voltage waveform for Problem 12.3.

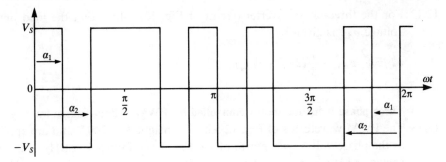

Fig. 12.21 PWM voltage waveform with two arbitrary switchings per quarter-cycle.

12.4 For the waveform of Fig. 12.20, calculate the values of α_1, and α_2 that will permit the 3rd and 5th harmonic components to be eliminated.

12.5 For the voltage waveform of Fig. 12.12 show that the Fourier coefficient b_n, in terms of the switching angles α_1, α_2, is given by

$$b_n = \frac{4V}{n\pi}[1 - \cos n\alpha_1 + \cos n\alpha_2]$$

12.6 Define relationships for the switching angles that need to be satisfied if the 3rd and 5th harmonic components are to be eliminated from the waveform of Fig. 12.21. Calculate appropriate values of α_1 and α_2.

12.7 A single-sided triangular carrier wave of peak height V_c contains six pulses per half-cycle and is modulated by a sine-wave $v_m(\omega t) = V_m \sin \omega t$ synchronised to the origin of a triangular pulse. Sketch waveforms of the resultant modulated wave if (i) $V_m = 0.5V_c$, (ii) $V_m = V_c$ and (iii) $V_m = 1.5V_c$. Which of these waveforms appears to contain the greatest fundamental (i.e. modulating frequency) value?

12.8 For the waveforms described in Problem 12.7 estimate, graphically, the values of ωt at which intersections occur between $v_c(\omega_c t)$ and $v_m(\omega t)$ when $V_m = V_c$. Use these to calculate values of the harmonics of the modulated wave up to $n = 21$ and thereby calculate the r.m.s. value.

12.9 The modulated voltage waveform described in Problem 12.7 is applied to a series R–L load in which $R = 25\,\Omega$ and $X_L = 50\,\Omega$ at 50 Hz. If the constant height of the PWM voltage wave is 400 V calculate the resulting current harmonics up to $n = 21$. Calculate the resultant r.m.s current. Compare the value of the current distortion factor CDF with the voltage distortion factor VDF.

12.10 Calculate the power dissipation in the R–L series circuit of Problem 12.9. Hence calculate the operating power factor.

12.11 The PWM voltage waveform $v_o(\omega t)$ of Fig. 12.4 is applied to the series R–L load, $R = 25\,\Omega$ and $X_L = 50\,\Omega$ at 50 Hz. If $V = 250\,V$ and $f = 50\,Hz$, deduce the waveform of the resulting current.

12.12 For the three-phase inverter circuit of Fig. 12.6 show that the load phase voltage v_{AN} is given by

$$v_{AN} = \frac{2}{3} v_{AN_s} - \frac{1}{3}(v_{BN_s} + v_{CN_s})$$

Three-phase induction motor controlled by PWM voltage source inverter

12.13 The PWM waveforms of Fig. 12.5 have a height $V = 200$ V and are applied as the phase voltage waveform of a three-phase, four-pole, 50 Hz, star-connected induction motor. The motor equivalent circuit parameters, referred to primary turns, are $R_1 = 0.32\,\Omega$, $R_2 = 0.18\,\Omega$, $X_1 = X_2 = 1.65\,\Omega$, $X_m = $ large. Calculate the motor r.m.s. current at 1440 r.p.m. What are the values of the main harmonic currents?

12.14 For the induction motor of Example 12.5 calculate the input power and hence the power factor for operation at 1440 r.p.m.

12.15 If only the fundamental current component results in useful torque production, calculate the efficiency of operation for the motor of Example 12.5 at 1440 r.p.m. (i) neglecting core losses and friction and windage, (ii) assuming that the core losses plus friction and windage are equal to the copper losses.

APPENDIX

General expressions for Fourier series

If a periodic function $e(\omega t)$, of any waveshape, is repetitive every 2π radians it may be expressed as a summation of harmonic terms:

$$e(\omega t) = \frac{a_0}{2} + \sum_{n=1}^{\infty}(a_n \cos n\omega t + b_n \sin n\omega t) \tag{A.1}$$

$$= \frac{a_0}{2} + \sum_{n=1}^{\infty} c_n \sin(n\omega t + \psi_n)$$

where

$$c_n = \sqrt{a_n^2 + b_n^2} = \text{peak value of } n\text{th harmonic} \tag{A.2}$$

$$\psi_n = \tan^{-1}\left(\frac{a_n}{b_n}\right) = \text{phase displacement of the } n\text{th harmonic} \tag{A.3}$$

Also,

$$a_n = n \sin \psi_n \tag{A.4}$$

$$b_n = n \cos \psi_n \tag{A.5}$$

The various coefficients in (A.1) are defined by the expressions

$$\frac{a_0}{2} = \frac{1}{2\pi}\int_0^{2\pi} e(\omega t)\, d\omega t \tag{A.6}$$

$$= \text{time average value} = \text{d.c. term value}$$

$$a_n = \frac{1}{\pi}\int_0^{2\pi} e(\omega t) \cos n\omega t \, d\omega t \tag{A.7}$$

General expressions for Fourier series

$$b_n = \frac{1}{\pi} \int_0^{2\pi} e(\omega t) \sin n\omega t \, d\omega t \qquad (A.8)$$

For the fundamental component

$$a_1 = \frac{1}{\pi} \int_0^{3\pi} e(\omega t) \cos \omega t \, d\omega t \qquad (A.9)$$

$$b_1 = \frac{1}{\pi} \int_0^{2\pi} e(\omega t) \sin \omega t \, d\omega t \qquad (A.10)$$

Note: The definitions of equations (A9), (A10) represent a sign convention. In some books the reverse definition is used but this does not affect the values of c_n and ψ_n.

Answers to problems

Chapter 1
1.7 $W = 60\,\text{W}$ (unaided), $C = 0.6\,\mu\text{F}$, $W = 1.9\,\text{W}$ (aided).
1.8 $ID_s = 75\,\text{W}$, $ID_{\text{on}} = 6\,\text{W}$, $\eta = 900/981 = 0.9$ p.u.
1.9 6.6 W
1.10 $R = 3\,\Omega$ (junction to case), $L = 3.7\,\text{cm}$, $T_{\text{case}} = 155°\text{C}$.
1.11 $R_{jSi} = 3\,\Omega$, $R_{Sa} = 0.5/L(\text{m})$, $L = 0.35\,\text{m}$, $T_s = 70°\text{C}$.
1.12 $R_{Sa} = 0.5/L(\text{m})$, $L = 0.133\,\text{m}$, $W_{\text{with}} = 4W$, $W_{\text{without}} = 0$, $W_s = 60\,\text{W}$
1.13 85 W
1.14 $0.09\,\Omega$
1.17 $R_\infty = 3\,\Omega$, $Z_{10} = 1/7\,\Omega$, $T_\infty = 100°\text{C}$, $T_{10} = 43°\text{C}$.
1.18 $L = 1/7\,\text{m}$, $T_{\text{sink}} = 160°\text{C}$.
1.19 Without aid, $L_{\min} = 0.124\,\text{m}$.
With aid, $ID_{\text{off}} = 0.42\,\text{W}$, $ID = 2.52\,\text{W}$, $T_c = 167.5°\text{C}$, $L_{\min} = 0.037\,\text{m}$.
1.20 $ID_s = 0.83\,\text{W}$, $ID_{\text{on}} = 2\,\text{W}$, $T_s = 111.7°\text{C}$, $R_{S_a} = 22.6\,\Omega$, $L = 4.4\,\text{cm}$.

Chapter 2
2.25 $22.5\,\Omega$
2.35 (a) bipolar (not FET), (b) SCR, (c) bipolar, (d) triac
2.36 (a) SCR, (b) triac, (c) bipolar, (d) GTO

Chapter 3
3.2 (ii) $50\,\mu\text{H}$, $1.25\,\mu\text{F}$
3.3 $L_{S_{\min}} = 0.33\,\mu\text{H}$.
3.4 $C_{S_{\min}} = 1\,\mu\text{F}$. Without the snubber, the product 'reapplied voltage times unevenly distributed uncombined charge' causes localised dissipation, hot-spots and probable destruction.
3.7 $0.65\,\mu\text{F}$
3.8 $3.5\,\text{nF}$
3.9 $0.25\,\mu\text{F}$, $35\,\mu\text{H}$
3.10 $20\,\mu\text{H}$

3.11 20 µF

3.12 Functions of turn-off snubber circuit: (ii) divert voltage stresses from switch, (ii) ensure switching path with SOA, (iii) reduce incidental dissipation.
$C_s = 75$ nF (choose 68 nF), $R_s = 53.3\,\Omega$ (choose 47 Ω)
$P = 2.27$ W (use 5W resistor)

3.13 RCD snubber circuits are concerned with switching loss minimisation. RC snubber circuits are concerned with dV/dt and transient overvoltage suppression.

Chapter 4

4.2 (a) S, (b), U, (c) U, (d) S, (e) S, (f) U

4.3 (a) In order for the motor to start, $b > d$ and

$$\sqrt{a^2 + 4c(b-d)} > 0$$

(b) $N = \dfrac{a + \sqrt{a^2 + 4c(b-d)}}{2c}$

(c) From (4.5), the criterion is $2cN > a$, which is true and therefore represents a stable point of operation.

4.4 (ii) see Fig. 4.5

Chapter 5

5.4 (i) $0.25V_{dc}$, $0.5V_{dc}$, $\sqrt{3}$, (ii) $0.75V_{dc}$, $0.866V_{dc}$, 0.577.
For $\gamma = \tfrac{3}{4}$, $C_1 = 0.45V_{dc}$, $C_2 = 0.318V_{dc}$, $C_3 = 0.15V_{dc}$, $V_{rms} \simeq 0.853V_{dc}$.

5.5 $V_{rms} = 0.866V_{dc}$.

5.6 (i) $0.25V_{dc}^2/R$, (ii) $0.75V_{dc}^2/R$.

5.7 $C_1 = 2V_{dc}/\pi$, $\psi_1 = 0$.

5.10 $V_{av}(\text{diode}) = -\gamma V_{dc}$, $V_{av}(\text{switch}) = (1-\gamma)V_{dc}$.

5.11 $V_{av} = 150$ v, $I_{av} = 7.5$ A, $I_{L1} = 0.5$ A, $I_{L_2} = 0.178$ A, $I_L = 7.52$ A, $P = 1131$ W.

5.12 $I_{min} = 6.5$ A, $I_{max} = 8.4$ A, $I_{S_{av}} = 5.75$ A, $P_{in} = 1150$ W.

5.13 $\gamma' = 0.651$, $\gamma = 0.678$, continuous current.

5.14 $I_{max} = 27.4$ A, $I_{av} = 20$ A.

5.15 $V_{av} = 33.33$ V, $I_{av} = 41.7$ A, $\eta = 70.4\%$, $I_1 = 20.56$ A, $I_2 = 5.16$ A, $I_3 = 0$, $I_L = 46.8$ A, $P = 1481$ W.

5.16 $V_{av} = 75$ V, $I_{av} = 25$ A, $\eta = 84.5\%$.

5.17 490 r.p.m.

5.18 $V_T = 250$ V, $I_T = 31.1$ A, $V_D = 250$ V, $I_d = 31.1$ A.

5.19 (c) $V_{av} = 0.75V_{dc}$, $V_{rms} = 0.866V_{dc}$, $RF = 0.577$, V_{rms}(approx) $= 0.853V_{dc}$.
(d) $V_{av}(\text{SCR}) = (1-\gamma)V_{dc}$, $V_{av}(\text{SCR})/V_{av} = (1-\gamma)/\gamma = 1$ for $\gamma = \tfrac{1}{2}$.

Chapter 6
6.2 A multiplier $E_m/2\pi$ applies to each element of the Table.

α	0°	30°	60°	90°
$n=1$	0.67	0.88	1.2	1.33
2	0.133	0.29	0.47	0.53
3	0.054	0.18	0.3	0.34
4	0	0.13	0.22	0.25
5	0	0.1	0.17	0.2
6	0	0.08	0.15	0.17

6.3 $V_{rms} = E_m/\sqrt{2}$ for all α.

α	0°	30°	60°	90°
V_{av}(p.u.)	1	0.866	0.5	1

6.4 87.1 A, 1629 W, 0.78.
6.5 discontinuous operation, $X = 191.5°$; 109.7 A, 23 014 W, 0.874.
6.6 972.4 r.p.m., 98.2 N m, 0.9, 93.8%, at full load; 41.3 N m, 0.87, 97.3%, at $\alpha = 15°$.
6.7 At $\alpha = 0°$ operation is (just) continuous; 11.94 N m, 4.88 A, $PF = 0.9$. At $\alpha = 30°$, $N = 693$ r.p.m.
6.8 48°, 43.6 mH.
6.15 $0 \leq \alpha \leq 43.4°$, $I = 181$ A (constant)
6.16 conduction remains continuous.
6.17 83.6%, 0.52.
6.18 318 V, 104.3 A.
6.19 88.3%, 0.955 at 1250 r.p.m. 93.8%, 0.45 at 6.25 r.p.m.

Chapter 7
7.2 (i) 2.81 A, 814.5 W, (ii) 1.62 A, 338 W, (iii) 0.43 A, 50 W.
7.6 (i) 2.1 A, (ii), 1.29 A, (iii) 0.32 A.
7.7 (i) 748 W, (ii) 310 W, (iii) 45.8 W.
7.13 (i) 45.6 µF, (ii) 45.6 µF, (iii) 15.1 µF. Note that these are maximum (not minimum) capacitance conditions.
7.14 (i) 22.8 µF, (ii) 22.8 µF, (iii) 7.5 µF.
7.15 (i) $PF = 0.84$, $PF_c = 0.93$, (ii) $PF = 0.542$, $PF_c = 0.722$, (iii) $PF = 0.21$, $PF_c = 0.254$.
7.18 (i) 2358 W, 2436 W, (ii) 786 W, 1011 W.
7.24 587 V, 3.24 A.

7.27 $PF_{30} = 0.827$, $PF_{60} = 0.477$, without compensation; $PF_{30} = 0.932$, $PF_{60} = 0.792$, when $R = V_c$; $\alpha = 33.25°$.

7.29 25.11 µF, $PF_{30} = 0.941$, $PF_{60} = 0.848$.

7.32 0, 12 051 W, 11 840 W, 10 133 W.

7.33 622 V, 37.24 A.

7.34 154°, $P_{165} = 0$, 52.6 kVA.

7.35 933 V, 32.5 A.

7.36 146.4°

7.37 0.675

Chapter 8

8.1 0.84 p.u., $-16.5°$.

8.2 70 V, 63.64 V, 31.82 V.

8.3 0.6 p.u.

8.4 (b) 465 W, (c) power is halved, (d) SCR in load branch, extinguished by natural commutation.

8.5 (b) 1152 W, (c) displacement factor = 0.843, distortion factor = 0.834, $PF = 0.707$.

8.6 (b) 0.897 p.u., (c) current will contain a d.c. component and even-order harmonics as well as odd-order harmonics of changed values.

8.7 (b) 0.8 p.u.

8.8 (b) $I_{av} = \dfrac{E_m}{\pi R}(1 + \cos\alpha)$, 6.73 A.

8.9 $\cos\psi_1 = 0.978$.

8.14 P versus α is in Fig. 8.7. Use (8.9), (8.34) for displacement factor and (8.38) for reactive voltamperes.

8.15 Distortion factor versus α is in Fig. 8.7.

8.16 18.8 µF, $PF_c = 8.74$, $PF = 0.814$.

8.18 (a) 1527.4 W, (b) $P = 0.663$ p.u., $PF = 0.814$, (c) $I_c = 2.82$ A, $C = 37.4$ µF.

8.19 While an SCR is on, I_1 and I_2 are very similar to the case without the transformer. While both SCRs are off, $I_2 = 0$ and I_1 draws its magnetising current, lagging V_1 by almost 90°.

8.22 215° (estimate), 221° (by iteration).

8.25 $X = 222°$, $P = 0.1$ p.u., $PF = 0.16$ lagging.

8.26 displacement factor = 0.179, distortion factor = 0.894, $PF = 0.16$.

8.27 $a_1 = -0.4$ p.u., $b_1 = 0.51$ p.u., $\psi_1 = -19.6°$.

8.28 0.716 p.u.

8.31 No. The supply current is not in time-phase with the voltage at every instant of the cycle.

8.36 All the waveforms have the same values of I_1, I_{rms}, P, PF and distortion factor. No apparent preference from an R load viewpoint. From the supply system viewpoint it is advisable to have minimum supply current interruption – use $N = 1$, $T = 2$.

8.37 $E_2 = 99.2\,\text{V}$, $E_3 = 100\,\text{V}$, $E_4 = 70.9\,\text{V}$ (peak values).

8.38 $\theta(t)$ is a triangle peaking at 4π. $I_{\text{dc}} = 0$. Load current contains both even and odd subharmonics (down to 1/4) and higher harmonics.

8.39 (a) $N = 17$, $T = 24$, $P = 0.70833$ p.u., (b) $I_n = 0.0213 E_m/R$, $f = 2.0833\,\text{Hz}$, (c) distortion factor $= 0.842 = PF$, displacement factor $= 1.0$.

If $R = X_c$, $\psi_1 = 54.7°$, displacement factor $= 0.58$ and the PF reduces!

Chapter 9

9.2 torque/ampere is halved.

9.3 $50° \leq \alpha \leq 110°$.

9.4 $50° \leq \alpha \leq 105°$.

9.5 111 A, 196 V (if star-connected), $T_{950} = 502\,\text{N m}$, $T_{750} = 313\,\text{N m}$, $36° \leq \alpha \leq 85°$ assuming constant phase-angle.

9.6 $T_{955} = 100\,\text{N m}$, $T_{500} = 27.4\,\text{N m}$, $I_{955} = 1.0$ p.u., $I_{500} = 0.61$ p.u., $18.8° \leq \alpha \leq 100°$.

9.7 111 A (r.m.s.), 70.7 A (mean) with star-connection, $V_{\max} = 240\sqrt{2} = 339\,\text{V}$ with delta-connection.

9.8 The no-load loss is represented by a resistor of $R = 240^2/403 = 48\,\Omega/\text{phase}$ at the terminals. $I_{950} = 0.36$ p.u., $\alpha_{950} = 110°$, $I_{500} = 0.175$ p.u., $\alpha_{500} = 125°$.

9.10 Ratio I/α is a function of α so that the system is functionally nonlinear. Linear system analysis does not apply.

9.12 8 rad/s

9.14 Requires the use of some form of closed-loop system.

9.15 'Concave' curve near origin suggests a nonlinear system of at least second order. Non-oscillatory small slope curve at high t suggests overdamping.

9.16 103.9 rad/s, $|G| = 0.343$, $K_{TG} \simeq 2$.

9.17 $30/(1 + 30 K_{TG})$.

9.21 $S_{T\max}(\alpha = 180°) = 4 S_{T\max}(\alpha = 0°)$.

Chapter 10

10.2 270 V

10.5 $\alpha = -114°$, 83.5%, 0.22.

10.6 $440\sqrt{2} = 623\,\text{V}$, $210/\sqrt{3} = 121.4\,\text{A}$.

10.7 4.2 mH. Insufficient information to calculate R_f.

10.8 At 1200 r.p.m.: 86%, 0.435. At 1420 r.p.m.: 88.4%, 0.572.

10.9 At 1200 r.p.m.: $PF = 0.879$. At 1420 r.p.m.: $PF = 0.938$. Transformer rating = inverter rating; at 1200 r.p.m. is 20 kVA nominal, 20.824 kVA from V_{dc} and 22.5 kVA from voltage and current.

10.10 Same as 10.6.

Answers to problems

10.11 At 1440 r.p.m., $\alpha = 90°$, $T = 663$ N m, $(N_{FL} - N_{NL})N_{NL} =$ slip $= 4\%$.

Chapter 11
11.2 At f_3, $T_3/T_1 = 0.69$.
11.6 $E_1 = 47.5$ V, $I_1 = 42.6 \angle -17.5°$.
11.7 0.945, 89.8%.
11.8 At 725 r.p.m., $I_2 = 10.175 \angle -1.4°$, $I_1 = 13.63 \angle -41.7°$, $P = 3125$ W, $PF = 0.634$, $\eta = 96.6\%$.
11.10 V_{AB_1} leads V_{AN_1} by 30°.
11.12 $V_{NO} = (V_{dc} - V_{AN})_0^\pi = (V_{dc} - V_{BN})_\pi^{2\pi}$
(a) V_{NO} is square-wave $\pm V_{dc}/3$ with three times supply frequency, (b) I_{NO} is square-wave $\pm V_{dc}/R$ with three times supply frequency.
11.13 $a_1 = 0$, $b_1 = 3E_m/\pi$, $V_1 = 3E_m/\pi\sqrt{2}$, $V_{av} = 2E_m/3$, $V_{av_1} = 6E_m/\pi^2$.
11.16 $a_1 = 0$, $b_1 = 4E_m/\pi = c_1$, $\cos\psi_1 = 2$, $E_{rms} = E$, distortion factor $= 0.9$.
11.18 $V_{rms} = 0.75$ V, distortion $= 0.99$.
11.19 4.35 A, $I_{dc} = 5.67$ A, 1135 W.
11.20 8.165 A, $I_{dc} = 20$ A, 400 W.
11.21 0.38 A, 8.72 A.

Chapter 12
12.2 (a) 1.03 V, (b) 0.9 V.
12.3 $V_n = \dfrac{4V}{n\pi}\left(1 - \cos\dfrac{n\pi}{6} + \cos\dfrac{n\pi}{3}\right)$,

$\delta = 120°$, $V_1 = 0.81$ V (cf. 1.1 V).
12.4 $\alpha_1 = 17.8°$, $\alpha_2 = 40°$.
12.6 $\alpha_1 = 23.6°$, $\alpha_2 = 33.3°$.
12.8 $b_1 = 0.99V$, $b_3 = 0.004V$, $b_5 = -0.001V$, $b_7 = -0.03V$, $b_9 = -0.21V$, $b_{11} = -0.184V$, $b_{13} = 0.11V$, $b_{15} = 0.14V$, $b_{17} = -0.02V$, $b_{19} = -0.12V$, $b_{21} = -0.01V$, $b_{rms} = 0.743V$.
12.9 $I_{rms} = 5.013$ A, CDF $= 0.999$, VDF $= 0.942$.
12.10 628 W, 0.42.
12.13 $I_{rms} = 12.67$ A, $I_1 = 12.5$ A, $I_7 = 1.59$ A, $I_{11} = 1.016$ A, $I_{17} = 0.65$ A, $I_{10} = 0.58$ A.
12.14 $P_{in} = 3265$ W, $VA_{in} = 3 \times 76.4 \times 15 = 3428$ VA, $PF = 0.95$.
12.15 (i) 92.3%, losses $= 251$ W, (ii) with same input current $\eta = 84.6\%$, with same load $\eta = 85.7\%$.

References and bibliography

(A) Books

(i) General

1. F. F. Mazda. *Thyristor Control*. Newnes–Butterworth, England, 1973.
2. R. S. Ramshaw. *Power Electronics*. Chapman & Hall, England, 1973.
3. F. Csaki, K. Gansky, I. Ipsits & S. Marti. *Power Electronics*. Academic Press, Budapest, Hungary, 1975.
4. S. B. Dewan & A. R. Straughen. *Power Semiconductor Circuits*. Wiley-Interscience, USA, 1975.
5. M. Ramamoorty. *Introduction to Thyristors and their Applications*. The Macmillan Press Ltd, India, 1977.
6. *General Electric SCR Manual*. GE, Schenectady, NY, USA, 6th edn, 1979.
7. R. K. Sugandhi & K. K. Sugandhi. *Thyristors – Theory and Applications*. J. Wiley & Sons, India, 1981.
8. G. K. Dubey, S. R. Doradla, A. Joslu, & R. M. K. Sinha. *Thyristor Power Controllers*. J. Wiley and Sons, New Delhi, India, 1986.
9. K. Thorborg. *Power Electronics*. Prentice-Hall (UK) Ltd, London, England, 1988.
10. N. Mohan, T. M. Undeland, W. P. Robbins. *Power Elecronics: Converters, Applications, and Design*. J. Wiley and Sons, USA, 1989.
11. J. G. Kassakian, M. F. Schlect & G. C. Verghese. *Principles of Power Electronics*. Addison-Wesley, USA, 1989.
12. M. J. Fisher. *Power Electronics*. PWS-Kent, USA, 1991.
13. B. W. Williams. *Power Electronics*, The Macmillan Press, England 2nd edn, 1992.

14. M. H. Rashid. *Power Electronics: Circuits, Devices and Applications.* Prentice-Hall, USA, 2nd edn, 1993.
15. C. W. Lander. *Power Electronics.* McGraw-Hill, England, 3rd edn, 1993.
16. B. M. Bird, K. G. King & D. A. G. Pedder. *An Introduction to Power Electronics.* J. Wiley and Sons, England, 1993.

(ii) Rectifiers and inverters

17. H. Rissik. *Mercury Arc Current Converters.* Sir Isaac Pitman and Sons, England, 1963.
18. E. W. Kimbark. *HVDC Transmission.* J. Wiley & Sons, USA, 1965.
19. J. Schaeffer. *Rectifier Circuits.* J. Wiley & Sons, USA, 1965.
20. B. Bedford & R. Hoft. *Principles of Inverter Circuits.* McGraw-Hill, USA, 1965.
21. R. M. Davis. *Power Diode and Thyristor Circuits.* Cambridge University Press, England, 1971.
22. B. R. Pelly. *Thyristor Phase Controlled Converters and Cycloconverters.* Wiley-Interscience, USA, 1971.
23. P. Wood. *Switching Power Converters.* Van Nostrand Reinhold, USA, 1981.
24. G. De. *Principles of Thyristorised Converters.* Oxford and IBH Publishing Co., India, 1982.
25. A. Kloss. *A Basic Guide to Power Electronics.* J. Wiley & Sons, England, 1984.
26. M. A. Slonim. *Theory of Static Converter Systems, Part A: Steady-State Processes.* Elsevier, USA, 1984.
27. G. Moltgen. *Converter Engineering* (translation from German). Siemens Aktiengesellschaft, Wiley, USA, 1984.
28. G. Seguier. *Power Electronic Converters, Vol. 1 – AC/DC Converters.* North Oxford Academic Press, England, 1986.
29. R. G. Hoft. *Semiconductor Power Electronics.* Van Nostrand Reinhold, USA, 1986.
30. C. Rombout, G. Seguier & R. Bausiere. *Power Electronic Converters, Vol. 2 – AC/AC Converters.* McGraw-Hill, England, 1987.
31. J. M. D. Murphy & F. G. Turnbull. *Power Electronic Control of AC Motors.* Pergamon Press, England, 1988.

(iii) Properties of waveforms

32. W. Shepherd. *Thyristor Control of AC Circuits.* Crosby Lockwood Staples, England, 1975.

33. W. Shepherd & P. Zand. *Energy Flow and Power Factor in Nonsinusoidal Circuits*. Cambridge University Press, England, 1979.

(iv) Harmonic control and electrical drives
34. A. Kusko. *Solid-state DC Motor Drives*. MIT Press, USA, 1960.
35. A. E. Fitzgerald & C. Kingsley. *Electric Machinery*. McGraw-Hill, USA, 2nd edn, 1961.
36. P. C. Sen. *Thyristor DC Drives*. J. Wiley and Sons, USA, 1981.
37. S. K. Pillai. *A First Course in Electrical Drives*. Wiley Eastern Ltd, New Delhi, India, 1982.
38. S. B. Dewan, S. R. Straughen & G. R. Slemon. *Power Semiconductor Drives*. Wiley-Interscience, USA, 1984.
39. W. Leonhard. *Control of Electrical Drives* (translation from German). Springer, West Germany, 1985.
40. G. K. Dubey. *Power Semiconductor Controlled Drives*. Prentice-Hall, USA, 1989.
41. C. B. Gray. *Electical Machines and Drive Systems*. Longman, England, 1989.

(v) Semiconductor physics and device properties
42. F. E. Gentry, F. W. Gutzwiller, N. Holonyak & E. E. Von Zastrow. *Semiconductor Controlled Rectifiers*. Prentice-Hall, USA, 1964.
43. P. E. Grey & C. L. Searle. *Electronic Principles*. J. Wiley & Sons, USA, 1967.
44. A. Blicher. *Thyristor Physics* (translation from German). Springer, West Germany, 1976.
45. J. M. Peter (Ed.). *The Power Transistor in its Environment*. Thomson-CSF-Semiconductor Division, Aix-en-Provence, France, 1978.
46. E. S. Oxner. *Power FETs and Their Applications*. Prentice-Hall, USA, 1982.
47. R. Sittig & P. Roggwiller. *Semiconductor Devices for Power Conditioning*. Plenum Press, USA, 1982.
48. P. D. Taylor. *Thyristor Design and Realisation*. J. Wiley & Sons, England, 1987.
49. B. J. Baliga. *Modern Power Devices*. J. Wiley and Sons, USA, 1987.
50. E. Ohno. *Introduction to Power Electronics*. Oxford Science Publications, Oxford, England, 1988.
51. M. Zambuto. *Semiconductor Devices*. McGraw-Hill International Editions, Singapore, 1989.
52. *Power Mosfet Transistor Data*, Motorola Inc., USA, 4th edn, 1989.

53. D. A. Grant & D. Gower. *Power MOSFETS – Theory and Applications*. J. Wiley and Sons, New York, USA, 1989.
54. R. S. Ramshaw. *Power Electronics Semiconductor Switches*. Chapman and Hall, London, England, 1993.

(B) Technical papers and review articles

TP1. E. R. Laithwaite. Electrical Variable-Speed Drives. *Engineers' Digest – Survey No. 3*, **25**, no. 10, 1964, pp. 115–65.

TP2. W. Shepherd & J. Stanway. An Experimental Closed-Loop Variable-Speed Drive Incorporating a Thyristor Driven Induction Motor. *Trans. IEEE*, **IGA-3**, no. 6, 1967, pp. 559–65.

TP3. W. Shepherd & J. Stanway. Slip Power Recovery in an Induction Motor by the use of a Three-Phase Thyristor Inverter. *Trans. IEEE*, **IGA-5**, no. 1, 1969, pp. 74–82.

TP4. D. E. Knight. Guidelines for Variable Speed Drive Choice. *Electrical Times, Issue 4274*, 28th March 1974, p. 5.

TP5. M. Ikamura, T. Nagano & T. Ogawa. Current Status of Power Gate Turn-off Switches. *IEEE Int. Semiconductor Power Conductor Conf.*, USA, 1977, pp. 39–49. (Included, with other relevant papers, in *Power Transistors: Device Design and Applications*. Eds. B. Jayant Baliga and Dan Y. Chen, IEEE Press, USA, 1984.)

TP6. B. R. Pelly. Power Semiconductors – A Status Review. *IEEE Int. Semiconductor Power Converter Conf.*, USA, 1982, pp. 1–19.

TP7. A. Woodworth & F. Burgum. Simple Rules for GTO Circuit Design. *Mullard Technical Publication M83-0137*, London, England, 1983.

TP8. D. W. Novotny & T. Lipo. Vector Control and Field Orientation. Chapter 11 of *Conf. On Dynamics and Control of A.C. Drives*, University of Wisconsin, Madison, USA, 1985. (17pp.)

TP9. A. J. Moyes & A. E. Murrell. Comparative Economics of Variable-Speed Drives – A User's Assessment. *IEE Conf. on Drives, Motors and Controls*, London, England, 1985, pp. 107–11.

TP10. N. Groves, G. Crayshaw, J. P. Ballard & I. C. Rohsler. *Solid State Electronic Devices for Power Switching*, ERA Technology, Leatherhead, Surrey, England, 1986.

TP11. J. D. Van Wyk, H. Ch. Skudelny & A. Muellen. Power Electronics Control of the Electromechanical Conversion Process and Some Applications. *Proc. IEE*, **133**, Part B, no. 6, 1986, pp. 369–99.

TP12. T. A. Lipo. Recent Progress in the Development of Solid-State AC Motor Drives. *Trans. IEEE*, **PE-3**, no. 2, 1988, pp. 105–17.

TP13. B. K. Bose. Power Electronics – an Emerging Technology. *Trans. IEEE*, **IE-36**, no. 3, 1989, pp. 403–12.
TP14. P. C. Sen. Electric Motor Drives and Control. *Trans. IEEE*, **IE-37**, no. 6, 1990, pp. 561–75.
TP15. *GTR Module (IGBT) Application Notes 3507D-A*, Toshiba Corp., Tokyo, Japan, 1991.
TP16. MOS Controlled Thyristor User's Guide. *Harris Semiconductor Data Booklet DB 307A*, undated.
TP17. B. K. Bose. Recent Advances in Power Electronics. *Trans. IEEE*, **PE-7**, no. 1, 1992, pp. 2–16.
TP18. K. Kamiyama, T. Ohmae & T. Sukegawa. Application Trends in AC Motor Drives. *Proc. IEEE–IECON '92*, San Diego, Cal, USA, 1992, pp. 31–6.
TP19. D. L. Blackburn. Status and Trends in Power Semiconductor Devices. *Proc. IEEE–IECON '93*, Hawaii, USA, Nov. 1993, pp. 619–25.
TP20. S. Tadakuma & M. Ehara. Historical and Predicted Trends of Industrial AC Drives. *Proc. IEEE–IECON '93*, Hawaii, USA, Nov. 1993, pp. 655–61.

Index

a.c. commutator motor, 147
adjustable speed drive, 121–51
  a.c. or d.c.? 147
  availability of supply, 137
  braking requirements, 135
  controllability, 134
  drive motors, 139
  effect of supply variation, 137
  efficiency, 130
  environment, 138
  load factor and duty cycle, 136
  loading of the supply, 137
  power factor, 136
  power/weight ratio, 136
  rating and capital cost, 130
  reliability, 135
  running costs, 138
  speed range, 130
  speed regulation, 134
  stability, 123–9
  torque–speed characteristics, 122
  trends in design, 149
air-gap flux, 152–6, 178, 346, 519
algorithm, 113, 145, 148
amplifier, 34, 116–17, 381, 384, 390, 392
anode, 58–60, 65–6, 69, 74, 78, 82
ANSI/IEEE Standard 591-1981, 138
apparent power (voltamperes), 289, 294
armature inductance, 155–7, 164–7, 172–3, 178–83, 189, 192, 195, 206–7, 214, 225, 230–5
armature resistance, 155–7, 164–6, 172–3, 178–84, 189, 192, 194–7, 201–2, 207, 212–14, 222, 230, 233–4
artificially (forced) commutated inverter, *see* current source inverter *and* voltage source inverter
ASIC (application specific, integrated circuit), 113
avalanching, 45, 51, 62, 74, 80
average power, *see* power, average

back e.m.f., 154–85, 192–211, 213–17
Baker clamp, 48, 52
base speed, 127

Bessel Function, 493
bipolar power transistor (BJT), 29, 36–48, 85-8, 115, 118, 149
  base current, 37
  base charging capacitance, 39
  complementary connection, 58–61
  construction, 41–3
  control gain ratio, 37–40
  current gain, 36–7
  extrinsic resistance, 39
  protection, 85
  rating, 43-6, 67
  saturation, 41, 45
  slew rate, 41, 84
  switching aids, 6–17
  switching characteristics 40–1
Bode diagram, 37, 38, 388
braking, 124, 126, 135, 366
branch-delta connection, 364, 368
breakdown voltage, 41, 43, 47, 61, 66
bridge rectifier, *see* single-phase bridge rectifier *and* three-phase bridge rectifier
brushless excitation system, 142–3, 191
brushless synchronous motor, 142–3
burst firing (integral cycle control), 323–36, 342–5

CAD (computer aided design), 113
capacitance, 6, 37–9, 44, 55
capacitor, 9–15, 18, 96–105, 171
capacitive compensation, 246–54, 259–65
carriers (semiconductor), 60–3, 74
cathode, 71, 77, 78, 82
circulating current, 222
closed-loop operation, 134, 143, 147, 380–93, 401–3, 518
collector, 41, 47, 53–4
commutation circuit, 81, 158, 445
complementary SCR, 61
controlled rectifier, *see* single-phase bridge rectifier *and* three-phase bridge rectifier
copper losses, 169, 202, 351, 393–4, 408, 417, 473–9
crane (hoist), 122–3
critical load inductance, 202, 217

cumulative feedback connected devices (thyristors), 57–63
current,
  average, 66, 194, 212, 254
  instantaneous, 3–5, 8–15, 95–105, 110, 152–5, 164–6, 170, 192, 195–6, 494
  maximum (non-recurrent) surge rating, 68
  maximum (recurrent) surge rating, 68
  mean forward (continuous) rating, 66
  peak, 3–5, 8–16, 33–4, 254
  positive sequence, 471
  negative sequence, 471
  rating, 41, 47–50, 61, 64–8, 70, 76
  ripple, 455–6
  root mean square (r.m.s.), 167–70, 242–4, 247–8, 254, 257, 286–8, 312–13, 327, 348–50, 404–33
current control loop, 518
current density, 43, 60, 66
current harmonics, 214, 246, 309–12, 416–18, 422–3, 472–6, 503–5
current limit control, 517–18
current source inverter (CSI), 516–18
current surge protection, 105
cycle selection (integral-cycle control), 323–36, 342–5
cycloconverter, 146–7, 445

Darlington connections, 36, 52–3, 106, 111, 115
d.c. chopper 152–89
  equations, 160–70
  class A performance, 170–1
  class B performance, 171–4
d.c. link current, 265–8, 407, 411–12
d.c. motor, 139, 152–7
  controlled by class A chopper, 158–70
  controlled by class B chopper, 171–4
  controlled by 1φ rectifier, 191–209, 233–4
  controlled by 3φ rectifier, 210–33, 234–5
  saturation, 153
depletion layer capacitance, 42–3
derating (of motor), 134, 474
describing function, 341
diac, 61, 77
diffusion, 42–3, 57, 60, 68–70
diode, 8–15, 45–8, 117, 158, 171–4, 192–3, 210, 220, 337, 396, 407, 460
displacement factor, 245–8, 258, 271, 291–2, 323–8
distortion factor, 245–8, 259, 328, 330–3, 455–7
distortion voltamperes, 293–6
doping level, 58, 75, 80
drain terminal, (of FET) 49–51
DSP (digital signal processor), 113
duty cycle, 47, 72, 136–7, 157, 172
dynamic braking, 135
dynamic stability (transient stability), 127–9
dynamometer wattmeter, 290

eddy current, 513
efficiency, 129–34, 179–85, 222–4, 351, 416–19, 476–9
electric generator, 122
electric (drive) motor, 139–50
electromagnetic interference, 106–7, 330

electrons, 61, 77
emitter, 39, 52–4
encoder, 143, 146, 384
Engineering Recommendation G5/3 (UK Electricity Council), 138
epitaxial deposition, 33, 36, 42, 43, 49, 53, 86,
extinction angle, 159, 163, 166, 200–3, 306–23, 366
extrinsic resistance, 39, 40, 43–4

feedback control system, 143, 379–80, 383–93, 518
field effect transistor (FET), *see* MOSFET
  characteristics, 49–52
  construction, 48–9
  rating, 67
field oriented (vector) control, 146
filter inductance, 267, 422–3, 427
filter resistance, 267
form factor, 287
four-quadrant operation, 125–7, 145, 150, 221
Fourier coefficients, 162–3, 192–9, 213–14, 245–6, 283–5, 309–14, 324–7, 450–4
freewheel diode, 192–3, 210–11, 220–1
frequency, 6, 37, 41, 43, 67, 107, 160, 288, 346–7, 445, 487, 490, 496, 514
friction and windage, *see* torque, friction and windage
fuse, 107, 137

gain coefficients (weighting factors), 58–60
gate control, 61, 70–2, 75–7, 80–1, 85–6, 115–18
gate turn-off (GTO) thyristor, 31, 36, 75–81, 86, 109, 115, 120, 150,
  construction, 75–6, 78
  control current gain, 79
  rating, 67
  turn-off, 77, 81
  turn-on, 76–7

Hall plate, 107
hard drive (of SCR), 111
harmonic reactance, 361–3, 471–2, 475
heating loads, 329–30
heat sink, 17–20, 27–8
holding current, 63, 74, 281
holes, 60–1
horsepower, 133–4, 210, 380, 393
hot-spot, 47, 64, 74, 77
hysteresis, 63, 513

IC (integrated circuit), 108, 112
IEEE standard, *see* ANSI/IEEE standard
impedance, 20, 198–202, 215–19, 305–6
incidental loss, 3, 15–17, 79, 83, 86–7
inductance, 9, 12–13, 15, 96–105, 171–4, 254, 349
induction motor, *see* three-phase induction motor
inertial torque, 124, 150, 156
insulated gate bipolar transistor (IGBT), 36, 53–7, 67, 86, 111–12, 118, 150
integral-cycle control, 323–36, 342–5
interdigitation, 43, 65, 70, 75, 79
inverter, *see* current source inverter (CSI),

Index   537

inverter (cont'd)
    naturally commutated inverter *and* voltage source inverter
iron losses (core losses), 169, 202, 474, 517
isolation circuitry 108–12

junction, 45, 58–63, 66, 74, 76
junction irradiation, 62
junction temperature, 19, 28, 30–1, 63

Kirchhoff's Law, 281

lamp flicker, 328–9
Laplace transform, 99, 341
latching current, 61
leakage current, 3, 7, 17, 45, 60, 63, 68
lighting control, 328–9
load torque, *see* torque load
low-pass filter, 288

magnetic saturation, 152–3, 156, 349, 435
mark–space ratio, *see also* duty cycle,
MCT (MOS controlled thyristor), 36, 82–6, 115, 150
microprocessor, 106, 112–3
Miller effect, 55–6
m.m.f. space harmonics, 358–61
m.m.f. time harmonics, 358–61
MOSFET, 6–7, 23, 48–52, 67, 86–7, 112, 118
moving-iron instrument, 287–8
multiple-pulse modulation, 487–90

natural commutation, 191
naturally commutated inverter, 265
natural sampling, 491–9
negative feedback, 383–93
negative sequence m.m.f., 360
negative sequence torque, 366, 475, 481–2
negative sequence voltage, 363–5
Nyquist diagram, 386, 389–92, 403

open-loop control, 134, 384, 517
opto-isolator, 111
overcurrent protection, 107–8
overlap, 415
overshoot, 128
overvoltage protection, 108

periodic time, 160
permanent magnet materials, 141–2, 149
permanent magnet synchronous motor, 141–2
phase-angle, 283, 304–5
phase-angle switching, 74, 116–17, 280–323, 362–403
plasma, 76, 77
positive sequence m.m.f., 360
positive sequence torque, 366, 475, 481–2
positive sequence voltage, 363–5
power,
    average (active), 155, 169, 202, 220, 259, 268, 288–91, 314–16, 327, 350–3, 416–18, 474, 518
    instantaneous, 314–5
    rating, 16, 21, 32–4, 70
    reactive, *see* reactive voltampores
power factor,
    adjustable speed drives, 129, 136
    d.c. motor control, 202–3, 220
    single-phase voltage controller, 291-6, 314–16, 327–8
    slip-energy recovery system, 418–19
    three-phase bridge rectifier, 245–8, 260–1
    three-phase induction motor, 356
    three-phase naturally commutated inverter, 271
power handling capability, 16, 32, 34, 57, 79
power/weight ratio, 129, 136
power transistor, *see* bipolar power transistor (BJT) *and* field effect transistor (FET)
Principle of Conservation of Energy, 289
Principle of Superposition, 361
pulley, 126
pulse transformer, 109–11
pulse-width modulation (PWM), 112–13, 149–50, 487–522

radio-frequency interference, *see* electromagnetic interference
reactive voltamperes, 268, 292–6
rectifier moving-coil instrument, 288
regenerative braking, 135
regenerative feedback, 58, 75
regular sampling, 499–501
reliability, 129, 135–6
resistance, 3–23, 37–41, 45, 50–1, 54, 71–2, 81, 95–101, 109–11, 155–6, 164–170, 192–7, 202, 212, 242–54, 256–61, 349–55, 363–4, 396–7, 407–16, 436–40, 445-55, 473–5, 495
reverse safe operating area (RBOSA), 47, 57
ripple factor, 161, 243
ripple frequency, 242, 267, 331

safe operating area (SOA), 33–4, 42–3, 47, 85
Schrage motor, 147
semiconductor switch, 2–3, 32, 191
semiconverter, *see* three-phase semiconverter
separately excited d.c. motor, 152–7
series connection (d.c. motor), 152–3
shaft encoder, 382–3
shaft speed, *see* speed
shunt connection (d.c. motor), 153–5
silicon controlled rectifier (SCR), 6, 28, 36, 63–73, 102–3, 108–9, 114–15, 119–20
    construction, 59–60, 68–70
    current gain, 58–60
    di/dt, 64–5
    dv/dt, 66, 95–103
    gate triggering, 61, 70–3, 116–17
    heat sink, 67–68
    ratings, 64–8, 70
    snubber circuits, 95–103
    turn-off, 63
    turn-on, 61
single-phase bridge rectifier, 191–210
single-phase voltage controller, 280–345
    with integral-cycle control, 323–36
    with phase-angle control, 280–323
        current harmonics, 281–5, 309–12
        power and power factor, 288–96, 314–16

Index 539

voltage harmonics, 281–5, 313–14
single pulse modulation, 487–9
sinusoidal modulation, 491–505
skin effect, 435, 473, 513
slip (of induction motor)
    fundamental frequency, 347–55, 404–17, 436–40, 471–9
    harmonic frequency, 363
    at peak torque, 355
slip-energy recovery (SER) systems, 404–34
snubber circuits, 87, 95–105
soft starting, 364
source terminal (of FET), 50–1
SPICE, 114
speed
    average, 155, 369–86
    instantaneous, 128, 154, 402
    ripple (oscillation), 329
spreading, 43, 80
steady-state performance, 123–7
step-wave inverter, *see* current source inverter (CSI) *and* voltage source inverter
stepper (stepping) motor, 139, 143–5
stray capacitance, 6, 51, 85
subharmonic, 324–7, 331
supply voltage dip, 330–1
surface passivation, 36, 68
switching aid, 6–17
switched reluctance drive, 139, 145
synchronous motor, 139–45
synchronous reluctance motor, 139, 142, 145

tachogenerator, 383–5
temperature rise, 17
THD (total harmonic distortion), 455–7
thermal resistivity (resistance), 20
three-phase bridge rectifier,
    controlled, d.c. motor load, 212–21
    controlled, high L load, 254–65
    controlled, R load, 236–53
    load side properties, 244
    supply side properties, 248
    uncontrolled, 396, 407
three-phase double converter, 221–2
three-phase induction motor, 146–9
    copper (winding) losses, 351, 408, 417, 473–9
    current control, 516–18
    equivalent circuits, 348–50, 363
    injected secondary voltage, 404–6
    iron losses, 474
    secondary resistance control, 393–8
    slip-energy recovery drive, 406–34
    transfer function, 381–93
    transformation ratio, 349–50, 419–22
    vector control, 146–7
    voltage control, nonsinusoidal, 362–78
    voltage control, sinusoidal, 346–58
    voltage control, 356–8
three-phase, naturally commutated inverter, 265–79, 406–8
three-phase semiconverter, 210–12
three-phase voltage controller, 362–78
thyristor, *see* cumulative feedback connected devices, gate turn-off (GTO) thyristor, MCT (MOS controlled thyristor) *and* silicon controlled rectifier (SCR) type
time constant, 6, 8–15, 164, 306, 380–1
time-cut strategies, 106
torque,
    average, 154–6, 187, 216–17, 350–3, 378–81, 412, 436, 475–6
    friction and windage, 124, 352
    harmonic, 475
    inertial, 124, 150, 156
    instantaneous, 129, 154
    load, 124
    negative sequence, 366, 475, 481–2
    peak, 355–6, 394, 412, 436–8
    positive sequence, 366, 475, 481–2
    ripple, 329, 476
torque–speed characteristic, 122–7, 154–6, 355–7, 370, 378, 385, 402, 406, 413, 478, 517
transfer function, 381–93
transient performance, 127–9, 380
transformer, 73, 349, 407, 420
transistor, *see* bipolar power transistor, field effect transistor *and* insulated gate bipolar transistor
travelling m.m.f. wave, 359–60
triac, 30, 73–5
triplen harmonics, 360, 493, 502
true r.m.s. instrument, 115, 288
turn-off aid, 9–15
turn-on aid, 8–9,

underdamped response, 128, 387
uniform sampling (regular sampling), 499–500

vector control, 146–7,
viscous friction, 124
voltage,
    average, 193, 212, 241–2, 244, 406–7, 420
    instantaneous, 3–5, 8–15, 95–105, 110, 152–5, 170, 192, 195–6, 211, 236, 267, 281, 457–8, 494–5
    rating, 43, 47–50, 70, 83
    r.m.s., 167–70, 242–4, 286–8, 313–14, 327, 404–33, 435–44, 455–7
voltage controller, *see* single-phase voltage controller *and* three-phase voltage controller
voltage harmonics, 160–3, 213–17, 281–5, 313–14, 450–4
voltage source inverter, 444–86, 487–510
    PWM, 512–16
    stepwave,
        induction motor load, 471–82
        R load, 444–58
        R–L load, 459–65

Ward–Leonard drive, 136, 147
weighting factor, 58–9

Zener diode, 50, 57, 83
zero sequence current, 360, 471
zero voltage switching (integral-cycle control), 281, 323–36